/ Advances in
IMMUNOLOGY

VOLUME **96**

Advances in IMMUNOLOGY

VOLUME **96**

Edited by
FREDERICK W. ALT
Howard Hughes Medical Institute, Boston, Massachusetts

Associate Editors
K. FRANK AUSTEN
Harvard Medical School, Boston, Massachusetts

TASUKU HONJO
Kyoto University, Kyoto, Japan

FRITZ MELCHERS
University of Basel, Basel, Switzerland

JONATHAN W. UHR
University of Texas, Dallas, Texas

EMIL R. UNANUE
Washington University, St. Louis, Missouri

AMSTERDAM • BOSTON • HEIDELBERG • LONDON
NEW YORK • OXFORD • PARIS • SAN DIEGO
SAN FRANCISCO • SINGAPORE • SYDNEY • TOKYO
Academic Press is an imprint of Elsevier

Academic Press is an imprint of Elsevier
84 Theobald's Road, London WC1X 8RR, UK
Radarweg 29, PO Box 211, 1000 AE Amsterdam, The Netherlands
Linacre House, Jordan Hill, Oxford OX2 8DP, UK
30 Corporate Drive, Suite 400, Burlington, MA 01803, USA
525 B Street, Suite 1900, San Diego, CA 92101-4495, USA

First edition 2007

Copyright © 2007 Elsevier Inc. All rights reserved

No part of this publication may be reproduced, stored in a retrieval system or transmitted in any form or by any means electronic, mechanical, photocopying, recording or otherwise without the prior written permission of the publisher

Permissions may be sought directly from Elsevier's Science & Technology Rights Department in Oxford, UK: phone (+44) (0) 1865 843830; fax (+44) (0) 1865 853333; email: permissions@elsevier.com. Alternatively you can submit your request online by visiting the Elsevier web site at http://elsevier.com/locate/permissions, and selecting *Obtaining permission to use Elsevier material*

Notice
No responsibility is assumed by the publisher for any injury and/or damage to persons or property as a matter of products liability, negligence or otherwise, or from any use or operation of any methods, products, instructions or ideas contained in the material herein. Because of rapid advances in the medical sciences, in particular, independent verification of diagnoses and drug dosages should be made

ISBN: 978-0-12-373709-0
ISSN: 0065-2776

For information on all Academic Press publications
visit our website at books.elsevier.com

Pinted and Bound in USA
07 08 09 10 11 10 9 8 7 6 5 4 3 2 1

Working together to grow
libraries in developing countries

www.elsevier.com | www.bookaid.org | www.sabre.org

ELSEVIER   BOOK AID International   Sabre Foundation

CONTENTS

Contributors ix

1. New Insights into Adaptive Immunity in Chronic Neuroinflammation   1

    Volker Siffrin, Alexander U. Brandt, Josephine Herz, and Frauke Zipp

    1. Introduction: Multiple Sclerosis Is a Heterogeneous Inflammatory Disease of the Nervous System   2
    2. Experimental Autoimmune Encephalomyelitis as a Model for MS   3
    3. Current Knowledge About Induction and Perseveration of Chronic Neuroinflammation   5
    4. Therapeutic Approaches to Chronic Neuroinflammation   20
    Acknowledgments   29
    References   29

2. Regulation of Interferon-γ During Innate and Adaptive Immune Responses   41

    Jamie R. Schoenborn and Christopher B. Wilson

    1. Introduction   43
    2. IFN-γ-Producing Cells   45
    3. Signaling Pathways Controlling IFN-γ Production by NK Cells   48
    4. Control of IFN-γ Production by NKT Cells   52
    5. Signaling Pathways in the Differentiation of CD4 and CD8 T Cells   53
    6. Transcription Factors Downstream of the TCR, Activating NK Receptors, and Cytokine Receptors   60
    7. Epigenetic Processes Govern Plasticity of Cell Fate Choices and Help to Identify Distal Regulatory Elements   72
    8. Transcriptional Regulatory Elements Within the *Ifng* Gene   74
    9. Functional Analysis of Candidate Distal Regulatory Elements in the *Ifng* Locus   82
    10. Conclusions and Future Directions   84
    Acknowledgments   85
    References   85

3. **The Expansion and Maintenance of Antigen-Selected CD8$^+$ T Cell Clones** — 103

   Douglas T. Fearon

   1. Background — 105
   2. The Behavior of the CD8$^+$ T Cell in Persistent Viral Infections — 111
   3. Clarifying the Role of IL-2 in the Clonal Expansion and Effector Differentiation of the CD8$^+$ T Cell — 115
   4. Coreceptors Mediating IL-2-Independent CD8$^+$ T Cell Clonal Expansion — 118
   5. Modifying the Antiproliferative Effects of Types I and II IFN — 120
   6. Transcriptional Control of Replicative Senescence: Bmi-1, Blimp-1, and BCL6/BCL6b — 123
   7. A Refined Model for CD8$^+$ T Cell Clonal Expansion: Sequential Phases of CD27-Dependent Self-Renewal and IL-2-Dependent Differentiation — 126
   8. Clinical Extensions of the TCR/CD27 Pathway: Adoptive CD8$^+$ T Cell Therapy — 127
   Acknowledgments — 129
   References — 129

4. **Inherited Complement Regulatory Protein Deficiency Predisposes to Human Disease in Acute Injury and Chronic Inflammatory States** — 141

   Anna Richards, David Kavanagh, and John P. Atkinson

   1. Altered Self Triggers Innate Immunity — 143
   2. Regulation of the Alternative Complement Pathway — 145
   3. Lessons from Homozygous Complement Regulatory Protein Deficiencies — 153
   4. Complement and Atypical Hemolytic Uremic Syndrome — 154
   5. Complement and Age-Related Macular Degeneration — 160
   6. Immunopathogenesis of aHUS and AMD — 163
   7. Treatment of aHUS and AMD — 167
   8. Conclusions: Lessons and Implications — 168
   References — 169

5. **Fc-Receptors as Regulators of Immunity** — 179

   Falk Nimmerjahn and Jeffrey V. Ravetch

   1. Introduction — 180
   2. The Family of Activating and Inhibitory FcRs — 181
   3. Activating and Inhibitory FcR Signaling Pathways — 183
   4. The Role of Activating and Inhibitory FcRs on Innate Immune Effector Cells — 185

5. Modulation of Antibody Activity	188
6. Activating and Inhibitory FcR Expression on DCs	190
7. FcγRIIB as a Master Regulator of Humoral Tolerance and Plasma Cell Survival	192
8. Summary and Outlook	195
Acknowledgments	196
References	196
Index	205
Content of Recent Volumes	211

See Color Plate Section in the back of this book

CONTRIBUTORS

Number in parentheses indicate the pages on which the authors' contributions begin.

John P. Atkinson (141)
Washington University School of Medicine, St. Louis, Missouri

Alexander U. Brandt (1)
Cecilie-Vogt-Clinic for Molecular Neurology, Charité-Universitaetsmedizin Berlin, 10117 Berlin, Germany; *and* Max-Delbrueck-Center for Molecular Medicine, 13125 Berlin, Germany

Douglas T. Fearon (103)
Wellcome Trust Immunology Unit, University of Cambridge, Medical Research Council Centre, Cambridge CB2 2QH, United Kingdom

Josephine Herz (1)
Cecilie-Vogt-Clinic for Molecular Neurology, Charité-Universitaetsmedizin Berlin, 10117 Berlin, Germany; *and* Max-Delbrueck-Center for Molecular Medicine, 13125 Berlin, Germany

David Kavanagh (141)
Washington University School of Medicine, St. Louis, Missouri

Falk Nimmerjahn (179)
Laboratory for Experimental Immunology and Immunotherapy, Nikolaus-Fiebiger-Center for Molecular Medicine, University of Erlangen-Nuremberg, Erlangen 91054, Germany; *and* Laboratory for Molecular Genetics and Immunology, Rockefeller University, New York, New York

Jeffrey V. Ravetch (179)
Laboratory for Molecular Genetics and Immunology, Rockefeller University, New York, New York

Anna Richards (141)
Washington University School of Medicine, St. Louis, Missouri

Jamie R. Schoenborn (41)
Molecular and Cellular Biology Graduate Program, University of Washington, Seattle, Washington

Volker Siffrin (1)
Cecilie-Vogt-Clinic for Molecular Neurology, Charité-Universitaetsmedizin Berlin, 10117 Berlin, Germany; *and* Max-Delbrueck-Center for Molecular Medicine, 13125 Berlin, Germany

Christopher B. Wilson (41)
Department of Immunology, University of Washington, Seattle, Washington

Frauke Zipp (1)
Cecilie-Vogt-Clinic for Molecular Neurology, Charité-Universitaetsmedizin Berlin, 10117 Berlin, Germany; *and* Max-Delbrueck-Center for Molecular Medicine, 13125 Berlin, Germany

CHAPTER 1

New Insights into Adaptive Immunity in Chronic Neuroinflammation

Volker Siffrin, Alexander U. Brandt, Josephine Herz, and **Frauke Zipp**

Contents

1. Introduction: Multiple Sclerosis Is a Heterogeneous Inflammatory Disease of the Nervous System — 2
2. Experimental Autoimmune Encephalomyelitis as a Model for MS — 3
3. Current Knowledge About Induction and Perseveration of Chronic Neuroinflammation — 5
   - 3.1. General considerations — 5
   - 3.2. $CD4^+$ T helper cells in chronic neuroinflammation — 5
   - 3.3. $CD8^+$ T cells in neuroinflammation: A never-ending controversy — 15
   - 3.4. B cells and antibody-mediated immune responses in chronic neuroinflammation — 17
4. Therapeutic Approaches to Chronic Neuroinflammation — 20
   - 4.1. General considerations — 20
   - 4.2. Current therapeutic concepts — 21
   - 4.3. New therapeutic concepts — 23
- Acknowledgments — 29
- References — 29

Cecilie-Vogt-Clinic for Molecular Neurology, Charité-Universitaetsmedizin Berlin, 10117 Berlin, Germany; and Max-Delbrueck-Center for Molecular Medicine, 13125 Berlin, Germany

Abstract

Understanding the immune response in the central nervous system (CNS) is crucial for the development of new therapeutic concepts in chronic neuroinflammation, which differs considerably from other autoimmune diseases. Special immunologic properties of inflammatory processes in the CNS, which is often referred to as an immune privileged site, imply distinct features of CNS autoimmune disease in terms of disease initiation, perpetuation, and therapeutic accessibility. Furthermore, the CNS is a stress-sensitive organ with a low capacity for self-renewal and is highly prone to bystander damage caused by CNS inflammation. This leads to neuronal degeneration that contributes considerably to the phenotype of the disease. In this chapter, we discuss recent findings emphasizing the predominant role of the adaptive immune system in the pathogenesis of chronic neuroinflammation, that is, multiple sclerosis (MS) in patients and experimental autoimmune encephalomyelitis (EAE) in rodents. In addition, we report on efforts to translate these findings into clinical practice with the aim of developing selective treatment regimens.

1. INTRODUCTION: MULTIPLE SCLEROSIS IS A HETEROGENEOUS INFLAMMATORY DISEASE OF THE NERVOUS SYSTEM

Multiple sclerosis (MS) is the most common chronic inflammatory disease of the central nervous system (CNS) in the western world, and leads to devastating disability in young adults with only limited treatment options available so far. Early in the disease course, most patients suffer from a relapsing-remitting course characterized by reversible neurological dysfunctions, such as impaired vision, paralysis, ataxia, and sensory deficits, and bladder, bowel, and sexual dysfunction. However, this phase of the disease is highly variable between different patients with regard to extent, duration, and clustering of the attacks. After ten years, about half of those patients will have entered a period of silent deterioration, less prominent attacks than in the first years after manifestation, and increasing cumulative disability. This gradual deterioration affects about 90% of MS patients after 20–25 years and typically comprises a decrease in lower extremity function and decline in ambulation. The clinical syndrome is caused by an autoimmune attack against the myelin sheath, histopathologically characterized by a complex picture of inflammation, demyelination, remyelination, axonal/neuronal damage (neurodegeneration) typically in subcortical but also cortical disseminated lesions. A similar disease can be induced in rodents by transferring myelin-specific lymphocytes. This, along with the fact that white matter plaques in the brain and spinal cord are its most obvious morphological signs, explains why for more than a century

MS was thought to be an inflammatory demyelinating disease. It only recently became evident that axons and their parent cell bodies, neurons, are also a major target in the CNS. In MS patients, early axonal pathology can be found, correlating with the number of infiltrating immune cells. Furthermore, in magnetic resonance imaging (MRI) in patients, "black holes" are the sign for complete tissue loss. Among other features, such as focal cortical thinning in the MRI and widespread gray matter involvement even at early disease stages, cortical lesions have been reported in patients, and these are reflected in frequently observed cognitive impairment. In principle, all aspects of immune reactions can be identified in MS lesions. Typical hallmarks of inflammatory plaques are CD4 and CD8 T cells, activated macrophages and microglia, and antibody and complement deposition. These static views derived from histopathology have led to extensive efforts to differentiate the crucial processes that initiate and perpetuate chronic neuroinflammation from those that are protective or potentially pure epiphenomena.

2. EXPERIMENTAL AUTOIMMUNE ENCEPHALOMYELITIS AS A MODEL FOR MS

To investigate the pathogenic mechanisms in MS, neuroimmunologists have generated disease models mimicking the human disease. The induction of experimental autoimmune encephalomyelitis (EAE) requires subcutaneous immunization with myelin proteins or peptides dissolved in proinflammatory adjuvants, derived from heat-inactivated *Mycobacterium tuberculosis*, and in some models additional intravenous injections with toxin from *Bordetella pertussis*. This strategy is called active EAE. Alternatively, encephalitogenic T cells derived from immunized animals can be isolated, *in vitro* expanded and transferred to naive animals, inducing a clinically and histopathologically similar disease called adoptive transfer or passive EAE.

While the priming of T cells in active EAE seems to be similar in all rodent EAE models, the effector phase in the target organ and the modes of chronification of the disease process seem to depend on strain- and species-specific properties. The Lewis rat, for instance, shows a monophasic disease with massive inflammation, but only a minor degree of demyelination. Several murine EAE models, mostly using the myelin basic protein (MBP) as inducing myelin protein, also show a monophasic but highly demyelinating EAE course (Nogai *et al.*, 2005) which resembles the human disease "acute disseminated encephalomyelitis" (ADEM), which has been shown to be associated with recent infections or vaccinations (although no controlled study exists for novel recombinant vaccines).

Both disease entities are alike in terms of histopathologic hallmarks such as short-lived degree of inflammation and extensive demyelination, but only minor axonal/neuronal degeneration and a high potential of complete resolution (Kawakami et al., 2004; Menge et al., 2005). The SJL/J mouse strain has found broad acceptance as being the closest model of the human situation in MS, as these mice develop a relapsing-remitting disease process upon induction with proteolipid protein (PLP) or adoptive transfer of encephalitogenic $CD4^+$ T cells recognizing the PLP peptide 139–151. As in MS, histopathologic findings include T cell and macrophage-dominated inflammation, extensive demyelination, axonal and neuronal damage (Aktas et al., 2005; Diestel et al., 2003) early on in the disease process (our unpublished data). The typical relapsing-remitting disease process has been shown to depend on epitope spreading, which means that the main T cell response in the relapse is directed against a different CNS myelin antigen in comparison to the one the animals were originally immunized with (McMahon et al., 2005). In support of these findings, Sercarz and colleagues showed that the observed immune dominance of certain antigens is mainly confined to the inductive phase of EAE, while so-called "cryptic" epitopes cannot easily induce the disease, but are associated with acute exacerbations (Lehmann et al., 1992). Furthermore, the priming of these relapse-inducing T cells is believed to occur in the CNS, as the earliest proliferation of these cells has been located *in vivo* in the CNS. This implies that naive T cells can gain access to the diseased CNS in EAE where efficient activation may lead to exacerbations. In this case, T cell priming has proved to be most efficient if performed by CNS derived "dendritic cell-like" antigen-presenting cells (APC) in the context of endogenously processed peptide in two distinct relapsing-remitting murine models (McMahon et al., 2005; McRae et al., 1995).

Another aspect of chronic neuroinflammation can be investigated in the C57Bl/6 mouse strain, in which immunization with the myelin oligodendrocyte glycoprotein (MOG) or its immunodominant epitope 35–55 induces a severe attack followed by incomplete recovery and a secondary-progressive stage of silent deterioration, as found in the later stages of MS (Mendel et al., 1995).

However, the main criticism of EAE models concerns the necessity of violent immunization. Considerable progress in this direction has been achieved by two competing groups who have recently published their development of a spontaneous opticospinal EAE in C57Bl/6 mice carrying both an MOG-specific T cell receptor and a MOG-specific B cell receptor. While the single-transgenic animals rarely develop spontaneous disease, double transgenics show a chronic progressive disease process (Bettelli et al., 2006a; Krishnamoorthy et al., 2006).

3. CURRENT KNOWLEDGE ABOUT INDUCTION AND PERSEVERATION OF CHRONIC NEUROINFLAMMATION

3.1. General considerations

There has been much debate about the importance of the different T cell subsets, as there is conflicting evidence about the numbers of $CD4^+$ and $CD8^+$ T cells and their ratio in the histopathology of active MS lesions (Babbe *et al.*, 2000; Sobel, 1989; Traugott *et al.*, 1983a; Wucherpfennig *et al.*, 1992). Influenced by the genetic association of MHC class II genes and MS (Haines *et al.*, 1996), and by the animal model for the human disease, $CD4^+$ T cells have been in the spotlight for many years. There is compelling evidence that $CD4^+$ T cells recognizing antigenic epitopes from the myelin sheath can, in many species, initiate a relapsing-remitting disease course that mimics the human disease both clinically and histologically. Findings in this and other autoimmune disease models led to the hypothesis that autoimmune disorders arise if autoaggressive immune responses generate self-recognizing T cells that attack the target organ. This concept traditionally centered around the hypothesis of self/nonself-discrimination originating in a highly specific T cell receptor recognition and thymic deletion of potentially self-reactive T lymphocytes. This rather simplistic approach suggested that autoimmune responses could arise if, in the context of genetic susceptibility, specific self-reactive T cells escaped deletion in the thymus and were reactivated by endogenous or foreign antigens via molecular mimicry. However, we are now beginning to understand the complexity and dynamics of immune responses as a tightly regulated system in which self-recognizing sentinels are part of the normal T cell repertoire that appears to be crucial for both self-tolerance and defense by immune modulatory functions, such as anergy and active suppression via regulatory T cells (Treg). In this concept, autoimmunity is regarded as an immune disregulation in which, despite redundant safety regulations, self-recognizing T cells are shifted toward a proinflammatory phenotype capable of initiating an autoimmune attack. Consequently, understanding T cell differentiation and regulation *in vivo* is now the key to gaining deeper insights into autoimmune phenomena.

3.2. $CD4^+$ T helper cells in chronic neuroinflammation

Upon induction of active EAE by immunization with the myelin peptides in adjuvant, the priming phase of T cells is followed by a sequential appearance of CNS-specific T helper cells (Th) in the secondary lymphoid organs and in extralymphoid tissues (as detected by IL-2/IFN-γ/IL-17-Elispot). Most strikingly, CNS-specific T cells are not detectable in the brain until (pre-)onset of clinical signs (Hofstetter *et al.*, 2005, 2007). In clinically ill animals, there is a close correlation of cell numbers with clinical EAE scores. Long-term monitoring of antigen-specific T cells suggests a peripheral pool of antigen-specific effector cells that are recruited to the brain and then decrease over a period of two to three months. The absolute number of neural antigen-specific T cells in the CNS therefore never exceeds the absolute number outside the brain, and these peripherally situated T cells have a similar potential to the CNS T cells for inducing the disease in adoptive transfer EAE. This refutes the hypothesis that myelin-specific T cells undergo a significant avidity or cytokine profile enrichment in the CNS. Interestingly, the highest frequencies of myelin-specific T cells can be detected outside the CNS before onset, culminating in the CNS at onset, and disappearing with complete resolution of clinical symptoms in the CNS. Focusing the ratio of antigen-specifically recruited Th cells versus unspecifically recruited cells in full-blown EAE, less than one in a thousand T cells seems to functionally recognize the disease-inducing antigen (determined by Elispot technique), as confirmed by other groups employing different methods (Steinman, 1996; Targoni *et al.*, 2001). These findings differ from results derived from adoptive transfer EAE models in which a much higher number of specific T cells was identified in the target organ. Moreover, there seems to be a highly synchronized pattern of T cell trafficking from the periphery into the target tissue (Flugel *et al.*, 2001). These differences to the active EAE model can be explained by the artificial *in vitro* expansion period before transfer and the lack of *de novo* priming in the secondary lymphoid organs of the host. The essence of Hofstetter's findings is the insight that a small number of antigen-specific Th cells seems to be sufficient to initiate and perpetuate full-blown EAE with resolution of clinical and histopathologic disease upon disappearance of these cells. These findings gave rise to the question of key features leading to the immune attack of myelin-specific Th cells against the CNS.

3.2.1. Th1 versus Th17

For many years, EAE was believed to depend on myelin-specific $CD4^+$ Th1 cells attacking the CNS, as targeted deletion of the interleukin 12 (IL-12) p40 gene or neutralization of the IL-12 p40 subunit with a monoclonal antibody protected against EAE. However, Cua and colleagues showed that IL-23 and not IL-12, as first suspected, was the critical factor in the development of EAE (Langrish *et al.*, 2005). This refueled the controversial debate over why there was normal or even increased disease susceptibility in the absence of other Th1-relevant elements, such as the IL-12 p35 subunit, interferon (IFN)-γ, IFN-γ receptor, or STAT1. Furthermore, IL-12-induced classic Th1 cells characteristically expressing IFN-γ demonstrated both proinflammatory and immunoregulatory functions in different models (Becher *et al.*, 2003; Feuerer *et al.*, 2006; Ivanov *et al.*, 2006; Park *et al.*, 2005).

The final resolution of this paradox was achieved by Cua and colleagues, who showed that IL-12 and IL-23, both heterodimers belonging to the IL-12 family, share the p40 subunit and are deficient in both IL-12 and IL-23. These researchers provided evidence that animals deficient in the IL-23 p19 subunit show a similar reduction in EAE susceptibility to IL-12/IL-23 p40 knockout mice. There was a complete absence of clinical and histological signs of EAE in these animals, while local application of IL-23 restored the potential to evolve EAE (Cua et al., 2003; Ivanov et al., 2006). The IL-23-dependent step in chronic neuroinflammation was found in the generation of the Th17 cell lineage. Th17 cells, originally discovered in the synovial fluid of patients with Lyme arthritis, have been shown to contribute to different autoimmune diseases by the highly proinflammatory cytokines IL-17, IL-22, GM-CSF, and tumor necrosis factor (TNF) (Infante-Duarte et al., 2000; Liang et al., 2006). IL-17 depletion by antibody-mediated neutralization or using knockout animals resulted in a significant reduction of EAE susceptibility. In line with these findings, Langrish et al. (2005) describe a much higher potential of Th17 CNS antigen-specific cells to induce EAE. Accordingly, different group report enhanced IL-17 and IL-23 expression by human PBMC and mononuclear cells in the cerebrospinal fluid (CSF) of patients suffering from MS, which strongly suggests involvement of these pathways in the human disease as well (Matusevicius et al., 1999; Vaknin-Dembinsky et al., 2006).

Further insight into the progeny of the Th17 lineage was added with the discovery that the pleiotropic cytokines IL-6 and TGF-β were crucial for commitment to the Th17 lineage, while IL-23 might be necessary to consolidate IL-17 production and effector function of memory Th cells. As with the Th1, Th2, and Treg lineages, a distinct key transcription factor, the orphan nuclear receptor RORγt, was identified to initiate the differentiation of the Th17 lineage. In mice with a targeted disruption of the RORγt gene, clinical symptoms of EAE were delayed and mild, contrasting with extensive inflammatory infiltrates residing in the spinal cord and expressing high amounts of IFN-γ (Ivanov et al., 2006). This is compelling evidence that clinical disease severity is not only dependent on the quantity of infiltrating cells and the amount of proinflammatory cytokines but also on the differentiation status of disease-initiating CNS-specific CD4$^+$ cells (Th17). These findings add important knowledge to our understanding of what qualities may turn a CNS antigen-reactive T cell encephalitogenic.

3.2.2. T cell trafficking to the CNS

For decades, it has been known that the production of proinflammatory cytokines by CNS antigen-specific CD4$^+$ cell clones is one of the conditions necessary to initiate neuroinflammation. Another crucial step is for the immune cells to actually reach the CNS by transmigrating from the blood into the target tissue. Generally, activated memory T lymphocytes are able to enter the CNS irrespective of their antigen specificity, whereas naive lymphocytes fail to enter the healthy CNS. T lymphoblasts rapidly appear in the CNS tissue upon active transfer regardless of MHC compatibility, T-cell phenotype, T cell receptor gene usage, or antigen specificity. In the rat, the peak of cell infiltration lies between 9 and 12 h after transfer and is about 100 times smaller in terms of cell numbers than in non-CNS sites (Hickey, 1999; Hickey et al., 1991). There are many components involved in this process of transmigration in health and disease (Fig. 1.1).

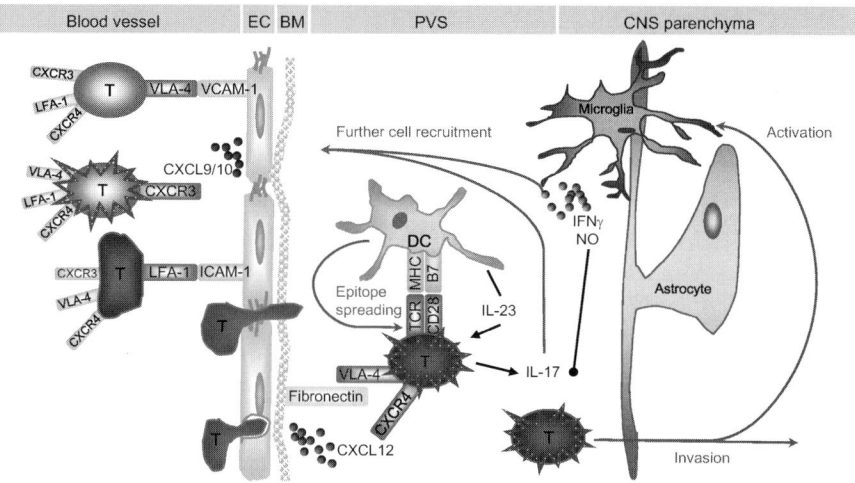

FIGURE 1.1 Multistep process of T (T) cell extravasation, perivascular restimulation, cell recruitment, and penetration of CNS parenchyma. Following profound functional changes in the periphery including upregulation of adhesion molecules and chemokine receptors, T cells adhere to and roll along the endothelium via adhesion molecules, such as VLA-4, get activated via chemokines (e.g., CXCL9 and 10), which induce G-protein–mediated promotion of further adhesion via ICAM-1–LFA-1-interactions leading to transmigration along a cytokine gradient via transcellular or paracellular diapedesis through the endothelium. Following transmigration T cells get first in the perivascular space (PVS) where interaction with dendritic cells (DC) via antigen recognition and costimulation allows penetration of the CNS parenchyma through the glia limitants. Additionally, cytokine release by DCs, particularly IL-23, supports maintenance of T cell differentiation status (Th17), resulting in the release of further proinflammatory effector cytokines such as IL-17. Additional chemotactic factors secreted by microglia activated by CNS infiltrated T cells promote further inflammatory cell recruitment and thus enhancement of inflammation. These processes are also compromised by counterregulatory processes, including IFN-γ and nitric oxide (NO) release by activated microglia. In addition to the release of soluble proinflammatory factors, DCs have been described as the preferred PVS cell population responsible for epitope spreading, an essential mechanism in inducing relapse phases of disease, suggesting that also naive T cells gain excess to the diseased CNS where efficient activation may lead to exacerbations. (See Plate 1 in Color Plate Section.)

3.2.2.1. Adhesion molecules A model for sequential extravasation in the postcapillary venules has been proposed consisting of tethering and rolling of lymphoblasts on the endothelium mediated by selectins (E-, P-, L-selectins) and Very Late Antigen-4 (VLA-4). These adhesion molecules interact with their carbohydrate ligands followed by chemokine receptor-induced activation of integrins (CAMs), which subsequently results in close adhesion to the endothelium. The last step consists of a chemokine driven diapedesis of the lymphoblast through the endothelium or in between two endothelial cells (Ransohoff *et al.*, 2003).

Interfering with selectin function has proved to have neither a beneficial nor a detrimental effect on T cell recruitment to the CNS in EAE (Engelhardt *et al.*, 1997, 2005), while the adhesion molecule VLA-4, a member of the α4-integrin family, has been defined as essential factor for T cells to enter the CNS and for CNS-specific T cell clones to transmit EAE (Keszthelyi *et al.*, 1996; Yednock *et al.*, 1992). Interestingly, some non-encephalitogenic T cell clones lacking VLA-4 have been rendered encephalitogenic by administering pertussis toxin, which is believed to "open the blood–brain barrier" but most probably has pleiotropic effects (Kuchroo *et al.*, 1993). Accordingly, blockade of VLA-4 by neutralizing nondepleting antibodies against the α4 integrin prevented CNS inflammation in different murine EAE models by drastically reducing the number of activated T cells reaching the target tissue; this was effective even if initiated in late stages of the disease. These findings led to the development of one of the most powerful but high-risk treatment strategies currently available, whose benefits and hazards for MS patients will be discussed in detail in the final chapter (Brocke *et al.*, 1999; Miller *et al.*, 2003; Niino *et al.*, 2006).

While the case of VLA-4 seems straightforward enough, the case of LFA-1–ICAM-1 (Leukocyte function-associated molecule-1, intercellular adhesion molecule-1) interaction is far less so. Functional neutralization of LFA-1 or ICAM-1 in EAE proved to be beneficial in some studies and detrimental in others, and even knockout animals revealed controversial results. Different activities of antibodies can sometimes be explained by distinct *in vivo* effects of the different clones, as partially agonistic, depleting, or differential steric properties may result from binding different epitopes. But opposing data in knockout animals are much more difficult to interpret and lead us to suspect that at least one of the mouse strains used was not a complete null mutation. The conclusion to be drawn at this point is that this interaction probably exceeds a pure binding process, leading also to some kind of activation and cell conditioning for diapedesis (Bullard *et al.*, 2007; Kobayashi *et al.*, 1995; Samoilova *et al.*, 1998; Welsh *et al.*, 1993).

3.2.2.2. Chemokines As mentioned above, T cells trying to enter the CNS require chemotaxis in order to leave the circulation. The interaction of chemokines with their chemokine receptors represents an important step between rolling/tethering and adhesion by activating the T cells and, most probably, offering them firm adhesion (Engelhardt and Ransohoff, 2005). Wekerle and coworkers showed that T cells require profound functional changes to allow them to infiltrate the CNS, which provides a reasonable explanation for the latency period after adoptive transfer of highly activated encephalitogenic T cells until clinical signs appear. The upregulation of different chemokines and chemokine receptors (CCR1/2/3/5/7, CXCR4) was particularly extensive before the recruitment of inflammatory cells to the CNS (Flugel et al., 2001). On the receptor side, T cells in noninflamed CSF and in MS predominantly express CCR7 and CXCR3, while monocytes in MS lesions express CCR1, CCR2, and CCR5 (Rebenko-Moll et al., 2006). Accordingly, studies in EAE have detected a multitude of chemokine attractants, for example, CXCL9–11 associated with CXCR3, CCL19/21 with CCR7, CCL2 with CCR2, and CXCL12 with CXCR4.

CCR1-deficient mice show an attenuated EAE course which corresponds with reduced monocyte infiltration in EAE lesions in keeping with findings of a specific CCR1 antagonist that reduced leukocyte infiltration and severity in EAE (Eltayeb et al., 2003; Rottman et al., 2000). Unfortunately, CCR1 blockade did not provide any significant beneficial effect in a Phase II clinical trial in patients suffering from relapsing-remitting MS, although it clearly modulated the activity of circulating monocytes (Zipp et al., 2006). These findings reflect the higher complexity of the human disease and the more chronic disease process in which redundancy of different chemokine receptors on effector monocytic cells (e.g., CCR2, CCR5) may develop as a consequence of ongoing T cell activation, impaired immune regulation, and disrupted feedback loops.

CXCR3 is preferentially expressed on Th1 cells generated *in vitro* and *in vivo* and has been shown to promote differential trafficking of IFN-γ-producing Th1 cells into sites of inflammation. Somewhat unexpectedly, CXCR3-deficient mice developed exaggerated EAE, refuting the hypothesis that CXCR3 deletion might lead to reduced transmigration of encephalitogenic T cells. Indeed, in knockout animals IFN-γ-producing T cells in the CNS were significantly lower in numbers than in wild-type controls without obvious differences in cell distribution or absolute numbers of infiltrating cells (Liu et al., 2006a). In the Th17 era, these findings could be a result of uninhibited Th17 action caused by reduced migration of IFN-γ-producing T cells into the CNS; this would certainly explain the association between enhanced clinical symptomatology, comparable infiltrates, and lower IFN-γ production. Furthermore, there is convincing evidence that *in vivo* chemotaxis is not as absolute and dichotomous as studies with *in vitro* differentiated Th1 and Th2 cells suggest. Hamann and colleagues have shown that *in vivo* generated cytokine-producing T cells exert tissue and infection-dependent differential chemotactic behavior. On the one hand, the CXCR3 ligands CXCL9 and CXCL10 robustly attracted IFN-γ-producing T cells that were generated by murine influenza infection in the lung. On the other hand, IL-10-producing and especially IL-10/IFN-γ expressing CD4$^+$CXCR3$^+$ T cells also strongly migrated toward their ligands in this Th1-mediated disease model (Debes *et al.*, 2006). Extrapolating these findings to Th1-mediated autoimmune diseases, it seems that not only proinflammatory Th1 cells but also the recruitment of potential Treg subsets might depend on Th1 cell-attracting chemokines, making interference with these complex processes unpredictable, especially in the human situation.

For pathological purposes, it is important to differentiate the chemotactic cues necessary for extravasation from those that are necessary in the target tissue for local homing to immune relevant sites in the CNS, which can also be interpreted as trapping cells in areas of interest. The homeostatic T cell chemokine CXCL12, together with its receptor CXCR4, is also believed to regulate baseline lymphocyte trafficking, but additionally plays a crucial role in fetal hematopoiesis as well as in cell proliferation and survival (Sallusto and Mackay, 2004). CXCL12 is one of the top candidates for homing of T cells into the CNS. This CXCR4 ligand is highly expressed by endothelial cells with strong polarization to the basolateral surface—the parenchymal side of the vessel. Blockade of CXCR4 by a specific small molecule antagonist led to enhanced EAE severity marked by more diffuse infiltration of inflammatory cells compared to typical vascular cuff forming in sham-treated animals without significant differences in absolute cell numbers. The clinical outcome can be explained by a bigger area of inflammation and also a higher degree of demyelination, though apart from a slight increase in the expression of proinflammatory cytokines in the tissue there were no gross differences in recruited cell populations. The authors interpret their data in terms of a tissue protective effect of CXCL12, which binds the receptor carrying proinflammatory cells to the area around the vessel (McCandless *et al.*, 2006). Hamann and coworkers showed that CXCL12-dependent migration seems to be nonselective for all types of effector/memory T cells in different organs of the naive mouse and also in different infectious diseases models. This suggests a vital role for CXCL12–CXCR4 driven chemotaxis in lymphocyte homing to immune relevant sites in different target organs, and presumably a central role for immune surveillance and antigen recognition *in vivo* (Debes *et al.*, 2006).

Finally, it should be kept in mind that chemokine encoding genes are switched on nonspecifically upon cell activation by different stimuli (Ubogu *et al.*, 2006), and chemokine receptors are not always involved in chemotaxis but can also deliver proapoptotic signals and sometimes function as scavenging receptors without any chemotactic function at all (D'Amico *et al.*, 2000; Lasagni *et al.*, 2003). It has been shown that the correlation of chemokine-driven migration may be variable in different inflammatory conditions. For example, IFN-γ-producing CD4$^+$ T cells migrated robustly toward the CXCR3 ligands in an *ex vivo* chemotaxis assay if isolated from influenza infected mice, but they responded poorly to their classic ligands if generated in parasitic infection (Debes *et al.*, 2006). In light of these limitations, caution should be exercised in transferring *in vitro* acquired knowledge to our autoimmune disease model, and especially to the human condition.

3.2.3. Containment of CNS inflammation by apoptosis

Apoptosis is, on the one hand, an important mechanism for the containment and reversal of CNS inflammation, but on the other hand, it has been shown to contribute to tissue damage, which is of high significance in low-proliferating tissue such as that of the CNS. In contrast to necrosis, apoptosis represents programmed cell death with a defined process of cell degradation, induced by specific death receptor/ligand systems without reactive inflammation. In chronic neuroinflammation, the research focus has been on the roles of the death ligand members of the TNF/NGF (nerve growth factor) superfamily, the CD95 (APO-1/Fas) ligand, and the recently characterized TNF-related apoptosis-inducing ligand (TRAIL).

CD95L, the ligand of CD95, is expressed on the surface of effector T lymphocytes, which enables them to target neighboring cells carrying the respective receptor. CD95 has a central position in preserving homeostasis of the immune system, leading to autoimmune phenomena on targeted deletion. In MS, CD95 has been shown to be an important factor for deregulated activation-induced cell death (AICD), which seems to be responsible for the survival of activated autoreactive T cells, most probably via enhanced levels of the soluble form of the apoptosis-inhibiting CD95 in the serum and CSF of MS patients (Bieganowska *et al.*, 1997; Zipp *et al.*, 1999). The importance of the CD95 system in the rodent model seems somewhat contradictory. On the one hand, different EAE models show a regulatory, suppressive effect of the CD95 system in EAE, not only leading to enhanced disease severity but also to absence of remission; on the other hand, these findings were concordant with other investigations showing CD95L expression by neurons which induced apoptosis of encephalitogenic T cells *in situ* (Aktas *et al.*, 2006; Flugel *et al.*, 2000). However, further investigations lend additional importance to CD95/CD95L by suggesting that animals deficient in CD95 in the CNS are protected from disease, most probably by the lack of apoptosis in oligodendrocytes, which might be one of the driving forces in the inflammatory cascade toward full-blown EAE (Hovelmeyer *et al.*, 2005).

This study also identified TNF-R1-mediated apoptosis in oligodendrocytes, supporting the idea of TNF-α as an ambiguous player in chronic neuroinflammation. Blockade of TNF-α is known as a prototypical proinflammatory cytokine, and blockade of TNF-α signaling has successfully been used to treat other autoimmune disorders. Even in EAE models, TNF-α blockers have had a protective effect, which led to the hope for a new therapeutic option in MS. Surprisingly, clinical trials with an antibody against TNF-α showed even exacerbated disease leading to the induction of relapses, and enhanced silent inflammatory activity in MRI (Anonymous, 1999). Indeed, further studies with TNF-α-deficient animals showed a chronic form of EAE with a delayed onset but pronounced accumulation of effector-memory Th cells, indicating an aborted contraction phase (Kassiotis and Kollias, 2001).

Another apoptotic mechanism involved in chronic neuroinflammation is TRAIL, which has been discussed as a protective factor against autoimmunity as it was shown to inhibit the proliferation of activated T cells and was associated with a better therapeutic response (Lunemann *et al.*, 2002). EAE studies supported these findings as the disease was significantly worse upon peripheral blockade of TRAIL. However, as in the CD95/95L system, the apoptotic effect of TRAIL was not restricted to the immune system, suggesting a death-inducing effect on neurons and oligodendrocytes if expressed on activated T cells in the CNS (Aktas *et al.*, 2005; Nitsch *et al.*, 2000).

3.2.4. Regulatory Th cells as key players in the control of chronic neuroinflammation

Immune tolerance is needed to preserve immune homeostasis and to prevent autoimmunity while ensuring effective host defense. Active mechanisms of immune regulation have been largely attributed to specialized T cell subsets termed Treg, which can be roughly divided into thymus derived and acquired Treg. Thymus derived "naturally occurring" $CD4^+CD25^+$ T cells (nTreg) are generated in the process of T cell maturation, and express the distinct lineage marker and transcription factor FoxP3. $CD4^+CD25^+FoxP3^+$ T cells have proved crucial for self-tolerance since targeted gene deletion leads to early fatal multiorgan lymphoproliferative autoimmune disease (Fontenot *et al.*, 2005). In EAE, nTreg have been associated with protection and recovery from clinical symptoms. Anderton and colleagues showed that IL-10 producing $CD4^+CD25^+FoxP3^+$ T cell accumulation in the CNS was associated with, and necessary for, clinical remission, which corresponded to a high *in vitro* regulatory potency of Treg recovered from the CNS (McGeachy *et al.*, 2005). Moreover, strain-specific EAE resistance in naive B10.S mice was associated with a higher ratio of CNS antigen-specific $CD4^+CD25^+$ Treg compared to the MHC class II-congenic EAE susceptible SJL/J mice, while absolute numbers of self-reactive $CD4^+$ T cells were similar. Accordingly, EAE resistance in the B10.S strain was overcome by depletion of $CD4^+CD25^+$ T cells (Reddy *et al.*, 2004). Furthermore, C57BL/6 animals immunized with MOG normally develop resistance to reimmunization after a first attack, but this was overcome by postrecovery depletion of $CD25^+$ T cells (McGeachy et al., 2005).

Although little is known about the progeny of the numerous $CD4^+CD25^+FoxP3^+$ Treg cells accumulating in the inflamed CNS, both expansion of nTreg and *de novo* differentiation generally seem possible (Kretschmer et al., 2005; Papiernik et al., 1998). As FoxP3 is inducible upon transgenic TGF-β-overexpression *in vitro*, the hypothesis has been formed that TGF-β production might be responsible for successful Treg induction from naive CD4 precursors, as well as its expansion and maintenance. This correlates well with *in vivo* findings, showing that TGF-β induces antigen-specific $CD4^+CD25^+FoxP3^+$ Treg expansion (Peng et al., 2004). Indeed, transfer of *in vitro* generated MOG-specific and TGF-β-producing $CD4^+$ cells protected recipients from clinical symptoms in actively induced MOG-EAE in C57BL/6 (Carrier et al., 2007), and TGF-β-neutralization abolished remission in active EAE in the SJL/J strain (Zhang et al., 2006). One recent *in vitro* coculture study suggests that neurons, in their cross talk with encephalitogenic T cell clones, might also be able to push the development of the Treg lineage, dependent on neuronal B7 costimulation and TGF-β expression (Liu et al., 2006b). In contrast with these TGF-β-hypotheses, which regard TGF-β as an anti-inflammatory cytokine and as the driving force for Treg development in EAE, there are several reports on ambivalent effects of TGF-β. It has been shown, for instance, that TGF-β, which is able to induce Treg *in vitro*, is also capable of inducing the highly pathogenic Th17 subset *in vitro* and *in vivo* in the context of an inflammatory milieu. This effect was dramatically illustrated by exacerbated EAE in C57BL/6 mice with a transgenic IL-2-promoter-dependent TGF-β-overexpression. Increased clinical outcome and mortality in this model were associated with an imbalance of $CD4^+CD25^+FoxP3^+$ Treg cells versus Th17 subset (Bettelli et al., 2006b). In support of these findings, $CD4^+$ T cell-restricted targeted deletion of the TGF-βRII led to resistance to EAE (Veldhoen et al., 2006b). In summary, TGF-β, while abrogating Th1 and Th2 development in inflamed and noninflamed environments, induces a regulatory phenotype only in the absence of inflammation and promotes the highly pathogenic Th17 subset in the presence of IL-6 (Veldhoen et al., 2006a). Therefore, any kind of therapeutic approach involving TGF-β or TGF-β-producing cell subsets should be carefully evaluated with regard to the potential generation of pathogenic Th17 cells in the realm of inflammation. In addition, effector mechanisms of Treg in the *in vivo* situation still remain unclear, despite the discussion of soluble factors, such as IL-10 and TGF-β, and despite the fact that immune modulatory surface molecules such as CTLA-4 (cytotoxic T lymphocyte antigen 4), membrane-bound TGF-β, and GITR (glucocorticoid-induced tumor necrosis factor receptor) seem to have a role in at least some models.

Rapid progress in understanding the central position of $CD4^+CD25^+FoxP3^+$ in immune modulation has meant that another Treg subset, characterized by high IL-10 production upon antigen stimulation, has lately been given short shrift in immunology. These acquired, antigen-driven Treg, called Tr1 cells, and arising during immune responses, were first described in bone marrow transplantation tolerance (Roncarolo et al., 1991) and were also successfully induced *in vitro* by complex protocols. Later, O'Garra and colleagues described the slightly distinct IL-10-Treg ($FoxP3^-$), which they derived from naive T cells by repetitive antigenic stimulation in the presence of immunosuppressive drugs. These IL-10-Treg were therapeutically effective in preventing EAE in an antigen-dependent manner at the site of inflammation (Barrat et al., 2002; Vieira et al., 2004). Additionally, our previous work indicates an induction of an IL-10-producing regulatory phenotype by application of the HMGCoA reductase inhibitor atorvastatin *in vitro* and *in vivo*, leading to a protective effect in different models of EAE (Aktas et al., 2003; Waiczies et al., 2005) (Fig. 1.2).

3.3. $CD8^+$ T cells in neuroinflammation: A never-ending controversy

The debate about the role of $CD8^+$ T cells in chronic neuroinflammation was started by histopathologic studies of active and chronic MS lesions, showing inconsistent results for both numbers and distribution in MS lesions. In some patients' biopsies, large numbers of oligoclonally expanded $CD8^+$ T cells prevailed within active demyelinating lesions (Babbe et al., 2000), while others presented with scarce, often only marginal infiltration of chronic active MS lesions by $CD8^+$ T cells (Traugott et al., 1983b). However, not only the numbers but also the significance of the presence of $CD8^+$ T cells in MS lesions has been heavily debated. On the one hand, there is evidence for a proinflammatory role from different passive transfer EAE models in which cytotoxic, myelin-specific $CD8^+$ T cells initiated severe CNS inflammation IFN-γ- and MHC-I dependently (Huseby et al., 2001; Sun et al., 2001). On the other hand, from a histological point of view, $CD8^+$-induced EAE differed significantly from conventional EAE as there were enhanced and prolonged meningeal involvement, extensive neutrophil recruitment, and signs of necrotic cell damage, all of which suggests a somewhat unspecific cytotoxic effect, as seen in other animal models of bystander damage (Banerjee et al., 2004). In sum, the case for $CD8^+$ T cells as the initiating force in chronic neuroinflammation does not seem very strong, but cytotoxic T cells might considerably contribute to axonal and neuronal damage. Furthermore, in Sun's study, absence of $CD4^+$ T cells in RAG-deficient animals delayed disease onset extremely, suggesting an important role

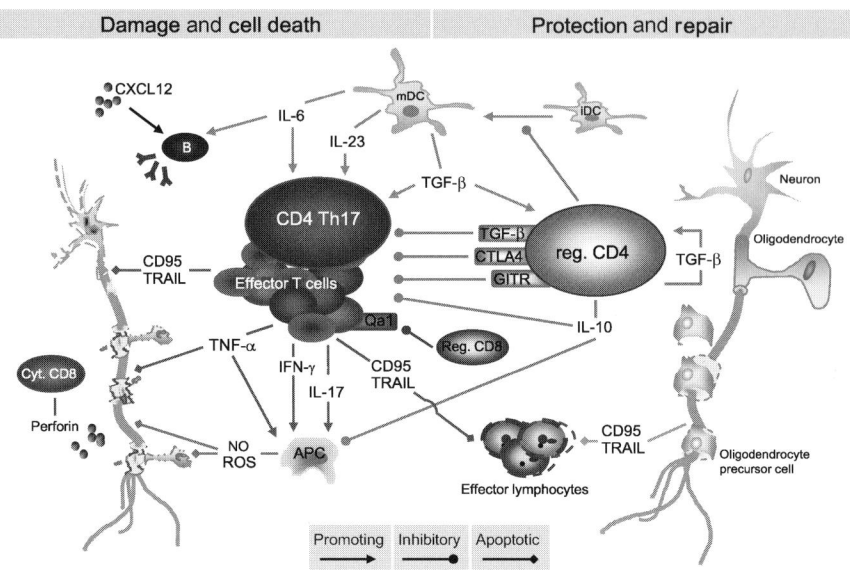

FIGURE 1.2 The pathogenesis of chronic neuroinflammation. Classically chronic neuroinflammation was regarded as CD4 Th1-mediated autoimmune disease. More recent data rather point to a disregulation of two dichotomous T cell subsets: the highly pathogenic CD4 Th17 cells and the recovery-mediating CD4 Treg. TGF-β, which has been associated with Treg for a long time, also promotes the differentiation of Th17 cells in the context of antigen-specific (re)activation by mature dendritic cells (mDC) secreting the proinflammatory cytokines IL-6 and IL-23. Uninhibited Th17 cells induce a massive recruitment of effector T cells and APC such as macrophages, B cells, and DC to the target organ. The release of proinflammatory cytokines such as IFN-γ, TNF-α, and IL-17 promotes CNS inflammation and tissue injury either by directly targeting neurons or indirectly via APC activation, which releases neurotoxic compounds like nitric oxide (NO) and reactive oxygen species (ROS). Further effector mechanisms involved in damage and cell death include perforin-mediated cytotoxicity and CD95 and TRAIL induced oligodendrocyte and neuronal cell death. However, CD95 and TRAIL-mediated apoptosis is also directed against effector lymphocytes, promoting reversal of CNS inflammation. Despite controversial discussions regarding the effector mechanisms of Treg, membrane-bound TGF-β, CTLA-4, and GITR as well as soluble released IL-10 and TGF-β are suggested to be involved in suppression of proinflammatory cells. Additional targets of Treg are DC in which they presumably induce or preserve an immature phenotype (iDC). Some reports also indicate a regulatory role for CD8 T cells via Qa1-mediated mechanisms. Due to dominance of either pathogenic CD4 Th17 or Treg in distinct disease phases, the balance is shifted periodically during the disease course resulting in phases of demyelination and neuronal/axonal degeneration but also phases of remyelination and regeneration. (See Plate 2 in Color Plate Section.)

for endogenous $CD4^+$ T cells in the initiation of clinical disease, which is not the case in a model of $CD8^+$ deficiency (Linker et al., 2005). Interestingly, this study suggested a beneficial role of $CD8^+$ T cells in conventional active EAE. Gold and coworkers describe a severe and mostly lethal EAE in β-2 microglobulin-deficient mice, which lack $CD8^+$ T cells and NK-T cells. They found typical EAE lesions with increased numbers of $CD4^+$ T cells associated with typical but also enhanced macrophage recruitment and equivalent demyelination and axonal damage, and this correlates well with "classic" but enhanced, $CD4^+$-affluent EAE pathology. This report supports a regulatory function of $CD8^+$ T cells on $CD4^+$ T cells which might be due to a Qa-1-dependent suppressor effect of $CD8^+$ T cells. Qa-1, a homologue of HLA-E in the human system and considered as atypical MHC, forms heterodimers with β-2 microglobulin, which is favorably expressed by activated $CD4^+$ T cells as well as B cells. Further evidence of a vital role for $CD8^+$ T cell transmitted $CD4^+$ T cell suppression was provided by Cantor and coworkers, who demonstrated elegantly that PLP-induced tolerance in C57Bl/6 could be overcome in Qa-1-deficient mice, while wild-type mice did not develop EAE. If Qa-1 was restituted in the $CD4^+$ T cells by retroviral transfer, those cells regained susceptibility to $CD8^+$-suppression. This mechanism was dependent on $CD8^+$ T cells being present during the $CD4^+$ priming process, and could not be surrogated by nonspecific, activated $CD8^+$ T cells; the specificity of the suppressor $CD8^+$ T cells involved has so far not been revealed (Hu et al., 2004).

3.4. B cells and antibody-mediated immune responses in chronic neuroinflammation

Persistent intrathecal immunoglobulin (Ig) synthesis is a key feature in the CSF of the majority of MS patients (Thompson et al., 1979) and contributes to diagnostic decision making. In fact, two distinct CSF parameters are important in the context of MS: the oligoclonal bands (OCB) and the Measles/Rubella/[Herpes] Zoster (MRZ) reaction. The OCB represent a distinct pattern of Ig in the CSF which cannot be found in the peripheral blood. They have a high sensitivity, being detectable in around 90% of MS patients. However, their presence is not restricted to MS, as they can be found transiently in many inflammatory CNS conditions as the expression of a humoral host response. For instance, Herpes simplex virus (HSV)-specific Ig can be detected in the CSF of patients suffering from HSV encephalitis, with the majority of antibodies directed against the pathogen (Vandvik et al., 1982). By contrast, OCB in MS have failed to display a predominant specificity inter- and intraindividually, as they are directed not only against many different pathogen-associated antigens but also against CNS structural epitopes. When compared to the corresponding ratio in blood, the MRZ-reaction describes an elevated ratio of intrathecally synthesized antibodies specific for MRZ (about 2% of all intrathecal IgG) versus all intrathecally synthesized IgG. It therefore has a higher specificity for MS than OCB, making it an attractive complement to the highly sensitive OCB (Reiber *et al.*, 1998). Although intrathecal antibodies are present in the human disease and also in some of the animal models, passive transfer studies and antibody depletion strategies failed to prove a clear-cut disease initiating or perpetuating effect (Antel and Bar-Or, 2006). In the animal model, some studies have shown a slightly exacerbating effect of MOG-specific antibodies on demyelination and clinical score in already established EAE (Urich *et al.*, 2006). Thus, in the context of T cell-mediated disease [or in the presence of a high number of autoreactive T cells (Bettelli *et al.*, 2006a; Krishnamoorthy *et al.*, 2006)], the disease can be modified by autoantibodies against myelin and other neural components, while systemic presence of these antibodies without T cell help has not proved to be pathogenic. Thus, intrathecal polyvalent antibody production in MS might in fact be an epiphenomenon rather than a disease promoting mechanism, and this can be utilized as diagnostic tool to discriminate MS from other CNS inflammatory conditions. This would also explain the broad array of specificities, which may reflect the individual patient's medical history of infections and CNS damage rather than an autoimmune response against an MS-relevant antigen. This concept of "bystander humoral response" has been supported by findings in basic research in B cell and plasma cell development. Radbruch and coworkers showed that migrating plasma blasts are capable of entering inflamed tissue by CXCR4- and CXCR3-mediated chemotaxis (which are also crucial factors for T cell migration to the CNS) irrespective of their antigen specificity. If they succeed in occupying a survival niche, which is defined by the availability of surviving factors, such as CXCL12 and other IL-6-induced B cell relevant factors, plasmablasts can develop into immobile, long-lived plasma cells, secreting large amounts of the antibodies (Radbruch *et al.*, 2006). Long-lived plasma cells are only present in the target organ during inflammation, disappearing after resolution of inflammation due to apoptosis, and the elimination of survival niches. This basic mechanism makes sense if there is an infectious attack against the CNS, as for instance in neuroborreliosis, in which bacteria attack the CNS, and the enrichment of peripherally generated plasma blasts leads to an enrichment of specific plasma blasts in the diseased target organ, and to a forceful antigen-specific local humoral response (Li *et al.*, 2006; Reiber and Peter, 2001). On the whole, this seems to be similar for MS in which chronically activated B cells, having undergone germinal center maturation, and plasma blasts are selectively enriched in the target tissue (Cepok *et al.*, 2005; Corcione *et al.*, 2004). In MS, however, there is no specific humoral immune response against a pathogen and thus no generation of specific plasma blasts in the bone marrow or spleen, which leads to an unselective attraction of traveling plasma blasts to the diseased target organ. This concept correlates well with low-level presence of B cells in chronic autoimmunity in the CNS (Esiri, 1977; Magliozzi et al., 2004). These findings lead us to hypothesize that intrathecal antibody synthesis in classic MS reflects a nonspecific recruitment of antibody-secreting cells into the CNS by inflammatory cues released in the process of T cell-mediated autoimmune inflammation. In situ, the provision of survival factors for plasma cells leads to long-term antibody production in the CNS. Indeed, myeloablative therapy in MS followed by successful autologous hematopoietic stem cell transplantation left OCB presence in the CSF unchanged, most likely resulting from the resistance of long-lived plasma cells to irradiation, cytostatics, and depletion by an anti-CD20 antibody (Rituximab, see below; Saiz et al., 2001).

A genuine autoimmune humoral response, as in the (NZB × NZW) F1 murine model for systemic lupus erythematodes (SLE), presents with stable and persistent autoantibodies in the serum as a hallmark of its pathology. These serum autoantibodies presumably derive from long-lived plasma cells in the bone marrow and are characteristically present already before onset of the disease (Hoyer et al., 2004). This is not the case for classic MS (Antel and Bar-Or, 2006), but some of these traits can be found in another type of chronic neuroinflammation called neuromyelitis optica (NMO, also Devic's syndrome), a demyelinating CNS disease with distinct clinical, therapeutic, and histopathologic features compared to MS. CNS infiltrates in NMO are marked by extensive eosinophil infiltration, complement activation, and necrotizing demyelination with prominent vascular hyalinization (Lucchinetti et al., 2002). Recent serological studies have identified a serum autoantibody called NMO-IgG, which binds to the abluminal face of CNS microvessels in different areas of the spinal cord and brain of murine CNS tissue, and which can be detected in the serum of most NMO patients while they are absent in healthy controls, MS, and other diseases (Lennon et al., 2004). This autoantibody presumably recognizes the aquaporin-4 water channel, which is highly expressed in astrocytic end feet along CNS microvasulature (Lennon et al., 2005). These findings make aquaporin-4 a likely suspect as an autoantigen which might be involved in initiating humoral autoimmune CNS disease. Another piece of evidence supporting this hypothesis is supplied by a first open trial in which patients were treated with the CD20-depleting antibody Rituximab. Results from this trial have been promising for this otherwise rapidly progressive disease, with six out of eight patients staying relapse-free over an observation period of one year (Cree et al., 2005). However, while serological studies, histopathologic results, and success of specific treatments are all crucial, it has yet to be proven that passive transferal of the antibody or serum from diseased animals is sufficient to induce the disease in healthy animals.

4. THERAPEUTIC APPROACHES TO CHRONIC NEUROINFLAMMATION

4.1. General considerations

Traditional therapy regimens aiming at broad immune modulation and suppression by steroids, Type I interferones, and cytostatics have only a limited impact on the disease progression, presumably reflecting the etiologic and pathological complexity of the clinical syndrome. This lack of long-term disease control might be due to the fact that our treatment strategies focus on the inflammatory processes outside the brain and on the attack against the myelin sheath, although even the earliest histopathologic descriptions from the end of the nineteenth century describe extensive gray matter and axonal pathology. Our ability to induce a similar disease in animals by injecting myelin components leading to pronounced demyelination and comparably minor axonal and neuronal damage in the early stages shifted the focus of inquiry from the search for therapeutic remedies to the investigation of the primary attack against the myelin sheath. Moreover, the visualization of CNS damage in MS with MRI is heavily biased toward the impressively obvious white matter demyelinating lesions, whereas gray matter lesions and neuronal degeneration can only be shown *in vivo* by intricate MRI spectroscopic methods which are not part of standard MRI protocols. However, beyond white matter demyelination, brains of patients show axonal pathology that correlates with immune cell infiltration and neuronal cell loss (Peterson *et al.*, 2001; Trapp *et al.*, 1998). Indeed, conventional MRI, being blind to neuronal damage, underreports the extent of the disease and neglects the neuropathologic heterogeneity in MS patients, which is very important in view of the fact that MRI is the preferential end point in clinical trials and contributes considerably to therapeutic decision making.

A recent paradigm shift has led to efforts by neuroimmunologists to acquire deeper insight into the neurodegenerative features of chronic neuroinflammation. Recently, proteolytic enzymes, cytokines, death ligands, such as TRAIL, oxidative products, such as 7-keto-cholesterol, and free radicals have been identified as potential contributors to neuronal damage (Zipp and Aktas, 2006). It has yet to be clarified what immune cells and what mechanisms initiate neurodegeneration in *in vivo* animal models and the human disease. In addition, inflammation may compromise energy metabolism and cause hypoxia and cytotoxicity, making neuroprotective drugs a likely candidate for future treatment strategies. The contribution of excitotoxicity was shown by Raine and coworkers, who influenced chronic neuronal damage in the EAE model via AMPA/kainate receptor antagonists (Pitt *et al.*, 2000). Unlike ischemic or traumatic models, in which dramatic metabolic failure leads to rapid irreversible neuronal injury, and where neuroprotectives have not had a significant effect, metabolic disturbances in relapsing-remitting MS seem to be much more subtle, and reversibility of neuronal impairment may give us the opportunity to save patients from silent deterioration. Accordingly, there have been some promising reports describing beneficial effects of Na^+-channel blockers on neurodegeneration in animal models of MS (Black et al., 2006).

4.2. Current therapeutic concepts

State-of-the-art therapy in MS rests upon two distinct pillars primarily aimed at reducing inflammatory activity in the CNS. On the one hand, high-dose steroid-pulse therapy is administered during acute attacks to induce remission, while on the other hand, continuous immune modulatory treatment with IFN-β and glatiramer acetate (GA) or suppressive treatment with Mitoxantrone have been shown to slow down the disease progress.

The mechanisms of both regimens seem to be pleiotropic; but as most of these results come from animal models or even *in vitro* investigations, it remains difficult to define the crucial mechanism for the *in vivo* human situation.

4.2.1. Steroids

Steroids can inhibit a broad spectrum of transcription factors involved in proinflammatory gene expression, which is what made them a universal tool in the treatment of chronic inflammatory diseases. Steroids given at high doses can induce apoptosis in CNS infiltrating T cells, leading to amelioration in adoptive transfer rat EAE models. Furthermore, even at low doses, transmigration of leukocytes seems to be significantly reduced mainly by modulating adhesion molecule expression, thus preventing blood–brain barrier breakdown. Moreover, steroids exert a downregulatory effect on antigen presenting cells (APC) by keeping down MHC expression and NO synthesis both in the CNS and in secondary lymphoid organs (Reichardt et al., 2006); they have also been shown to induce strongly IL-10-biased Treg, preventing EAE if generated in the presence of vitamin D_3 (Barrat et al., 2002).

Even if steroids have been proven to be of therapeutic value in inducing remission in acute MS attacks, there are a few caveats to keep in mind. First, there is no evidence at all for a beneficial effect of continuous steroid treatment, and long-term disability can certainly not be influenced by this therapeutic option. Furthermore, results from a toxin-induced demyelination model suggest that steroid treatment might interfere with successful remyelination, possibly by reducing debris removal and thus impairing myelin-induced oligodendrocyte precursor cells. However, this needs further investigation, as the toxin-induced demyelination model is quite different from EAE and MS, in which there is much more recruited inflammation and resulting axonal damage and neurodegeneration (Chari *et al.*, 2006; Miller, 1999).

4.2.2. Interferon-β

In contrast to steroids, immune modulatory drugs such as IFN-β or GA have a beneficial long-term outcome reflected in a reduction of relapse rate and lower disability progression. Under physiological circumstances, IFN-β, a Type I IFN, is a cytokine produced by different cell types upon Toll-like receptor (TLR) activation, which is involved in defense against different infections via recognition of conserved pathogen-associated pattern. IFN-β signals through the common Type I IFN receptor (IFN-α/β-R), inducing diverse immune modulatory effects as well as multiple antiviral effector mechanisms. On the one hand, there is reduced allover activation of T cells and reduced expression of transmigration-relevant adhesion molecules in treated MS patients, and on the other hand, IFN-β has a beneficial effect in EAE by shifting the IL-12/-23/IL-10 balance in favor of the regulatory cytokine. Another putative mechanism consists in altered antigen presentation by reduced costimulation and decreased MHC class II expression (Hartung *et al.*, 2004).

4.2.3. Glatiramer acetate

GA, a synthetic amino acid copolymer with a certain cross-priming reactivity of MBP-specific immune responses, has been shown to be effective in suppressing and treating rodent EAE models induced with different myelin proteins, and also in large trials in the human disease. In the light of *in vitro* data from human T cell lines, the effector mechanism was assumed to be a Th1–Th2 shift of GA/MBP-specific immune responses, which suggested a model in which peripherally GA-activated Th2 cell clones entered the CNS and exerted bystander suppression by MBP activation. However, this *in vitro* effect on human T cell lines has never been verified *ex vivo* in MS patients treated with GA. On the other hand, there is *in vitro* data to suggest a blockade of the MHC binding site which might result in reduced T cell activation in the CNS, though conclusive *ex vivo* human data is lacking. Recently, there have been some investigations suggesting a neuroprotective effect by induction of brain-derived neurotrophic factor (BDNF), but this might simply be an unspecific effect of T cell activation, as neuroprotective factors have been shown to be released in CNS injury models and *in vitro* activated T and B cells as well. Taken together, all of these uncertainties raise more questions about the mechanism of action for GA than have been answered so far, even if the clinical effect of GA on disease progression in MS has been clearly proven (Arnon *et al.*, 1996).

4.2.4. Mitoxantrone

Escalating MS treatment beyond immune modulation with the immunosuppressive anthracendione Mitoxantrone has been shown to be effective in EAE and in different MS subtypes. The mode of action is quite straightforward, since Mitoxantrone, as a cell cycle-independent cytotoxic agent, has profound inhibitory and also apoptotic effects on all hematopoietic cells. B cell function, in particular, is severely impaired in MS patients treated with Mitoxantrone, but T cell activation, proliferation and cytokine production also plummet, and APC function seems severely impaired. Unfortunately, because of severe side effects such as dose-dependent cardiotoxicity, this regimen is reserved for specific cases and even then cannot be administered on a lifetime basis (Fox, 2004).

4.3. New therapeutic concepts

Current research on new therapeutic regimens focuses on precise target mechanisms rather than unspecific immunosuppression, as applied to patients with autoimmune diseases in the past. However, it should be taken into account that highly specific treatment approaches, such as selected altered autoantigens or selected T cell receptor peptides, have so far not resulted in clinical therapies (Bielekova *et al.*, 2000). As in a number of autoimmune diseases, several autoantigens seem to be responsible. Furthermore, epitope spreading and promiscuous recognition by the T cell receptor may (among other factors) explain the lack of an antigen/HLA/T cell receptor-specific therapy despite extensive research. Another important issue is the establishment of effective therapies that can be given orally since most of the currently established therapeutic agents must be administered intramuscularly, subcutaneously, or intravenously. Other therapeutic developments in MS aim to directly prevent the ultimate damaging processes or to induce repair mechanisms of neural cells. However, modern clinical research into better treatment options approaches modulation of lymphocyte function in many different ways—a result of our deeper understanding of the underlying neuroimmunological concepts of MS, which we presented in the previous chapters. Below, we will outline some of the most encouraging potential therapies.

4.3.1. Interference with immune cell migration

4.3.1.1. Blocking VLA-4 One of the most promising new therapies which have already entered clinical practice consists in interfering with integrin function via the monoclonal antibody Natalizumab. This antibody against the VLA-4 α4-subchain, which is expressed on most mononuclear cells upon activation, blocks adhesion to VCAM-1 on endothelial cells and dramatically limits lymphocyte trafficking across the blood–brain barrier (Sheremata et al., 2005). Its beneficial effect in MS patients was shown in two Phase III clinical trials, both as a monotherapy (Polman et al., 2006) and as an add-on therapy to IFN-β-1a (Rudick et al., 2006). In both trials, Natalizumab treatment led to a substantial and significant decrease of relapse rate compared to placebo and a substantially lower proportion of disease progression, which corresponded to a dramatically reduced presence of lymphocytes in the CSF. However, two MS patients in the trials and one patient treated with Natalizumab in an independent study with Morbus Crohn developed progressive multifocal leukoencephalopathy (PML) during treatment (Kleinschmidt-DeMasters and Tyler, 2005; Langer-Gould et al., 2005; Van Assche et al., 2005). PML is caused by the reactivation and/or *de novo* infection of oligodendrocytes with JC virus, which leads to a virus-induced progressive demyelination (Zurhein and Chou, 1965) almost exclusively in severely immunosuppressed individuals (Berger et al., 1987). The JC virus is usually acquired in childhood, leading to positive serum antibody levels in adults in more than 80% of cases. It hibernates mainly in epithelial cells of the kidneys (Chesters et al., 1983). The hypothesis has been formed that the absence of immune surveillance produced by Natalizumab allows the JC virus to attack oligodendrocytes, leading to a devastating clinical outcome. In support of this hypothesis, one of the patients, diagnosed in time, was saved simply by removal of the drug. Two of the three patients suffering from PML had a history of another immunosuppressive therapy simultaneous to therapy with Natalizumab, which means that the magnitude of immune restraint was probably enhanced due to combination-immunosuppressive therapy or a previously altered immune system. Although initially withdrawn from the market, Natalizumab was reapproved by the FDA and EMEA. The dramatic clinical improvement of MS might outweigh the risk of PML in selected, rapidly progressing MS patients, thereby justifying Natalizumab treatment, especially if given as a monotherapy.

4.3.1.2. Blocking sphingosine-1-phosphate receptors Sphingosine-1-phosphate (S1P) is a lysophospholipid that exerts a variety of effects on survival and migratory behavior of many cell types (Rosen and Goetzl, 2005). It acts via a family of G-protein–coupled receptors, of which several subtypes—namely lysophophatidic acid (LPA)(1) (LPA), LPA(2), and LPA(4)—are predominantly expressed by B- and T cells. Fingolimod, or FTY720, is a derivative of myriocin, an ingredient from fungus *Isaria sinclairii* used in traditional Chinese medicine. It was only recently discovered that fingolimod binds to S1P receptors, and preferentially to the LPA(1) receptor after phosphorylation, hence allowing specific inhibition of lymphocytic systemic migration (Brinkmann et al., 2002). The rationale for a therapeutic use in MS lies in the attempt to trap lymphocytes in systemic lymphatic organs, thus blocking lymphocyte migration into the CNS. Several studies with experimental autoimmune encephalitis showed promising results, with substantial decreases in disease severity when animals were treated with fingolimod (Fujino *et al.*, 2003; Kataoka *et al.*, 2005; Webb *et al.*, 2004). The beneficial effect of fingolimod treatment compared to placebo was recently shown in a Phase II clinical trial. For both the primary end point Gadolinium-enhancing MRI lesions and the annual relapse rate, a significant and substantial reduction over placebo was shown (Kappos *et al.*, 2006). Following these promising results, a Phase III clinical trial is currently being conducted. The fact that this therapy is orally administered makes fingolimod one of the most eagerly anticipated new options for MS treatment.

4.3.2. Interference with lymphocyte activation

4.3.2.1. Blocking IL-2 receptor IL-2 plays a key role in T cell activation and proliferation at the site of inflammation. Daclizumab is a humanized antibody against the α-chain of the IL-2 receptor (CD25) and limits T cell expansion by blocking IL-2 signaling in T cells. Its potential in immunomodulatory therapy was proven by its treatment of acute renal rejection after transplantation (Vincenti *et al.*, 1998). After promising results in EAE (Engelhardt *et al.*, 1989; Hayosh and Swanborg, 1987; Rose *et al.*, 1991), a Phase II trial, designed as a pilot trial with a baseline-to-treatment design, showed profound reduction of disease activity in MRI and significantly reduced clinical disease severity (Bielekova *et al.*, 2004). In a second trial, similarly beneficial results were observed (Rose *et al.*, 2004). Daclizumab is currently being investigated in a Phase III clinical trial.

4.3.2.2. Inhibiting T Cell cycle progression Blockade of the β-3-hydroxy-3-methylglutaryl coenzyme A (HMG-CoA) reductase (HMGCR) by inhibitors (HMGCRI), also known as statins, results in interference with T cell cycle progression and plays a beneficial role in chronic neuroinflammation (Aktas *et al.*, 2003; Waiczies *et al.*, 2005). The HMGCR pathway utilizes several key enzymes to convert intermediary metabolites via a series of sequential organic reactions (rearrangements, condensation reactions, etc.) that finally lead to cholesterol synthesis. This pathway is a source of hydrophobic molecules important for a wide range of inter- and intracellular functions involving hormonal communication, protein synthesis, electron transport, protein lipid modification for membrane anchoring and intracellular signaling (via isoprenoids), and cell membrane maintenance (via cholesterol). Previous studies demonstrating the anti-inflammatory nature of statins revealed the possible benefits of employing these agents for the treatment of inflammatory autoimmune disorders such as MS or rheumatoid arthritis (RA). According to our data, and those of other groups, *in vivo* statin treatment prevented and reversed disease progression in murine EAE (Aktas *et al.*, 2003), collagen-induced arthritis (CIA) (Leung *et al.*, 2003), and systemic lupus erythematosus (SLE) (Lawman *et al.*, 2004). Leung and colleagues showed a marked decrease in development of CIA following simvastatin therapy, using doses that were unable to significantly alter cholesterol concentrations *in vivo*. Together with Youssef *et al.* (2002), our group has reported that atorvastatin prevents and reverses chronic and relapsing paralysis in murine EAE by targeting Th1 cells. It has been proposed that statins are therapeutically active in EAE by inducing the secretion of Th2 anti-inflammatory cytokines (IL-4, IL-5, and IL-10) and transforming growth factor (TGF)-β, and, conversely, by suppressing Th1 proinflammatory cytokines (IL-2, IL-12, IFN-γ, and TNF-α) (Aktas *et al.*, 2003; Nath *et al.*, 2004; Youssef *et al.*, 2002). Additionally, adoptive transfer of induced Th2 cells protected recipient mice from induction of EAE. In general, *ex vivo* analysis of the diverse autoimmune animal models revealed a significant suppression of autoantigen-specific Th1 humoral and cellular immune responses following statin treatment. Based on these observations, and the fact that statins are generally well-tolerated orally administered drugs (Shepherd *et al.*, 2004), a number of pilot clinical trials have been carried out with patients suffering from autoimmune disorders. McCarey *et al.* reported that in patients suffering with RA, atorvastatin reduced the rate of the inflammatory variables C-reactive protein and erythrocyte sedimentation by 50% and 28%, respectively, with a modest influence on clinical disease manifestations (McCarey *et al.*, 2004). *In vivo*, specifically in animal models of autoimmune disease, HMGR inhibitors have been reported to skew proinflammatory cytokines toward a regulatory/anti-inflammatory phenotype. Atorvastatin promoted a cytokine shift toward a Th2 phenotype, demonstrated in the expression of STAT6 phosphorylation and secreted Th2 cytokines (IL-4, IL-5, IL-10, and TGF-β) in treated animals following *ex vivo* antigen stimulation. By contrast, STAT4 phosphorylation was inhibited and secretion of Th1 cytokines (IFN-γ, IL-2, IL-12, and TNF-α) was suppressed in these animals (Aktas *et al.*, 2003; Nath *et al.*, 2004; Youssef *et al.*, 2002). Since this experiment was *in vitro*, we do not observe the dramatic skewing of Th subpopulations observed in EAE (unpublished observation), and it is not yet clear whether this therapeutic effect of HMGR inhibitors in the mouse can be extrapolated to the human system. Importantly, adoptive transfer of Th cells from atorvastatin-treated animals into recipients prevented the development of EAE (Youssef *et al.*, 2002).

It has also been reported that HMGCRI reduce inflammation and Th1 responses by inhibiting NF-κB activation (Leung *et al.*, 2003). Moreover, certain HMGCRI have been shown to inhibit T cell stimulation independently of HMGCR, by interacting directly and allosterically with LFA-1

(Weitz-Schmidt *et al.*, 2001). In APC, namely B cells and macrophages, HMGCRI inhibit the expression of HLA class II antigens in response to IFN by suppressing the inducible promoter IV of the transactivator CIITA (Kwak *et al.*, 2000).

We have shown that HMGCRI decrease T cell proliferation by direct TCR engagement (Aktas *et al.*, 2003) and interference with cell cycle regulation. This was represented by a downregulation of cyclin-dependent kinase (CDK)-4 and upregulation of $p27^{kip1}$, which had previously been reported as the mechanism of action of statins only in mesangial cells (Danesh *et al.*, 2002). This view has been confirmed in initial clinical trials demonstrating the benefit of statin therapy in MS (Vollmer *et al.*, 2004). Currently, several clinical trials are under way to further explore the treatment effect of orally administered statins in MS, one of them in our own laboratory utilizing atorvastatin.

4.3.2.3. Inhibition of pyrimidine de novo synthesis Teriflunomide, a leflunomide metabolite for oral therapy, inhibits the *de novo* synthesis of pyrimidines (O'Connor *et al.*, 2006) and thus limits T- and B cell proliferation. Several other effects on the immune system were reported that can be attributed to an additional inhibition of tyrosine kinase by leflunomide (Herrmann *et al.*, 2000), but their relevance for the therapeutic effect of leflunomide is unclear. Inhibition of pyrimidine *de novo* synthesis seems to favor blockade of Th1 cell over Th2 cell proliferation, which could potentially be of value in MS therapy (Dimitrova *et al.*, 2002). In experimental immune neuritis and EAE, a beneficial effect of leflunomide and teriflunomide was shown (Korn *et al.*, 2001; Styren *et al.*, 2004). Oral therapy with teriflunomide in a randomized, double-blinded and placebo-controlled Phase II clinical trial led to a significant reduction in MRI activity in MS patients, and to positive effects on relapse rate and disease progression (O'Connor *et al.*, 2006). A Phase III clinical trial is currently being conducted to test efficiency and adverse effects on a larger clinical scale.

4.3.2.4. Blocking the CD40 costimulatory pathway Blocking the costimulatory pathway CD40L(CD154)-CD40 can inhibit stimulation and activation of proinflammatory T cells in the CNS (Howard and Miller, 2001; Howard *et al.*, 1999). The monoclonal antibody against CD40L/CD154 has shown very promising effects in EAE (Grewal *et al.*, 1996; Howard *et al.*, 2002). However, thrombembolic complications in first trials in humans with this antibody (Kawai *et al.*, 2000) delayed its introduction into clinical practice. This effect had not been observed in EAE, as CD40L is not expressed on murine but on human platelets. As a result, new trials with coadministration of heparin were initiated and are still running.

4.3.3. Depletion of defined cell subsets

4.3.3.1. Depleting CD20 positive cells of the B cell lineage CD20 is a transmembrane phosphoprotein that is expressed on B cells and pre-B cells but not on long-lived plasma cells (Tedder *et al.*, 1988). The monoclonal antibody Rituximab, which is widely used in treating neoplastic B cell diseases (Grillo-Lopez *et al.*, 2002), leads to a depletion of CD20 expressing cells via complement activation and probably cell-mediated cytotoxicity (Kennedy *et al.*, 2004) for a period of 6–12 months (Cree *et al.*, 2005). As autoantibody-mediated mechanisms are still under debate for MS patients, particularly when unresponsive or only partially responsive to standard treatment regimens, the rationale for the use of Rituximab in MS therapy is to induce a potentially beneficial effect, especially in progressive MS, with some Phase II clinical trials still currently using Rituximab as an add-on. Apart from the predominant role B cells play in humoral immunity, they also contribute to the priming and modulation of T cell-dependent autoimmune responses. Indeed, in terms of antigen presentation, it has been suggested that B cell cytokines might modulate autoimmune processes, as selective depletion of IL-10 in B cells led to the absence of remission in EAE (Fillatreau *et al.*, 2002). This should be kept in mind when interfering with B cell presence in the CNS.

4.3.3.2. Depletion of CD52 positive cells CD52 is expressed on the vast majority of differentiated lymphocytes, monocytes, and macrophages. Alemtuzumab, a humanized antibody against CD52 based on CAMPATH-1 (Hale *et al.*, 1988), and utilized for lymphoma therapy, leads to a transient but profound depletion of CD52 expressing cells. Interestingly, the analysis of effects on different lymphocyte populations revealed a profound depletion of $CD4^+$ and $CD8^+$ T cells, whereas B cell counts increased. One clinical trial initially showed a positive effect on MRI activity in treated MS patients (Moreau *et al.*, 1994), but one-third of patients developed antibody-mediated autoimmune thyroid disease (Coles *et al.*, 1999). Further monitoring of patients treated with this antibody revealed progressive brain and spinal cord atrophy associated with clinical deterioration, though reduction of inflammatory activity in MRI was sustained (Paolillo *et al.*, 1999). In keeping with these findings, patients in early stages of disease with frequent attacks seemed to benefit from therapy, whereas patients in later stages of silent deterioration showed virtually no benefit (Coles *et al.*, 2006).

4.3.4. Expanding regulatory T cells

One very promising approach was the development of a superagonistic CD28 antibody that could activate T cells without ligation of the T cell receptor (Luhder *et al.*, 2003). In animal models, it was shown that this approach preferentially stimulated and expanded regulatory $CD4^+CD25^+$ T cells, proposing a beneficial effect for autoimmune disease, in this case MS (Rodriguez-Palmero *et al.*, 1999). Subsequently, it was shown that a CD28 superagonist can be an effective treatment for EAE by enhancing and promoting Treg function. However, when the humanized version of the antibody (TGN1412) was tested recently in a Phase I study, it led to extremely severe adverse events. Shortly after the first administration of the drug serum, virtually all cytokines shot up to immense levels in all six of the volunteers that received TGN1412. This cytokine storm was followed by severe lymphocyte and monocyte depletion, leaving the patients with multiorgan failure and permanent damage (Suntharalingam *et al.*, 2006). So far, the detailed mechanisms underlying this event remain unclear, leaving us with a stark reminder of the caution with which immunologic research data from animal models must be translated, and the difficulties in advancing translational medicine.

Nevertheless, we think that the expansion, and also the functional enhancement of Treg, is one of the most promising strategies in treating autoimmunity. Even in an unimpaired immune system, autoreactive T cells are part of the normal repertoire, but different types of Treg subsets prevent autoimmune processes in health and disease. The greatest therapeutic challenge consists in selectively modulating these cells *in vitro* with the aim of restoring the immunologic balance *in vivo*. One of the major problems of this approach lies in the necessity of antigen-specific reactivation of these cells at the site of inflammation, or at least in proximity, to exert their suppressor function. However, as in a number of autoimmune diseases, no predominant autoantigen(s) have been identified, thus complicating this approach. Finally, among other factors, epitope spreading and promiscuous recognition by the T cell receptor may prove to be further challenges along the way to the successful adoptive transfer, or *in vivo* expansion, of effective Treg therapy for a sustained improvement of chronic autoimmune disorders.

ACKNOWLEDGMENTS

The authors wish to acknowledge funding from the *Deutsche Forschungsgemeinschaft* (*DFG*) and the *Bundesministerium für Bildung und Forschung* (*BMBF*) of Germany. The authors thank Andrew Mason for reading the manuscript as a native English speaker, and Carmen Infante-Duarte and Sonia Waiczies for valuable discussion.

REFERENCES

Aktas, O., Waiczies, S., Smorodchenko, A., Dorr, J., Seeger, B., Prozorovski, T., Sallach, S., Endres, M., Brocke, S., Nitsch, R., and Zipp, F. (2003). Treatment of relapsing paralysis in experimental encephalomyelitis by targeting Th1 cells through atorvastatin. *J. Exp. Med.* **197,** 725–733.

Aktas, O., Smorodchenko, A., Brocke, S., Infante-Duarte, C., Topphoff, U. S., Vogt, J., Prozorovski, T., Meier, S., Osmanova, V., Pohl, E., Bechmann, I., Nitsch, R., *et al.* (2005). Neuronal damage in autoimmune neuroinflammation mediated by the death ligand TRAIL. *Neuron* **46,** 421–432.

Aktas, O., Prozorovski, T., and Zipp, F. (2006). Death ligands and autoimmune demyelination. *Neuroscientist* **12,** 305–316.

Anonymous (1999). TNF neutralization in MS: Results of a randomized, placebo-controlled multicenter study. The Lenercept Multiple Sclerosis Study Group and The University of British Columbia MS/MRI Analysis Group. *Neurology* **53,** 457–465.

Antel, J., and Bar-Or, A. (2006). Roles of immunoglobulins and B cells in multiple sclerosis: From pathogenesis to treatment. *J. Neuroimmunol.* **180,** 3–8.

Arnon, R., Sela, M., and Teitelbaum, D. (1996). New insights into the mechanism of action of copolymer 1 in experimental allergic encephalomyelitis and multiple sclerosis. *J. Neurol.* **243,** S8–S13.

Babbe, H., Roers, A., Waisman, A., Lassmann, H., Goebels, N., Hohlfeld, R., Friese, M., Schroder, R., Deckert, M., Schmidt, S., Ravid, R., and Rajewsky, K. (2000). Clonal expansions of CD8(+) T cells dominate the T cell infiltrate in active multiple sclerosis lesions as shown by micromanipulation and single cell polymerase chain reaction. *J. Exp. Med.* **192,** 393–404.

Banerjee, K., Biswas, P. S., Kumaraguru, U., Schoenberger, S. P., and Rouse, B. T. (2004). Protective and pathological roles of virus-specific and bystander CD8+ T cells in herpetic stromal keratitis. *J. Immunol.* **173,** 7575–7583.

Barrat, F. J., Cua, D. J., Boonstra, A., Richards, D. F., Crain, C., Savelkoul, H. F., Waal-Malefyt, R., Coffman, R. L., Hawrylowicz, C. M., and O'Garra, A. (2002). In vitro generation of interleukin 10-producing regulatory CD4(+) T cells is induced by immunosuppressive drugs and inhibited by T helper type 1 (Th1)- and Th2-inducing cytokines. *J. Exp. Med.* **195,** 603–616.

Becher, B., Durell, B. G., and Noelle, R. J. (2003). IL-23 produced by CNS-resident cells controls T cell encephalitogenicity during the effector phase of experimental autoimmune encephalomyelitis. *J. Clin. Invest.* **112,** 1186–1191.

Berger, J. R., Kaszovitz, B., Post, M. J., and Dickinson, G. (1987). Progressive multifocal leukoencephalopathy associated with human immunodeficiency virus infection. A review of the literature with a report of sixteen cases. *Ann. Intern. Med.* **107,** 78–87.

Bettelli, E., Baeten, D., Jager, A., Sobel, R. A., and Kuchroo, V. K. (2006a). Myelin oligodendrocyte glycoprotein-specific T and B cells cooperate to induce a Devic-like disease in mice. *J. Clin. Invest.* **116,** 2393–2402.

Bettelli, E., Carrier, Y., Gao, W., Korn, T., Strom, T. B., Oukka, M., Weiner, H. L., and Kuchroo, V. K. (2006b). Reciprocal developmental pathways for the generation of pathogenic effector TH17 and regulatory T cells. *Nature* **441,** 235–238.

Bieganowska, K. D., Ausubel, L. J., Modabber, Y., Slovik, E., Messersmith, W., and Hafler, D. A. (1997). Direct *ex vivo* analysis of activated, Fas-sensitive autoreactive T cells in human autoimmune disease. *J. Exp. Med.* **185,** 1585–1594.

Bielekova, B., Goodwin, B., Richert, N., Cortese, I., Kondo, T., Afshar, G., Gran, B., Eaton, J., Antel, J., Frank, J. A., McFarland, H. F., and Martin, R. (2000). Encephalitogenic potential of the myelin basic protein peptide (amino acids 83–99) in multiple sclerosis: Results of a phase II clinical trial with an altered peptide ligand. *Nat. Med.* **6,** 1167–1175.

Bielekova, B., Richert, N., Howard, T., Blevins, G., Markovic-Plese, S., McCartin, J., Frank, J. A., Wurfel, J., Ohayon, J., Waldmann, T. A., McFarland, H. F., and Martin, R. (2004). Humanized anti-CD25 (daclizumab) inhibits disease activity in multiple sclerosis patients failing to respond to interferon beta. *Proc. Natl. Acad. Sci. USA* **101,** 8705–8708.

Black, J. A., Liu, S., Hains, B. C., Saab, C. Y., and Waxman, S. G. (2006). Long-term protection of central axons with phenytoin in monophasic and chronic-relapsing EAE. *Brain* **129**, 3196–3208.

Brinkmann, V., Davis, M. D., Heise, C. E., Albert, R., Cottens, S., Hof, R., Bruns, C., Prieschl, E., Baumruker, T., Hiestand, P., Foster, C. A., Zollinger, M., *et al.* (2002). The immune modulator FTY720 targets sphingosine 1-phosphate receptors. *J. Biol. Chem.* **277**, 21453–21457.

Brocke, S., Piercy, C., Steinman, L., Weissman, I. L., and Veromaa, T. (1999). Antibodies to CD44 and integrin alpha4, but not L-selectin, prevent central nervous system inflammation and experimental encephalomyelitis by blocking secondary leukocyte recruitment. *Proc. Natl. Acad. Sci. USA* **96**, 6896–6901.

Bullard, D. C., Hu, X., Schoeb, T. R., Collins, R. G., Beaudet, A. L., and Barnum, S. R. (2007). Intercellular adhesion molecule-1 expression is required on multiple cell types for the development of experimental autoimmune encephalomyelitis. *J. Immunol.* **178**, 851–857.

Carrier, Y., Yuan, J., Kuchroo, V. K., and Weiner, H. L. (2007). Th3 cells in peripheral tolerance. I. Induction of Foxp3-positive regulatory T cells by Th3 cells derived from TGF-beta T cell-transgenic mice. *J. Immunol.* **178**, 179–185.

Cepok, S., Rosche, B., Grummel, V., Vogel, F., Zhou, D., Sayn, J., Sommer, N., Hartung, H. P., and Hemmer, B. (2005). Short-lived plasma blasts are the main B cell effector subset during the course of multiple sclerosis. *Brain* **128**, 1667–1676.

Chari, D. M., Zhao, C., Kotter, M. R., Blakemore, W. F., and Franklin, R. J. (2006). Corticosteroids delay remyelination of experimental demyelination in the rodent central nervous system. *J. Neurosci. Res.* **83**, 594–605.

Chesters, P. M., Heritage, J., and McCance, D. J. (1983). Persistence of DNA sequences of BK virus and JC virus in normal human tissues and in diseased tissues. *J. Infect. Dis.* **147**, 676–684.

Coles, A. J., Wing, M., Smith, S., Coraddu, F., Greer, S., Taylor, C., Weetman, A., Hale, G., Chatterjee, V. K., Waldmann, H., and Compston, A. (1999). Pulsed monoclonal antibody treatment and autoimmune thyroid disease in multiple sclerosis. *Lancet* **354**, 1691–1695.

Coles, A. J., Cox, A., Le Page, E., Jones, J., Trip, S. A., Deans, J., Seaman, S., Miller, D. H., Hale, G., Waldmann, H., and Compston, D. A. (2006). The window of therapeutic opportunity in multiple sclerosis: Evidence from monoclonal antibody therapy. *J. Neurol.* **253**, 98–108.

Corcione, A., Casazza, S., Ferretti, E., Giunti, D., Zappia, E., Pistorio, A., Gambini, C., Mancardi, G. L., Uccelli, A., and Pistoia, V. (2004). Recapitulation of B cell differentiation in the central nervous system of patients with multiple sclerosis. *Proc. Natl. Acad. Sci. USA* **101**, 11064–11069.

Cree, B. A., Lamb, S., Morgan, K., Chen, A., Waubant, E., and Genain, C. (2005). An open label study of the effects of rituximab in neuromyelitis optica. *Neurology* **64**, 1270–1272.

Cua, D. J., Sherlock, J., Chen, Y., Murphy, C. A., Joyce, B., Seymour, B., Lucian, L., To, W., Kwan, S., Churakova, T., Zurawski, S., Wiekowski, M., *et al.* (2003). Interleukin-23 rather than interleukin-12 is the critical cytokine for autoimmune inflammation of the brain. *Nature* **421**, 744–748.

D'Amico, G., Frascaroli, G., Bianchi, G., Transidico, P., Doni, A., Vecchi, A., Sozzani, S., Allavena, P., and Mantovani, A. (2000). Uncoupling of inflammatory chemokine receptors by IL-10: Generation of functional decoys. *Nat. Immunol.* **1**, 387–391.

Danesh, F. R., Sadeghi, M. M., Amro, N., Philips, C., Zeng, L., Lin, S., Sahai, A., and Kanwar, Y. S. (2002). 3-Hydroxy-3-methylglutaryl CoA reductase inhibitors prevent high glucose-induced proliferation of mesangial cells via modulation of Rho GTPase/p21 signaling pathway: Implications for diabetic nephropathy. *Proc. Natl. Acad. Sci. USA* **99**, 8301–8305.

Debes, G. F., Dahl, M. E., Mahiny, A. J., Bonhagen, K., Campbell, D. J., Siegmund, K., Erb, K. J., Lewis, D. B., Kamradt, T., and Hamann, A. (2006). Chemotactic responses of IL-4-, IL-10-, and IFN-gamma-producing CD4+ T cells depend on tissue origin and microbial stimulus. *J. Immunol.* **176,** 557–566.

Diestel, A., Aktas, O., Hackel, D., Hake, I., Meier, S., Raine, C. S., Nitsch, R., Zipp, F., and Ullrich, O. (2003). Activation of microglial poly(ADP-ribose)-polymerase-1 by cholesterol breakdown products during neuroinflammation: A link between demyelination and neuronal damage. *J. Exp. Med.* **198,** 1729–1740.

Dimitrova, P., Skapenko, A., Herrmann, M. L., Schleyerbach, R., Kalden, J. R., and Schulze-Koops, H. (2002). Restriction of de novo pyrimidine biosynthesis inhibits Th1 cell activation and promotes Th2 cell differentiation. *J. Immunol.* **169,** 3392–3399.

Eltayeb, S., Sunnemark, D., Berg, A. L., Nordvall, G., Malmberg, A., Lassmann, H., Wallstrom, E., Olsson, T., and Ericsson-Dahlstrand, A. (2003). Effector stage CC chemokine receptor-1 selective antagonism reduces multiple sclerosis-like rat disease. *J. Neuroimmunol.* **142,** 75–85.

Engelhardt, B., and Ransohoff, R. M. (2005). The ins and outs of T-lymphocyte trafficking to the CNS: Anatomical sites and molecular mechanisms. *Trends Immunol.* **26,** 485–495.

Engelhardt, B., Diamantstein, T., and Wekerle, H. (1989). Immunotherapy of experimental autoimmune encephalomyelitis (EAE): Differential effect of anti-IL-2 receptor antibody therapy on actively induced and T-line mediated EAE of the Lewis rat. *J. Autoimmun.* **2,** 61–73.

Engelhardt, B., Vestweber, D., Hallmann, R., and Schulz, M. (1997). E- and P-selectin are not involved in the recruitment of inflammatory cells across the blood-brain barrier in experimental autoimmune encephalomyelitis. *Blood* **90,** 4459–4472.

Engelhardt, B., Kempe, B., Merfeld-Clauss, S., Laschinger, M., Furie, B., Wild, M. K., and Vestweber, D. (2005). P-selectin glycoprotein ligand 1 is not required for the development of experimental autoimmune encephalomyelitis in SJL and C57BL/6 mice. *J. Immunol.* **175,** 1267–1275.

Esiri, M. M. (1977). Immunoglobulin-containing cells in multiple-sclerosis plaques. *Lancet* **2,** 478.

Feuerer, M., Eulenburg, K., Loddenkemper, C., Hamann, A., and Huehn, J. (2006). Self-limitation of Th1-mediated inflammation by IFN-gamma. *J. Immunol.* **176,** 2857–2863.

Fillatreau, S., Sweenie, C. H., McGeachy, M. J., Gray, D., and Anderton, S. M. (2002). B cells regulate autoimmunity by provision of IL-10. *Nat. Immunol.* **3,** 944–950.

Flugel, A., Schwaiger, F. W., Neumann, H., Medana, I., Willem, M., Wekerle, H., Kreutzberg, G. W., and Graeber, M. B. (2000). Neuronal FasL induces cell death of encephalitogenic T lymphocytes. *Brain Pathol.* **10,** 353–364.

Flugel, A., Berkowicz, T., Ritter, T., Labeur, M., Jenne, D. E., Li, Z., Ellwart, J. W., Willem, M., Lassmann, H., and Wekerle, H. (2001). Migratory activity and functional changes of green fluorescent effector cells before and during experimental autoimmune encephalomyelitis. *Immunity* **14,** 547–560.

Fontenot, J. D., Rasmussen, J. P., Williams, L. M., Dooley, J. L., Farr, A. G., and Rudensky, A. Y. (2005). Regulatory T cell lineage specification by the forkhead transcription factor foxp3. *Immunity* **22,** 329–341.

Fox, E. J. (2004). Mechanism of action of mitoxantrone. *Neurology* **63,** S15–S18.

Fujino, M., Funeshima, N., Kitazawa, Y., Kimura, H., Amemiya, H., Suzuki, S., and Li, X. K. (2003). Amelioration of experimental autoimmune encephalomyelitis in Lewis rats by FTY720 treatment. *J. Pharmacol. Exp. Ther.* **305,** 70–77.

Grewal, I. S., Foellmer, H. G., Grewal, K. D., Xu, J., Hardardottir, F., Baron, J. L., Janeway, C. A., Jr., and Flavell, R. A. (1996). Requirement for CD40 ligand in costimulation induction, T cell activation, and experimental allergic encephalomyelitis. *Science* **273,** 1864–1867.

Grillo-Lopez, A. J., Hedrick, E., Rashford, M., and Benyunes, M. (2002). Rituximab: Ongoing and future clinical development. *Semin. Oncol.* **29,** 105–112.
Haines, J. L., Ter Minassian, M., Bazyk, A., Gusella, J. F., Kim, D. J., Terwedow, H., Pericak-Vance, M. A., Rimmler, J. B., Haynes, C. S., Roses, A. D., Lee, A., Shaner, B., *et al.* (1996). A complete genomic screen for multiple sclerosis underscores a role for the major histocompatability complex. The Multiple Sclerosis Genetics Group. *Nat. Genet.* **13,** 469–471.
Hale, G., Dyer, M. J., Clark, M. R., Phillips, J. M., Marcus, R., Riechmann, L., Winter, G., and Waldmann, H. (1988). Remission induction in non-Hodgkin lymphoma with reshaped human monoclonal antibody CAMPATH-1H. *Lancet* **2,** 1394–1399.
Hartung, H. P., Bar-Or, A., and Zoukos, Y. (2004). What do we know about the mechanism of action of disease-modifying treatments in MS? *J. Neurol.* **251**(Suppl. 5), v12–v29.
Hayosh, N. S., and Swanborg, R. H. (1987). Autoimmune effector cells. IX. Inhibition of adoptive transfer of autoimmune encephalomyelitis with a monoclonal antibody specific for interleukin 2 receptors. *J. Immunol.* **138,** 3771–3775.
Herrmann, M. L., Schleyerbach, R., and Kirschbaum, B. J. (2000). Leflunomide: An immunomodulatory drug for the treatment of rheumatoid arthritis and other autoimmune diseases. *Immunopharmacology* **47,** 273–289.
Hickey, W. F. (1999). Leukocyte traffic in the central nervous system: The participants and their roles. *Semin. Immunol.* **11,** 125–137.
Hickey, W. F., Hsu, B. L., and Kimura, H. (1991). T-lymphocyte entry into the central nervous system. *J. Neurosci. Res.* **28,** 254–260.
Hofstetter, H. H., Targoni, O. S., Karulin, A. Y., Forsthuber, T. G., Tary-Lehmann, M., and Lehmann, P. V. (2005). Does the frequency and avidity spectrum of the neuroantigen-specific T cells in the blood mirror the autoimmune process in the central nervous system of mice undergoing experimental allergic encephalomyelitis? *J. Immunol.* **174,** 4598–4605.
Hofstetter, H. H., Toyka, K. V., Tary-Lehmann, M., and Lehmann, P. V. (2007). Kinetics and organ distribution of IL-17-producing CD4 cells in proteolipid protein 139–151 peptide-induced experimental autoimmune encephalomyelitis of SJL mice. *J. Immunol.* **178,** 1372–1378.
Hovelmeyer, N., Hao, Z., Kranidioti, K., Kassiotis, G., Buch, T., Frommer, F., von Hoch, L., Kramer, D., Minichiello, L., Kollias, G., Lassmann, H., and Waisman, A. (2005). Apoptosis of oligodendrocytes via Fas and TNF-R1 is a key event in the induction of experimental autoimmune encephalomyelitis. *J. Immunol.* **175,** 5875–5884.
Howard, L. M., and Miller, S. D. (2001). Autoimmune intervention by CD154 blockade prevents T cell retention and effector function in the target organ. *J. Immunol.* **166,** 1547–1553.
Howard, L. M., Miga, A. J., Vanderlugt, C. L., Dal Canto, M. C., Laman, J. D., Noelle, R. J., and Miller, S. D. (1999). Mechanisms of immunotherapeutic intervention by anti-CD40L (CD154) antibody in an animal model of multiple sclerosis. *J. Clin. Invest.* **103,** 281–290.
Howard, L. M., Ostrovidov, S., Smith, C. E., Dal Canto, M. C., and Miller, S. D. (2002). Normal Th1 development following long-term therapeutic blockade of CD154-CD40 in experimental autoimmune encephalomyelitis. *J. Clin. Invest.* **109,** 233–241.
Hoyer, B. F., Moser, K., Hauser, A. E., Peddinghaus, A., Voigt, C., Eilat, D., Radbruch, A., Hiepe, F., and Manz, R. A. (2004). Short-lived plasmablasts and long-lived plasma cells contribute to chronic humoral autoimmunity in NZB/W mice. *J. Exp. Med.* **199,** 1577–1584.
Hu, D., Ikizawa, K., Lu, L., Sanchirico, M. E., Shinohara, M. L., and Cantor, H. (2004). Analysis of regulatory CD8 T cells in Qa-1-deficient mice. *Nat. Immunol.* **5,** 516–523.
Huseby, E. S., Liggitt, D., Brabb, T., Schnabel, B., Ohlen, C., and Goverman, J. (2001). A pathogenic role for myelin-specific CD8(+) T cells in a model for multiple sclerosis. *J. Exp. Med.* **194,** 669–676.

Infante-Duarte, C., Horton, H. F., Byrne, M. C., and Kamradt, T. (2000). Microbial lipopeptides induce the production of IL-17 in Th cells. *J. Immunol.* **165**, 6107–6115.

Ivanov, I. I., McKenzie, B. S., Zhou, L., Tadokoro, C. E., Lepelley, A., Lafaille, J. J., Cua, D. J., and Littman, D. R. (2006). The orphan nuclear receptor RORgammat directs the differentiation program of proinflammatory IL-17+ T helper cells. *Cell* **126**, 1121–1133.

Kappos, L., Antel, J., Comi, G., Montalban, X., O'Connor, P., Polman, C. H., Haas, T., Korn, A. A., Karlsson, G., and Radue, E. W. (2006). Oral fingolimod (FTY720) for relapsing multiple sclerosis. *N. Engl. J. Med.* **355**, 1124–1140.

Kassiotis, G., and Kollias, G. (2001). TNF and receptors in organ-specific autoimmune disease: Multi-layered functioning mirrored in animal models. *J. Clin. Invest.* **107**, 1507–1508.

Kataoka, H., Sugahara, K., Shimano, K., Teshima, K., Koyama, M., Fukunari, A., and Chiba, K. (2005). FTY720, sphingosine 1-phosphate receptor modulator, ameliorates experimental autoimmune encephalomyelitis by inhibition of T cell infiltration. *Cell Mol. Immunol.* **2**, 439–448.

Kawai, T., Andrews, D., Colvin, R. B., Sachs, D. H., and Cosimi, A. B. (2000). Thromboembolic complications after treatment with monoclonal antibody against CD40 ligand. *Nat. Med.* **6**, 114.

Kawakami, N., Lassmann, S., Li, Z., Odoardi, F., Ritter, T., Ziemssen, T., Klinkert, W. E., Ellwart, J. W., Bradl, M., Krivacic, K., Lassmann, H., Ransohoff, R. M., *et al.* (2004). The activation status of neuroantigen-specific T cells in the target organ determines the clinical outcome of autoimmune encephalomyelitis. *J. Exp. Med.* **199**, 185–197.

Kennedy, A. D., Beum, P. V., Solga, M. D., DiLillo, D. J., Lindorfer, M. A., Hess, C. E., Densmore, J. J., Williams, M. E., and Taylor, R. P. (2004). Rituximab infusion promotes rapid complement depletion and acute CD20 loss in chronic lymphocytic leukemia. *J. Immunol.* **172**, 3280–3288.

Keszthelyi, E., Karlik, S., Hyduk, S., Rice, G. P., Gordon, G., Yednock, T., and Horner, H. (1996). Evidence for a prolonged role of alpha 4 integrin throughout active experimental allergic encephalomyelitis. *Neurology* **47**, 1053–1059.

Kleinschmidt-DeMasters, B. K., and Tyler, K. L. (2005). Progressive multifocal leukoencephalopathy complicating treatment with natalizumab and interferon beta-1a for multiple sclerosis. *N. Engl. J. Med.* **353**, 369–374.

Kobayashi, Y., Kawai, K., Honda, H., Tomida, S., Niimi, N., Tamatani, T., Miyasaka, M., and Yoshikai, Y. (1995). Antibodies against leukocyte function-associated antigen-1 and against intercellular adhesion molecule-1 together suppress the progression of experimental allergic encephalomyelitis. *Cell Immunol.* **164**, 295–305.

Korn, T., Toyka, K., Hartung, H. P., and Jung, S. (2001). Suppression of experimental autoimmune neuritis by leflunomide. *Brain* **124**, 1791–1802.

Kretschmer, K., Apostolou, I., Hawiger, D., Khazaie, K., Nussenzweig, M. C., and von Boehmer, H. (2005). Inducing and expanding regulatory T cell populations by foreign antigen. *Nat. Immunol.* **6**, 1219–1227.

Krishnamoorthy, G., Lassmann, H., Wekerle, H., and Holz, A. (2006). Spontaneous opticospinal encephalomyelitis in a double-transgenic mouse model of autoimmune T cell/B cell cooperation. *J. Clin. Invest.* **116**, 2385–2392.

Kuchroo, V. K., Martin, C. A., Greer, J. M., Ju, S. T., Sobel, R. A., and Dorf, M. E. (1993). Cytokines and adhesion molecules contribute to the ability of myelin proteolipid protein-specific T cell clones to mediate experimental allergic encephalomyelitis. *J. Immunol.* **151**, 4371–4382.

Kwak, B., Mulhaupt, F., Myit, S., and Mach, F. (2000). Statins as a newly recognized type of immunomodulator. *Nat. Med.* **6**, 1399–1402.

Langer-Gould, A., Atlas, S. W., Green, A. J., Bollen, A. W., and Pelletier, D. (2005). Progressive multifocal leukoencephalopathy in a patient treated with natalizumab. *N. Engl. J. Med.* **353,** 375–381.

Langrish, C. L., Chen, Y., Blumenschein, W. M., Mattson, J., Basham, B., Sedgwick, J. D., McClanahan, T., Kastelein, R. A., and Cua, D. J. (2005). IL-23 drives a pathogenic T cell population that induces autoimmune inflammation. *J. Exp. Med.* **201,** 233–240.

Lasagni, L., Francalanci, M., Annunziato, F., Lazzeri, E., Giannini, S., Cosmi, L., Sagrinati, C., Mazzinghi, B., Orlando, C., Maggi, E., Marra, F., Romagnani, S., *et al.* (2003). An alternatively spliced variant of CXCR3 mediates the inhibition of endothelial cell growth induced by IP-10, Mig, and I-TAC, and acts as functional receptor for platelet factor 4. *J. Exp. Med.* **197,** 1537–1549.

Lawman, S., Mauri, C., Jury, E. C., Cook, H. T., and Ehrenstein, M. R. (2004). Atorvastatin inhibits autoreactive B cell activation and delays lupus development in New Zealand black/white F1 mice. *J. Immunol.* **173,** 7641–7646.

Lehmann, P. V., Forsthuber, T., Miller, A., and Sercarz, E. E. (1992). Spreading of T-cell autoimmunity to cryptic determinants of an autoantigen. *Nature* **358,** 155–157.

Lennon, V. A., Wingerchuk, D. M., Kryzer, T. J., Pittock, S. J., Lucchinetti, C. F., Fujihara, K., Nakashima, I., and Weinshenker, B. G. (2004). A serum autoantibody marker of neuromyelitis optica: Distinction from multiple sclerosis. *Lancet* **364,** 2106–2112.

Lennon, V. A., Kryzer, T. J., Pittock, S. J., Verkman, A. S., and Hinson, S. R. (2005). IgG marker of optic-spinal multiple sclerosis binds to the aquaporin-4 water channel. *J. Exp. Med.* **202,** 473–477.

Leung, B. P., Sattar, N., Crilly, A., Prach, M., McCarey, D. W., Payne, H., Madhok, R., Campbell, C., Gracie, J. A., Liew, F. Y., and McInnes, I. B. (2003). A novel anti-inflammatory role for simvastatin in inflammatory arthritis. *J. Immunol.* **170,** 1524–1530.

Li, L., Narayan, K., Pak, E., and Pachner, A. R. (2006). Intrathecal antibody production in a mouse model of Lyme neuroborreliosis. *J. Neuroimmunol.* **173,** 56–68.

Liang, S. C., Tan, X. Y., Luxenberg, D. P., Karim, R., Dunussi-Joannopoulos, K., Collins, M., and Fouser, L. A. (2006). Interleukin (IL)-22 and IL-17 are coexpressed by Th17 cells and cooperatively enhance expression of antimicrobial peptides. *J. Exp. Med.* **203,** 2271–2279.

Linker, R. A., Rott, E., Hofstetter, H. H., Hanke, T., Toyka, K. V., and Gold, R. (2005). EAE in beta-2 microglobulin-deficient mice: Axonal damage is not dependent on MHC-I restricted immune responses. *Neurobiol. Dis.* **19,** 218–228.

Liu, L., Huang, D., Matsui, M., He, T. T., Hu, T., Demartino, J., Lu, B., Gerard, C., and Ransohoff, R. M. (2006a). Severe disease, unaltered leukocyte migration, and reduced IFN-gamma production in CXCR3−/− mice with experimental autoimmune encephalomyelitis. *J. Immunol.* **176,** 4399–4409.

Liu, Y., Teige, I., Birnir, B., and Issazadeh-Navikas, S. (2006b). Neuron-mediated generation of regulatory T cells from encephalitogenic T cells suppresses EAE. *Nat. Med.* **12,** 518–525.

Lucchinetti, C. F., Mandler, R. N., McGavern, D., Bruck, W., Gleich, G., Ransohoff, R. M., Trebst, C., Weinshenker, B., Wingerchuk, D., Parisi, J. E., and Lassmann, H. (2002). A role for humoral mechanisms in the pathogenesis of Devic's neuromyelitis optica. *Brain* **125,** 1450–1461.

Luhder, F., Huang, Y., Dennehy, K. M., Guntermann, C., Muller, I., Winkler, E., Kerkau, T., Ikemizu, S., Davis, S. J., Hanke, T., and Hunig, T. (2003). Topological requirements and signaling properties of T cell-activating, anti-CD28 antibody superagonists. *J. Exp. Med.* **197,** 955.

Lunemann, J. D., Waiczies, S., Ehrlich, S., Wendling, U., Seeger, B., Kamradt, T., and Zipp, F. (2002). Death ligand TRAIL induces no apoptosis but inhibits activation of human (auto) antigen-specific T cells. *J. Immunol.* **168,** 4881–4888.

Magliozzi, R., Columba-Cabezas, S., Serafini, B., and Aloisi, F. (2004). Intracerebral expression of CXCL13 and BAFF is accompanied by formation of lymphoid follicle-like structures in the meninges of mice with relapsing experimental autoimmune encephalomyelitis. *J. Neuroimmunol.* **148,** 11–23.

Matusevicius, D., Kivisakk, P., He, B., Kostulas, N., Ozenci, V., Fredrikson, S., and Link, H. (1999). Interleukin-17 mRNA expression in blood and CSF mononuclear cells is augmented in multiple sclerosis. *Mult. Scler.* **5,** 101–104.

McCandless, E. E., Wang, Q., Woerner, B. M., Harper, J. M., and Klein, R. S. (2006). CXCL12 limits inflammation by localizing mononuclear infiltrates to the perivascular space during experimental autoimmune encephalomyelitis. *J. Immunol.* **177,** 8053–8064.

McCarey, D. W., McInnes, I. B., Madhok, R., Hampson, R., Scherbakov, O., Ford, I., Capell, H. A., and Sattar, N. (2004). Trial of Atorvastatin in Rheumatoid Arthritis (TARA). double-blind, randomised placebo-controlled trial. *Lancet* **363,** 2015–2021.

McGeachy, M. J., Stephens, L. A., and Anderton, S. M. (2005). Natural recovery and protection from autoimmune encephalomyelitis: Contribution of CD4+CD25+ regulatory cells within the central nervous system. *J. Immunol.* **175,** 3025–3032.

McMahon, E. J., Bailey, S. L., Castenada, C. V., Waldner, H., and Miller, S. D. (2005). Epitope spreading initiates in the CNS in two mouse models of multiple sclerosis. *Nat. Med.* **11,** 335–339.

McRae, B. L., Vanderlugt, C. L., Dal Canto, M. C., and Miller, S. D. (1995). Functional evidence for epitope spreading in the relapsing pathology of experimental autoimmune encephalomyelitis. *J. Exp. Med.* **182,** 75–85.

Mendel, I., Kerlero, D. R., and Ben Nun, A. (1995). A myelin oligodendrocyte glycoprotein peptide induces typical chronic experimental autoimmune encephalomyelitis in H-2b mice: Fine specificity and T cell receptor V beta expression of encephalitogenic T cells. *Eur. J. Immunol.* **25,** 1951–1959.

Menge, T., Hemmer, B., Nessler, S., Wiendl, H., Neuhaus, O., Hartung, H. P., Kieseier, B. C., and Stuve, O. (2005). Acute disseminated encephalomyelitis: An update. *Arch. Neurol.* **62,** 1673–1680.

Miller, D. H., Khan, O. A., Sheremata, W. A., Blumhardt, L. D., Rice, G. P., Libonati, M. A., Willmer-Hulme, A. J., Dalton, C. M., Miszkiel, K. A., and O'Connor, P. W. (2003). A controlled trial of natalizumab for relapsing multiple sclerosis. *N. Engl. J. Med.* **348,** 15–23.

Miller, R. H. (1999). Contact with central nervous system myelin inhibits oligodendrocyte progenitor maturation. *Dev. Biol.* **216,** 359–368.

Moreau, T., Thorpe, J., Miller, D., Moseley, I., Hale, G., Waldmann, H., Clayton, D., Wing, M., Scolding, N., and Compston, A. (1994). Preliminary evidence from magnetic resonance imaging for reduction in disease activity after lymphocyte depletion in multiple sclerosis. *Lancet* **344,** 298–301.

Nath, N., Giri, S., Prasad, R., Singh, A. K., and Singh, I. (2004). Potential targets of 3-hydroxy-3-methylglutaryl coenzyme A reductase inhibitor for multiple sclerosis therapy. *J. Immunol.* **172,** 1273–1286.

Niino, M., Bodner, C., Simard, M. L., Alatab, S., Gano, D., Kim, H. J., Trigueiro, M., Racicot, D., Guerette, C., Antel, J. P., Fournier, A., Grand'Maison, F., *et al.* (2006). Natalizumab effects on immune cell responses in multiple sclerosis. *Ann. Neurol.* **59,** 748–754.

Nitsch, R., Bechmann, I., Deisz, R. A., Haas, D., Lehmann, T. N., Wendling, U., and Zipp, F. (2000). Human brain-cell death induced by tumour-necrosis-factor-related apoptosis-inducing ligand (TRAIL). *Lancet* **356,** 827–828.

Nogai, A., Siffrin, V., Bonhagen, K., Pfueller, C. F., Hohnstein, T., Volkmer-Engert, R., Bruck, W., Stadelmann, C., and Kamradt, T. (2005). Lipopolysaccharide injection induces relapses of experimental autoimmune encephalomyelitis in nontransgenic mice via bystander activation of autoreactive CD4+ cells. *J. Immunol.* **175,** 959–966.

O'Connor, P. W., Li, D., Freedman, M. S., Bar-Or, A., Rice, G. P., Confavreux, C., Paty, D. W., Stewart, J. A., and Scheyer, R. (2006). A Phase II study of the safety and efficacy of teriflunomide in multiple sclerosis with relapses. *Neurology* **66,** 894–900.

Paolillo, A., Coles, A. J., Molyneux, P. D., Gawne-Cain, M., MacManus, D., Barker, G. J., Compston, D. A., and Miller, D. H. (1999). Quantitative MRI in patients with secondary progressive MS treated with monoclonal antibody Campath 1H. *Neurology* **53,** 751–757.

Papiernik, M., de Moraes, M. L., Pontoux, C., Vasseur, F., and Penit, C. (1998). Regulatory CD4 T cells: Expression of IL-2R alpha chain, resistance to clonal deletion and IL-2 dependency. *Int. Immunol.* **10,** 371–378.

Park, H., Li, Z., Yang, X. O., Chang, S. H., Nurieva, R., Wang, Y. H., Wang, Y., Hood, L., Zhu, Z., Tian, Q., and Dong, C. (2005). A distinct lineage of CD4 T cells regulates tissue inflammation by producing interleukin 17. *Nat. Immunol.* **6,** 1133–1141.

Peng, Y., Laouar, Y., Li, M. O., Green, E. A., and Flavell, R. A. (2004). TGF-beta regulates in vivo expansion of Foxp3-expressing CD4+CD25+ regulatory T cells responsible for protection against diabetes. *Proc. Natl. Acad. Sci. USA* **101,** 4572–4577.

Peterson, J. W., Bo, L., Mork, S., Chang, A., and Trapp, B. D. (2001). Transected neurites, apoptotic neurons, and reduced inflammation in cortical multiple sclerosis lesions. *Ann. Neurol.* **50,** 389–400.

Pitt, D., Werner, P., and Raine, C. S. (2000). Glutamate excitotoxicity in a model of multiple sclerosis. *Nat. Med.* **6,** 67–70.

Polman, C. H., O'Connor, P. W., Havrdova, E., Hutchinson, M., Kappos, L., Miller, D. H., Phillips, J. T., Lublin, F. D., Giovannoni, G., Wajgt, A., Toal, M., Lynn, F., *et al.* (2006). A randomized, placebo-controlled trial of natalizumab for relapsing multiple sclerosis. *N. Engl. J. Med.* **354,** 899–910.

Radbruch, A., Muehlinghaus, G., Luger, E. O., Inamine, A., Smith, K. G., Dorner, T., and Hiepe, F. (2006). Competence and competition: The challenge of becoming a long-lived plasma cell. *Nat. Rev. Immunol.* **6,** 741–750.

Ransohoff, R. M., Kivisakk, P., and Kidd, G. (2003). Three or more routes for leukocyte migration into the central nervous system. *Nat. Rev. Immunol.* **3,** 569–581.

Rebenko-Moll, N. M., Liu, L., Cardona, A., and Ransohoff, R. M. (2006). Chemokines, mononuclear cells and the nervous system: Heaven (or hell) is in the details. *Curr. Opin. Immunol.* **18,** 683–689.

Reddy, J., Illes, Z., Zhang, X., Encinas, J., Pyrdol, J., Nicholson, L., Sobel, R. A., Wucherpfennig, K. W., and Kuchroo, V. K. (2004). Myelin proteolipid protein-specific CD4+CD25+ regulatory cells mediate genetic resistance to experimental autoimmune encephalomyelitis. *Proc. Natl. Acad. Sci. USA* **101**(43), 15434–15439. 26–10–2004. (GENERIC).

Reiber, H., and Peter, J. B. (2001). Cerebrospinal fluid analysis: Disease-related data patterns and evaluation programs. *J. Neurol. Sci.* **184,** 101–122.

Reiber, H., Ungefehr, S., and Jacobi, C. (1998). The intrathecal, polyspecific and oligoclonal immune response in multiple sclerosis. *Mult. Scler.* **4,** 111–117.

Reichardt, H. M., Gold, R., and Luhder, F. (2006). Glucocorticoids in multiple sclerosis and experimental autoimmune encephalomyelitis. *Expert. Rev. Neurother.* **6,** 1657–1670.

Rodriguez-Palmero, M., Hara, T., Thumbs, A., and Hunig, T. (1999). Triggering of T cell proliferation through CD28 induces GATA-3 and promotes T helper type 2 differentiation *in vitro* and *in vivo*. *Eur. J. Immunol.* **29,** 3914–3924.

Roncarolo, M. G., Bacchetta, R., Bigler, M., Touraine, J. L., de Vries, J. E., and Spits, H. (1991). A SCID patient reconstituted with HLA-incompatible fetal stem cells as a model for studying transplantation tolerance. *Blood Cells* **17,** 391–402.

Rose, J. W., Lorberboum-Galski, H., Fitzgerald, D., McCarron, R., Hill, K. E., Townsend, J. J., and Pastan, I. (1991). Chimeric cytotoxin IL2-PE40 inhibits relapsing experimental allergic encephalomyelitis. *J. Neuroimmunol.* **32,** 209–217.

Rose, J. W., Watt, H. E., White, A. T., and Carlson, N. G. (2004). Treatment of multiple sclerosis with an anti-interleukin-2 receptor monoclonal antibody. *Ann. Neurol.* **56,** 864–867.

Rosen, H., and Goetzl, E. J. (2005). Sphingosine 1-phosphate and its receptors: An autocrine and paracrine network. *Nat. Rev. Immunol.* **5,** 560–570.

Rottman, J. B., Slavin, A. J., Silva, R., Weiner, H. L., Gerard, C. G., and Hancock, W. W. (2000). Leukocyte recruitment during onset of experimental allergic encephalomyelitis is CCR1 dependent. *Eur. J. Immunol.* **30,** 2372–2377.

Rudick, R. A., Stuart, W. H., Calabresi, P. A., Confavreux, C., Galetta, S. L., Radue, E. W., Lublin, F. D., Weinstock-Guttman, B., Wynn, D. R., Lynn, F., Panzara, M. A., and Sandrock, A.W (2006). Natalizumab plus interferon beta-1a for relapsing multiple sclerosis. *N. Engl. J. Med.* **354,** 911–923.

Saiz, A., Carreras, E., Berenguer, J., Yague, J., Martinez, C., Marin, P., Rovira, M., Pujol, T., Arbizu, T., and Graus, F. (2001). MRI and CSF oligoclonal bands after autologous hematopoietic stem cell transplantation in MS. *Neurology* **56,** 1084–1089.

Sallusto, F., and Mackay, C. R. (2004). Chemoattractants and their receptors in homeostasis and inflammation. *Curr. Opin. Immunol.* **16,** 724–731.

Samoilova, E. B., Horton, J. L., and Chen, Y. (1998). Experimental autoimmune encephalomyelitis in intercellular adhesion molecule-1-deficient mice. *Cell Immunol.* **190,** 83–89.

Shepherd, J., Hunninghake, D. B., Stein, E. A., Kastelein, J. J., Harris, S., Pears, J., and Hutchinson, H. G. (2004). Safety of rosuvastatin. *Am. J. Cardiol.* **94,** 882–888.

Sheremata, W. A., Minagar, A., Alexander, J. S., and Vollmer, T. (2005). The role of alpha-4 integrin in the aetiology of multiple sclerosis: Current knowledge and therapeutic implications. *CNS Drugs* **19,** 909–922.

Sobel, R. A. (1989). T-lymphocyte subsets in the multiple sclerosis lesion. *Res. Immunol.* **140,** 208–211.

Steinman, L. (1996). A few autoreactive cells in an autoimmune infiltrate control a vast population of nonspecific cells: A tale of smart bombs and the infantry. *Proc. Natl. Acad. Sci. USA* **93,** 2253–2256.

Styren, S. D., Barbier, A. J., Selk, D. E., and Wettstein, J. G. (2004). Beneficial effects of teriflunomide in experimental allergic encephalomyelitis. Washington, DC: Society for Neuroscience, 2004 Abstract Program No. 344. 5, Online. 2004. (GENERIC).

Sun, D., Whitaker, J. N., Huang, Z., Liu, D., Coleclough, C., Wekerle, H., and Raine, C. S. (2001). Myelin antigen-specific CD8+ T cells are encephalitogenic and produce severe disease in C57BL/6 mice. *J. Immunol.* **166,** 7579–7587.

Suntharalingam, G., Perry, M. R., Ward, S., Brett, S. J., Castello-Cortes, A., Brunner, M. D., and Panoskaltsis, N. (2006). Cytokine storm in a phase 1 trial of the anti-CD28 monoclonal antibody TGN1412. *N. Engl. J. Med.* **355,** 1018–1028.

Targoni, O. S., Baus, J., Hofstetter, H. H., Hesse, M. D., Karulin, A. Y., Boehm, B. O., Forsthuber, T. G., and Lehmann, P. V. (2001). Frequencies of neuroantigen-specific T cells in the central nervous system versus the immune periphery during the course of experimental allergic encephalomyelitis. *J. Immunol.* **166,** 4757–4764.

Tedder, T. F., McIntyre, G., and Schlossman, S. F. (1988). Heterogeneity in the B1 (CD20) cell surface molecule expressed by human B-lymphocytes. *Mol. Immunol.* **25,** 1321–1330.

Thompson, E. J., Kaufmann, P., Shortman, R. C., Rudge, P., and McDonald, W. I. (1979). Oligoclonal immunoglobulins and plasma cells in spinal fluid of patients with multiple sclerosis. *Br. Med. J.* **1,** 16–17.

Trapp, B. D., Peterson, J., Ransohoff, R. M., Rudick, R., Mork, S., and Bo, L. (1998). Axonal transection in the lesions of multiple sclerosis. *N. Engl. J. Med.* **338,** 278–285.

Traugott, U., Reinherz, E. L., and Raine, C. S. (1983a). Multiple sclerosis. Distribution of T cells, T cell subsets and Ia-positive macrophages in lesions of different ages. *J. Neuroimmunol.* **4,** 201–221.

Traugott, U., Reinherz, E. L., and Raine, C. S. (1983b). Multiple sclerosis: Distribution of T cell subsets within active chronic lesions. *Science* **219**, 308–310.
Ubogu, E. E., Cossoy, M. B., and Ransohoff, R. M. (2006). The expression and function of chemokines involved in CNS inflammation. *Trends Pharmacol. Sci.* **27**, 48–55.
Urich, E., Gutcher, I., Prinz, M., and Becher, B. (2006). Autoantibody-mediated demyelination depends on complement activation but not activatory Fc-receptors. *Proc. Natl. Acad. Sci. USA* **103**, 18697–18702.
Vaknin-Dembinsky, A., Balashov, K., and Weiner, H. L. (2006). IL-23 is increased in dendritic cells in multiple sclerosis and down-regulation of IL-23 by antisense oligos increases dendritic cell IL-10 production. *J. Immunol.* **176**, 7768–7774.
Van Assche, G., Van Ranst, M., Sciot, R., Dubois, B., Vermeire, S., Noman, M., Verbeeck, J., Geboes, K., Robberecht, W., and Rutgeerts, P. (2005). Progressive multifocal leukoencephalopathy after natalizumab therapy for Crohn's disease. *N. Engl. J. Med.* **353**, 362–368.
Vandvik, B., Vartdal, F., and Norrby, E. (1982). Herpes simplex virus encephalitis: Intrathecal synthesis of oligoclonal virus-specific IgG, IgA and IgM antibodies. *J. Neurol.* **228**, 25–38.
Veldhoen, M., Hocking, R. J., Atkins, C. J., Locksley, R. M., and Stockinger, B. (2006a). TGFbeta in the context of an inflammatory cytokine milieu supports de novo differentiation of IL-17-producing T cells. *Immunity* **24**, 179–189.
Veldhoen, M., Hocking, R. J., Flavell, R. A., and Stockinger, B. (2006b). Signals mediated by transforming growth factor-beta initiate autoimmune encephalomyelitis, but chronic inflammation is needed to sustain disease. *Nat. Immunol.* **7**, 1151–1156.
Vieira, P. L., Christensen, J. R., Minaee, S., O'Neill, E. J., Barrat, F. J., Boonstra, A., Barthlott, T., Stockinger, B., Wraith, D. C., and O'Garra, A. (2004). IL-10-secreting regulatory T cells do not express Foxp3 but have comparable regulatory function to naturally occurring CD4+CD25+ regulatory T cells. *J. Immunol.* **172**, 5986–5993.
Vincenti, F., Kirkman, R., Light, S., Bumgardner, G., Pescovitz, M., Halloran, P., Neylan, J., Wilkinson, A., Ekberg, H., Gaston, R., Backman, L., and Burdick, J. (1998). Interleukin-2-receptor blockade with daclizumab to prevent acute rejection in renal transplantation. Daclizumab Triple Therapy Study Group. *N. Engl. J. Med.* **338**, 161–165.
Vollmer, T., Key, L., Durkalski, V., Tyor, W., Corboy, J., Markovic-Plese, S., Preiningerova, J., Rizzo, M., and Singh, I. (2004). Oral simvastatin treatment in relapsing-remitting multiple sclerosis. *Lancet* **363**, 1607–1608.
Waiczies, S., Prozorovski, T., Infante-Duarte, C., Hahner, A., Aktas, O., Ullrich, O., and Zipp, F. (2005). Atorvastatin induces T cell anergy via phosphorylation of ERK1. *J. Immunol.* **174**, 5630–5635.
Webb, M., Tham, C. S., Lin, F. F., Lariosa-Willingham, K., Yu, N., Hale, J., Mandala, S., Chun, J., and Rao, T. S. (2004). Sphingosine 1-phosphate receptor agonists attenuate relapsing-remitting experimental autoimmune encephalitis in SJL mice. *J. Neuroimmunol.* **153**, 108–121.
Weitz-Schmidt, G., Welzenbach, K., Brinkmann, V., Kamata, T., Kallen, J., Bruns, C., Cottens, S., Takada, Y., and Hommel, U. (2001). Statins selectively inhibit leukocyte function antigen-1 by binding to a novel regulatory integrin site. *Nat. Med.* **7**, 687–692.
Welsh, C. T., Rose, J. W., Hill, K. E., and Townsend, J. J. (1993). Augmentation of adoptively transferred experimental allergic encephalomyelitis by administration of a monoclonal antibody specific for LFA-1 alpha. *J. Neuroimmunol.* **43**, 161–167.
Wucherpfennig, K. W., Newcombe, J., Li, H., Keddy, C., Cuzner, M. L., and Hafler, D. A. (1992). T cell receptor V alpha-V beta repertoire and cytokine gene expression in active multiple sclerosis lesions. *J. Exp. Med.* **175**, 993–1002.
Yednock, T. A., Cannon, C., Fritz, L. C., Sanchez-Madrid, F., Steinman, L., and Karin, N. (1992). Prevention of experimental autoimmune encephalomyelitis by antibodies against alpha 4 beta 1 integrin. *Nature* **356**, 63–66.

Youssef, S., Stuve, O., Patarroyo, J. C., Ruiz, P. J., Radosevich, J. L., Hur, E. M., Bravo, M., Mitchell, D. J., Sobel, R. A., Steinman, L., and Zamvil, S. S. (2002). The HMG-CoA reductase inhibitor, atorvastatin, promotes a Th2 bias and reverses paralysis in central nervous system autoimmune disease. *Nature* **420,** 78–84.

Zhang, X., Reddy, J., Ochi, H., Frenkel, D., Kuchroo, V. K., and Weiner, H. L. (2006). Recovery from experimental allergic encephalomyelitis is TGF-beta dependent and associated with increases in CD4+LAP+ and CD4+CD25+ T cells. *Int. Immunol.* **18,** 495–503.

Zipp, F., and Aktas, O. (2006). The brain as a target of inflammation: Common pathways link inflammatory and neurodegenerative diseases. *Trends Neurosci.* **29,** 518–527.

Zipp, F., Krammer, P. H., and Weller, M. (1999). Immune (dys)regulation in multiple sclerosis: Role of the CD95-CD95 ligand system. *Immunol. Today* **20,** 550–554.

Zipp, F., Hartung, H. P., Hillert, J., Schimrigk, S., Trebst, C., Stangel, M., Infante-Duarte, C., Jakobs, P., Wolf, C., Sandbrink, R., Pohl, C., and Filippi, M. (2006). Blockade of chemokine signaling in patients with multiple sclerosis. *Neurology* **67,** 1880–1883.

Zurhein, G., and Chou, S. M. (1965). Particles pesembling Papova Viruses in human cerebral demyelinating disease. *Science* **148,** 1477–1479.

CHAPTER 2

Regulation of Interferon-γ During Innate and Adaptive Immune Responses

Jamie R. Schoenborn* and **Christopher B. Wilson**[†]

Contents			
	1.	Introduction	43
	2.	IFN-γ-Producing Cells	45
		2.1. NK cells	45
		2.2. NKT cells	45
		2.3. CD8 T cells	46
		2.4. CD4 T cells play multiple roles in adaptive immunity	46
	3.	Signaling Pathways Controlling IFN-γ Production by NK Cells	48
		3.1. NK receptors provide a dynamic rheostat to control NK cell responses	48
		3.2. IL-12 is a potent activator of IFN-γ production in NK cells	49
		3.3. IL-15 and IL-2 regulate NK cell development and contribute to IFN-γ production	51
		3.4. TGF-β is a negative regulator of IFN-γ production and NK cell development	51
	4.	Control of IFN-γ Production by NKT Cells	52
	5.	Signaling Pathways in the Differentiation of CD4 and CD8 T Cells	53
		5.1. Naive T cells require antigen stimulation for proliferation and effector commitment	53
		5.2. T cells require cytokine signals for sustained IFN-γ expression	54

* Molecular and Cellular Biology Graduate Program, University of Washington, Seattle, Washington
[†] Department of Immunology, University of Washington, Seattle, Washington

5.3. Other factors influencing
　　　　　Th1 lineage commitment　　　　　　　　　　　57
　　　5.4. TGF-β and IL-6 negatively regulate IFN-γ
　　　　　production and Th1 development　　　　　　58
　　　5.5. IFN-γ production by memory T cells in response
　　　　　to cytokine stimulation　　　　　　　　　　　59
6. Transcription Factors Downstream of the TCR,
　　Activating NK Receptors, and Cytokine Receptors　60
　　　6.1. Factors downstream of the TCR, costimulatory,
　　　　　and activating NK receptors　　　　　　　　60
　　　6.2. STATs are activated in response to cytokine
　　　　　receptor signaling　　　　　　　　　　　　68
　　　6.3. T-box family members are crucial for IFN-γ
　　　　　secretion　　　　　　　　　　　　　　　　70
7. Epigenetic Processes Govern Plasticity of Cell
　　Fate Choices and Help to Identify Distal
　　Regulatory Elements　　　　　　　　　　　　　　72
8. Transcriptional Regulatory Elements Within the
　　Ifng Gene　　　　　　　　　　　　　　　　　　74
　　　8.1. Regulatory elements within the *Ifng* promoter
　　　　　and gene　　　　　　　　　　　　　　　　74
　　　8.2. Identification of candidate distal regulatory
　　　　　elements in the *Ifng* locus　　　　　　　　77
9. Functional Analysis of Candidate Distal Regulatory
　　Elements in the *Ifng* Locus　　　　　　　　　　82
10. Conclusions and Future Directions　　　　　　　　84
　　Acknowledgments　　　　　　　　　　　　　　　　85
　　References　　　　　　　　　　　　　　　　　　85

Abstract　　Interferon-γ (IFN-γ) is crucial for immunity against intracellular pathogens and for tumor control. However, aberrant IFN-γ expression has been associated with a number of autoinflammatory and autoimmune diseases. This cytokine is produced predominantly by natural killer (NK) and natural killer T (NKT) cells as part of the innate immune response, and by Th1 CD4 and CD8 cytotoxic T lymphocyte (CTL) effector T cells once antigen-specific immunity develops. Herein, we briefly review the functions of IFN-γ, the cells that produce it, the cell extrinsic signals that induce its production and influence the differentiation of naïve T cells into IFN-γ-producing effector T cells, and the signaling pathways and transcription factors that facilitate, induce, or repress production of this cytokine. We then review and discuss recent insights regarding the molecular regulation of IFN-γ, focusing on work that has led to the identification and characterization of distal regulatory elements and epigenetic modifications with the IFN-γ locus (*Ifng*) that govern its expression. The epigenetic modifications and three-dimensional structure of the *Ifng* locus in naive CD4 T cells, and the modifications they undergo as these cells differentiate into effector T cells, suggest a model whereby the chromatin architecture of *Ifng* is poised to facilitate either rapid opening or silencing during Th1 or Th2 differentiation, respectively.

1. INTRODUCTION

The canonical Th1 cytokine, interferon-γ (IFN-γ), is critical for innate and adaptive immunity against viral and intracellular bacterial infections. In humans, genetic deficiencies in the interleukin (IL)-12/IL-23/IFN-γ pathways that result in decreased IFN-γ induction or signaling are associated with strikingly increased susceptibility to mycobacterial infections (Filipe-Santos *et al.*, 2006). Susceptibility to what are normally weakly pathogenic mycobacterial strains is greatly increased in such patients, whereas susceptibility to the more pathogenic mycobacteria that cause leprosy and tuberculosis has been observed less frequently and primarily in individuals with incomplete loss-of-function mutations in these pathways (Casanova and Abel, 2002). Systemic infections with *Salmonella* are also more common (de Jong *et al.*, 1998), but, unlike the risk for mycobacterial infection, are most often observed in those with defects in IL-12/IL-23 production or signaling rather than IFN-γ signaling; this difference suggests that risk for *Salmonella* infection may result both from a defect in IFN-γ production and the production of other cytokines, like IL-17 (MacLennan *et al.*, 2004).

IFN-γ is also involved in tumor control (Ikeda *et al.*, 2002; Rosenzweig and Holland, 2005). IFN-γ directly enhances the immunogenicity of tumor cells and stimulates the immune response against transformed cells. Human tumors can evade this form of control by becoming unresponsive to IFN-γ (Kaplan *et al.*, 1998).

Mice with targeted genetic deficiencies resulting in the loss of IFN-γ induction, production, or responsiveness are highly susceptible to infections due to intracellular bacteria, including mycobacteria, *Salmonella* (John *et al.*, 2002), Listeria (Harty and Bevan, 1995), intracellular protozoans (including Toxoplasma and Leishmania), and certain viruses (Dalton *et al.*, 1993; Huang *et al.*, 1993; Jouanguy *et al.*, 1999). Such mice also display a greater range, number, and aggressiveness of naturally occurring and induced tumors (Kaplan *et al.*, 1998).

The importance of IFN-γ in the immune system stems in part from its ability to inhibit viral replication directly, but most importantly derives from its immunostimulatory and immunomodulatory effects. IFN-γ, either directly or indirectly, upregulates both major histocompatibility complex (MHC) class I and class II antigen presentation by increasing expression of subunits of MHC class I and II molecules, TAP1/2, invariant chain, and the expression and activity of the proteasome. IFN-γ also contributes to macrophage activation by increasing phagocytosis and priming the production of proinflammatory cytokines and potent antimicrobials, including superoxide radicals, nitric oxide, and hydrogen peroxide (Boehm et al., 1997). As described below, IFN-γ also controls the differentiation of naive CD4 T cells into Th1 effectors, which mediate cellular immunity against viral and intracellular bacterial infections. Although necessary for clearing many types of infections, excess IFN-γ has been associated with a pathogenic role in chronic autoimmune and autoinflammatory diseases, including inflammatory bowel disease, multiple sclerosis, and diabetes mellitus (Bouma and Strober, 2003; Neurath et al., 2002; Skurkovich and Skurkovich, 2003). IFN-γ enhances lymphocyte recruitment and prolonged activation in tissues (Hill and Sarvetnick, 2002; Savinov et al., 2001). Thus, the induction, duration, and amount of IFN-γ produced must be closely controlled and delicately balanced for optimum host wellness.

The primary sources of IFN-γ are natural killer (NK) cells and natural killer T (NKT) cells, which are effectors of the innate immune response, and CD8 and CD4 Th1 effector T cells of the adaptive immune system. NK and NKT cells constitutively express IFN-γ mRNA, which allows for rapid induction and secretion of IFN-γ on infection. In contrast to NK and NKT cells, naive CD4 and CD8 T cells produce little IFN-γ immediately following their initial activation. However, naive CD4 and CD8 T cells can gain the ability to efficiently transcribe the gene encoding IFN-γ (*IFNG* in humans and *Ifng* in mice) over several days in a process that is dependent on their proliferation, differentiation, upregulation of IFN-γ-promoting transcription factors, and remodeling of chromatin within the *Ifng* locus. Naive CD8 T cells are programmed to differentiate into IFN-γ-producing cytotoxic effectors by default, whereas CD4 T cells can differentiate into a number of effector lineages, of which only Th1 CD4 effector T cells produce substantial amounts of IFN-γ. The process of effector differentiation in CD4 T cells, and to a lesser extent in CD8 T cells, is influenced by the nature of the infecting pathogen and the cytokine milieu emanating from the innate immune system in response to the pathogen. These differences in priming conditions in turn can result in stable changes to the chromatin structure of the gene encoding IFN-γ, either facilitating high-level expression in Th1 CD4 and CD8 effector T cells or silencing expression in other effector lineages.

2. IFN-γ-PRODUCING CELLS

2.1. NK cells

NK cells are a key component of the innate immune system providing early cellular defense against viruses and other intracellular pathogens, and contributing to the early detection and destruction of transformed host cells. NK cells develop in the bone marrow from a common lymphoid progenitor that also gives rise to B and T cells. NK cells express inhibitory receptors that recognize MHC class I molecules, which are expressed by all nucleated cells, serve as a marker for "self," and inhibit NK cell activation. Because antigen presentation via MHC class I is key to the CD8 T cell response, pathogens frequently downregulate MHC class I in an attempt to elude elimination by the CD8 T cell response; transformed cells often have reduced or absent MHC class I as well. Loss of MHC class I, in combination with binding of activating receptors to ligands on infected or transformed cells, leads to NK cell activation. Such activation causes NK cells to release cytotoxic granules containing perforin and granzymes, which induce programmed cell death in the target cell. Activated NK cells are also the primary rapid and potent producers of IFN-γ during the early innate immune response, which contributes directly to the innate immune response and also shapes the type and quality of the adaptive immune response that is subsequently elicited. NK cells transcribe *Ifng* at the earliest stages of NK cell development in the bone marrow. *Ifng* transcription increases during NK maturation and is rapidly and potently induced by NK activation, which also triggers translation and secretion of IFN-γ (Stetson et al., 2003).

2.2. NKT cells

NKT cells share characteristics of both NK and T cell lineages in that they express several activating and inhibitory receptors of the NK repertoire, as well as a rearranged T cell receptor (TCR) in association with the CD3 signaling complex. Many NKT cells express the CD4 coreceptor and surface markers associated with an activated or memory T cell phenotype: $CD44^{hi}$, $CD62L^{lo}$, $CD69^+$, and $IL-25^{hi}$. The largest and best-characterized population of NKT cells includes those expressing an invariant T cell receptor (TCR) consisting of a rearranged Vα14/Jα18 in the mouse and Vα24/Jα18 in humans. Invariant NKT cells recognize lipid antigens in the context of the nonclassical MHC class I molecule CD1d. These cells are derived from $CD4^+$ $CD8^+$ double-positive thymocytes (Egawa et al., 2005) that have been positively selected in the thymus by endogenous CD1d molecules presenting host-derived lipids (Gapin et al., 2001; Zhou et al., 2004a). Mature invariant NKT cells are strongly reactive to the marine sponge-derived glycolipid α-galactosyl ceramide (α-GalCer) presented by CD1d, and as such α-GalCer has been used as an immunostimulatory molecule in mouse tumor and infection models (Gansert *et al.*, 2003; Kawano *et al.*, 1997). NKT cells can be positively or negatively regulated by an array of cellular and microbial lipids presented by CD1d (Brutkiewicz, 2006), including phospholipids from *Leishmania* (Amprey *et al.*, 2004), mycobacterial lipids (Fischer *et al.*, 2004), and lysosomal glycosphingolipids (Zhou *et al.*, 2004a). Invariant NKT cells contribute to innate or preadaptive immunity to tumors and a range of infections, and help to maintain self-tolerance and to prevent autoimmunity. They do so by the rapid secretion of cytokines, including IFN-γ, tumor necrosis factor (TNF)-α, IL-4, IL-5, IL-13, granulocyte-macrophage colony-stimulating factor (GM-CSF), and IL-2. Furthermore, NKT cells are capable of Fas ligand- and perforin-dependent cytotoxic killing on TCR stimulation (Kronenberg, 2005; Kronenberg and Gapin, 2002).

2.3. CD8 T cells

Effector and memory CD8 T cells—often referred to as CTL—are important in the control of infection with viruses and certain other intracellular pathogens, and transformed host cells (Glimcher *et al.*, 2004; Harty *et al.*, 2000; Williams *et al.*, 2006). CD8 CTL are activated by signals from the TCR and costimulatory molecules in response to cognate MHC class I: peptide complexes presented by infected cells. Similar to NK cells, CTL mediate the destruction of target cells via the directed release of cytotoxic granules containing perforin and granzymes onto an infected cell. Activated CD8 CTL also mediate target cell killing by the interaction of Fas ligand with Fas on target cells. CD8 CTLs also produce copious amounts of IFN-γ in response to activation via the TCR or in response to IL-12 and IL-18 (Glimcher *et al.*, 2004). IFN-γ in turn increases expression of MHC class I, thus making infected cells more readily recognizable by CD8 CTLs.

2.4. CD4 T cells play multiple roles in adaptive immunity

In contrast to CD8 T cells that are hardwired for cytolytic function and IFN-γ production, CD4 T cells can adopt a variety of functional effector responses. At present, four distinct CD4 effector lineages have been described: Th1, Th2, Th17, and regulatory T cells (Tregs), of which only Th1 CD4 effector T cells produce large amounts of IFN-γ (Dong, 2006; Harrington *et al.*, 2006; Murphy and Reiner, 2002; Weaver *et al.*, 2006). The contributions of Th1 and Th2 CD4 T cells to cell-mediated or antiparasitic and humoral immunity, respectively, have been extensively studied

(Murphy and Reiner, 2002). Th1 cells combat infection by intracellular pathogens by producing IFN-γ and IL-2 to stimulate and sustain an effective cellular immune response. Th2 cells, on the other hand, produce the cytokines IL-4, IL-5, and IL-13 that promote clearance of infection by multicellular helminths. In mice, IL-4 and IL-5 secretion induces antibody (Ab) class switching in activated B cells from IgM and IgD to IgG1 and IgE and augments IgA production, respectively. IgG1 and IgA are functionally important for neutralization and targeting of extracellular pathogens for phagocytosis and killing by macrophages and neutrophils. IgE targets helminths for attack by eosinophils and triggers the activation of mast cells, thereby inducing mucus secretion, smooth muscle contraction, and vasodilatation to facilitate expulsion of helminths, while also playing a predominant role in asthma and allergy (Bischoff, 2007; Grimbaldeston *et al.*, 2006). CD4 Th17 cells are thought to protect against extracellular bacteria, particularly in the gut, by the secretion of the cytokines IL-17a, IL-17f, and IL-22 (Liang *et al.*, 2006). Th17 cells may also be primary mediators of experimental autoimmune encephalomyelitis (EAE), a mouse model of multiple sclerosis, and of psoriasis and inflammatory bowel disease. The fourth lineage of CD4 T cells, the Treg lineage, inhibits self-reactive adaptive responses that cause autoimmunity. Natural Tregs appear to arise as a separate lineage of CD4 T cells during thymic development (Sakaguchi, 2005). In addition to these natural Tregs, other Tregs can also be induced from naive CD4 T cell precursors as part of the adaptive immune response (Lohr *et al.*, 2006). Tregs dampen the immune response and prevent autoimmunity by direct suppression via cell–cell contact of effector T cells as well as by secretion of cytokines. Interestingly, Th17 cells share overlapping developmental factors with Treg cells (Weaver *et al.*, 2006), suggesting that the generation of these two CD4 T cell lineages must be finely balanced to provide protective immunity without undue risk for autoimmunity.

As described more fully below, the lineage choice between Th1, Th2, and Th17 is strongly influenced by the cytokine milieu produced by dendritic cells (DCs) and other antigen-presenting cells (APCs) at the time of T cell priming. Along with signals via the TCR and costimulatory receptors, cytokine receptor signaling pathways then induce or upregulate expression of the lineage-specific transcription factors, T-box expressed in T cells (T-bet), GATA-binding protein-3 (GATA-3), and retinoid orphan receptor-γ T (ROR-γT), in developing Th1, Th2, and Th17 cells, respectively; these transcription factors appear to be the "master regulators" of these respective cell fates. By contrast, Treg development, commitment, and/or survival require the forkhead family transcription factor, FoxP3 (Fontenot *et al.*, 2005), though the mechanisms by which this transcription factor acts are incompletely understood.

3. SIGNALING PATHWAYS CONTROLLING IFN-γ PRODUCTION BY NK CELLS

3.1. NK receptors provide a dynamic rheostat to control NK cell responses

The cytoplasmic storage of lytic granules and continual transcription of effector cytokines, including *Ifng*, position NK cells to respond within minutes to hours on activation, thereby contributing to early stages of immunity to infection. A fine balance of activating and inhibitory receptors are involved in the control of NK cell responses, and several of the receptors and downstream signaling events are shared with receptors found in other cells such as T cells (reviewed by Lanier, 2005 and Vivier *et al.*, 2004). Activating receptors recognize stress-induced ligands expressed on infected, damaged, or transformed cells and are balanced by the function of inhibitory receptors, which bind classical and nonclassical MHC class I molecules to prevent NK cell destruction of healthy host cells. Damage, transformation, or infection can downregulate the expression of MHC class I molecules, reducing the strength of inhibitory signals and contributing to NK cell activation. Inhibitory receptors contain immunoreceptor tyrosine-based inhibitory motifs (ITIMs) that signal via phosphorylated Src homology 2 domain tyrosine phosphatase-1 (SHP-1), SHP-2, or SHIP to prevent sustained calcium signaling and decrease the phosphorylation of a number of intracellular signaling molecules including Syk, phospholipase C-γ (PLC-γ), Shc, Vav, zeta chain-associated protein kinase of 70 kDa (ZAP70), SH2-domain-containing leukocyte protein of 76 kDa (SLP-76), and linker for activation of T cells (LAT). By contrast, activating receptors include the Fcγ receptor CD16, CD94/natural-killer group 2, member C (NKG2C), CD94/NKG2E, and NKG2D, in addition to Ly49H and Ly49D in the mouse, and in humans the short form killer immunoglobulin receptors (KIRs). These activating receptors couple to signaling adaptor molecules FcRγ, CD3ζ, or DAP12 to transmit activating signals into the cytoplasm via phosphorylation of immunoreceptor tyrosine-based activating motifs (ITAMs), resulting in the activation of protein tyrosine kinases (PTKs) of the Src family. Src PTKs then phosphorylate tyrosines on secondary PTKs of the Syk family, resulting in the recruitment of a number of cytosolic adaptor molecules including SLP-76, 3BP2, and LAT, which in turn lead to the activation of downstream signaling cascades, including mitogen-activated protein kinase (MAPK), PLC-γ, and the Son of sevenless (Sos)/Ras pathways. A fourth signaling adaptor, DAP10 functions as a costimulatory molecule similar to CD28 on the surface of T cells to activate Akt. Together, these signaling pathways downstream of activating receptors lead to the induction of NK cell effector mechanisms, including the rapid release of cytotoxic molecules and secretion of cytokines including IFN-γ.

The signaling pathways leading to IFN-γ secretion have been defined *in vitro* using a number of pharmacological inhibitors and by Ab cross-linking of activating receptors. NK cells appear to use distinct signaling pathways to induce cytotoxicity and cytokine secretion effector mechanisms, with the latter being more heavily dependent on MAPK signaling. Inhibition of early signaling events by blocking activation of spleen tyrosine kinase (SYK) and Src kinases, or downstream events including phosphoinositol 3-kinase (PI3K) and MAPK pathways, greatly decrease the ability of NK cells to produce IFN-γ. More specifically, activation of MAPK pathways involving extracellular signal-regulated kinase (ERK) and p38 kinases leads to cytokine secretion, in part through the activation of Fos and Jun transcription factors. Furthermore, *Syk−/− Zap70−/−* NK cells fail to produce IFN-γ when stimulated *in vitro* through a number of different activating receptors. IFN-γ production in response to NKG2D activation specifically requires DAP12/Syk and stimulates Janus kinase 2 (JAK2), signal transducers and activators of transcription 5 (STAT5), PI3K, and MAPK pathways to induce secretion of IFN-γ and other cytokines. Finally, calcium flux is necessary for high-level IFN-γ production in human NK cells in response to CD16 stimulation (Tassi and Colonna, 2005; Tassi *et al.*, 2005).

3.2. IL-12 is a potent activator of IFN-γ production in NK cells

In addition to NK cell activating receptors, a number of soluble and contact-dependent signals contribute to the activation of NK cells (reviewed by Newman and Riley, 2007), of which secretion of cytokines by infected cells and APCs are most well characterized. Among the cytokines that activate NK cells are IL-12, IL-18, IL-2, IL-15, and type I IFNs (IFN-α and IFN-β) (reviewed by Lieberman and Hunter, 2002).

IL-12, originally identified based on its ability to enhance NK cell cytotoxic killing, has long been recognized for its ability to stimulate IFN-γ secretion by NK cells. Binding of IL-12 to its receptor activates the transcription factor STAT4, which can facilitate IFN-γ expression both directly and indirectly by upregulating the expression of CD44 and CD28 (Kobayashi *et al.*, 1989). CD80/CD86 ligation of the CD28 molecule on the surface of NK cells can activate the transcription factor NFκB and affect several NK cell functions, including enhancing IFN-γ production *in vitro* as well as in response to infection *in vivo* (Fitzgerald *et al.*, 2000; Ghosh *et al.*, 1993).

The ability of IL-12 to stimulate NK cell production of IFN-γ can be further increased by the presence of other soluble and cell-bound ligands, including type I IFNs, TNF-α, and IL-18. Many of the synergistic stimuli that enhance IL-12 activation of IFN-γ production by NK cells share the ability to activate the transcription factor NFκB or STAT4. Binding of

IL-18 to its receptor activates NFκB and MAPK (Strengell et al., 2003), and markedly augments IL-12-induced IFN-γ production by NK cells (Hoshino et al., 1999). IFN-γ secreted by NK cells can bind to its receptor, IFN-γR, thereby activating STAT1 and upregulating T-bet expression (Afkarian et al., 2002). While this positive feedback loop is important for Th1 effector differentiation, it only modestly influences subsequent IFN-γ secretion by NK cells (Lee et al., 2000). In addition to the IFN-γR, the receptor for type I IFNs (IFNAR), signals through STAT1. *STAT1*−/− mice show impaired resistance to viral infection and tumors and decreased NK cytolytic function, but the ability of NK cells to secrete IFN-γ *ex vivo* in response to IL-12 stimulation is only marginally diminished (Lee et al., 2000).

On viral infection, most nucleated cells produce type I IFNs (IFN-α and IFN-β), which are distinct from the type II interferon, IFN-γ. Type I IFNs bind IFNAR on infected and neighboring cells, activating STAT1, STAT2, and STAT4, which results in the inhibition of host and viral protein synthesis, resistance to viral replication, and increased presentation of antigens by MHC class I (Decker et al., 2005; Lieberman and Hunter, 2002). NK cell cytotoxicity is largely dependent on type I IFNs (Nguyen et al., 2002a). IFN-β strongly enhances the cytotoxicity of human (Gerosa et al., 2005; Hanabuchi et al., 2006; Marshall et al., 2006) and murine NK cells (Hunter et al., 1997), but only modestly enhances IFN-γ production when used in combination with other stimuli and does not enhance on its own. IFN-α treatment of primary NK cells or the human NK-92 cell line induces low level, transient transcription of *IFNG*, which peaks around 1 h after stimulation, whereas IL-12 stimulation results in substantially longer and more robust *IFNG* transcription (Matikainen et al., 2001). While neither IFN-α nor IL-18 alone can stimulate substantial IFN-γ secretion by human or mouse NK cells, the combination of these two cytokines stimulates IFN-γ production in amounts similar to IL-12 alone (Matikainen et al., 2001). These data suggest that IFN-α can activate STAT4 in an IL-12-independent and transient manner that requires additional stimuli, either IL-12 or IL-18 to fully induce IFN-γ secretion from NK cells (Hunter et al., 1997).

Human plasmacytoid dendritic cells (pDCs) and myeloid DCs (mDCs) can respond to microbial stimuli, resulting in the release of cytokines that influence NK cell viability and function. Furthermore, pDCs and mDCs can provide accessory ligands that mediate NK cell activation. Activation of pDCs via Toll-like receptor 9 (TLR9) by various CpGs oligonucleotides stimulates the secretion of IFN-α and TNF-α. In human NK cells, IFN-α and TNF-α can synergize to induce IFN-γ secretion (Marshall et al., 2006). In response to poly (I:C), mDCs induce IFN-γ production by NK cells through cell–cell contact and IL-12-dependent mechanisms (Gerosa et al., 2005). Resting human NK cells, as well as T cells and macrophages, express the glucocortoid-induced TNF receptor (GITR), stimulation of which promotes NK cell lytic function and IFN-γ secretion in synergy with IL-2, IFN-α, and NKG2D signaling (Hanabuchi et al., 2006). The expression of the GITR ligand (GITRL) on the surface of activated pDCs (Hanabuchi et al., 2006) provides additional evidence that NK cell activation relies on accessory signals from DCs.

3.3. IL-15 and IL-2 regulate NK cell development and contribute to IFN-γ production

The development of mature NK cells requires IL-15 production from the bone marrow stromal cells and monocytes (Carson et al., 1997; Puzanov et al., 1996; Vosshenrich et al., 2005). In the secondary lymphoid tissues, IL-15 is thought to be produced and trans-presented to NK cells by APCs, thereby providing survival and/or proliferative signals necessary for homeostatic maintenance of mature, peripheral NK cells, and priming these cells for cytolytic activity and the production of IFN-γ (Lucas et al., 2007; Ma et al., 2006; Williams et al., 1998). Transfer of mature NK cells into *IL-15−/−* or *IL-15Rα−/−* mice results in a dramatic loss of the transferred NK cells (Cooper et al., 2002; Koka et al., 2003), further demonstrating the need of NK cells for IL-15-induced survival signals. Like IL-12 and IFN-α, IL-15 acts in concert with IL-18 to stimulate IFN-γ production (Strengell et al., 2003). IL-2, which shares the common β and γ chains of its receptor with IL-15, can also promote NK cell growth, differentiation, cytolytic activity, and IFN-γ production *in vitro* (Fujii et al., 1998). However, unlike IL-15, IL-2 is dispensable for NK cell development and function *in vivo*. IL-2 does not directly affect IL-12Rβ2 expression (Wang et al., 1999) and only minimally affects IFN-γ secretion by mouse NK cells (Chang and Aune, 2005), but exposure of NK cells to IL-2 prior to stimulation with IL-12 increases expression of the IL-12Rβ2, suggesting that IL-2 may enhance NK response to IL-12 *in vivo* (Wang et al., 2000).

3.4. TGF-β is a negative regulator of IFN-γ production and NK cell development

While IL-2, IL-15, IL-12, IL-18, and type I IFNs promote IFN-γ production by NK cells, excess or prolonged production of IFN-γ can lead to protracted inflammation and immune activation, resulting in inflammatory and autoimmune pathologies. The immunoregulatory cytokine transforming growth factor-β (TGF-β) inhibits the expression of IFN-γ by NK cells (Li et al., 2006b). TGF-β induces the phosphorylation of SMAD2 and/or SMAD3 signaling proteins, which then bind with a common SMAD4 partner, translocate to the nucleus and bind to the promoters of target genes and recruit activating or repressive complexes (Massague, 1998). As shown by chromatin immunoprecipitation (ChIP), SMAD3/4 heterodimers can bind to the *Ifng* promoter and repress transcription (Yu *et al.*, 2006). In addition to directly inhibiting *Ifng* transcription, TGF-β inhibits the expression of T-bet, STAT4, and IL-12Rβ2 (Gorelik and Flavell, 2000; Lin *et al.*, 2005), all of which are important for IFN-γ expression. In addition to blocking expression of IFN-γ, TGF-β also has selective effects on NK cell development and homeostasis. Mice with deficiencies in TGF-β signaling in NK cells develop increased numbers of NK cells as early as 3 weeks after birth (Laouar *et al.*, 2005). dnTGF-βRII transgenic mice also have higher frequencies of NK cells that are sustained much longer after *L. major* infection than do control mice (Laouar *et al.*, 2005). IL-12, IL-15, and IL-18 decrease the ability of NK cells to respond to TGF-β by decreasing transcription and surface expression of the receptor for TGF-β and the SMAD2 and SMAD3 signaling molecules, and by partially rescuing expression of the transcription factor T-bet. Furthermore, enforced expression of T-bet in TGF-β-treated NK cells maintains IFN-γ production (Gorelik and Flavell, 2000; Yu *et al.*, 2006).

4. CONTROL OF IFN-γ PRODUCTION BY NKT CELLS

NKT cells are activated on TCR recognition of lipid antigens presented by CD1d. The ability of NKT cells to produce large amounts of both Th1 and Th2 cytokines is well documented and contrasts sharply with conventional T cells, which produce one or the other. Nearly all unstimulated NKT cells transcribe *Ifng* and *Il4* (the gene encoding IL-4) and well over half transcribe message for both cytokines (Matsuda *et al.*, 2003; Stetson *et al.*, 2003), indicating that the transcription factors and epigenetic status necessary for high-level expression of these cytokines are already in place prior to stimulation of NKT cells. Nonetheless, exposure to IL-12 plus IL-18 selectively induces the secretion of IFN-γ in NKT cells (Nagarajan and Kronenberg, 2007) as it does in NK cells and memory/effector Th1 and CD8 T cells. However, NKT cells from mice deficient for either the IL-4R or IL-12R produce IFN-γ and IL-4 in amounts similar to controls in response to α-GalCer stimulation, which is not the case for conventional T cells. As IL-12 and IL-4 play little role in influencing the types of cytokines produced by NKT cells, it is not surprising that attempts to "polarize" NKT cells by varying the dose, route, or timing of antigen has failed to alter the immediate cytokine profile (Matsuda *et al.*, 2003). Thus, it has been proposed that NKT cells are relatively fixed in their ability to simultaneously express both *Ifng* and *Il4*.

The mechanisms that allow both classes of cytokines to be produced by NKT cells remain poorly understood (Kronenberg, 2005). One possibility is that only a fraction of NKT cells secretes IL-4 and another fraction secretes IFN-γ. In this model, the choice of cytokines secreted by an individual NKT cell could either be random, be determined by the differential activation of NK receptors on the surface of an individual NKT cell, or be determined by differences in the structure of the lipopeptides recognized by NKT cells, which in turn would result in qualitatively or quantitatively different signals via the TCR. For example, reduced aliphatic chain length on derivatives of α-GalCer led to increased Th2 cytokine secretion but impaired IFN-γ secretion as a result of reduced activation of the NFκB transcription factor c-Rel (Oki *et al.*, 2004), whereas an analog of α-GalCer containing a carbon replacement resulted in increased IFN-γ secretion (Schmieg *et al.*, 2003). Alternatively, there could be subsets of NKT cells more poised toward the production of IFN-γ or IL-4. Finally, NKT cells may secrete both IFN-γ and IL-4, as the findings of Stetson *et al.* (2003) and Matsuda *et al.* (2003) suggest, and the resulting modulation of the immune response may be in part due to the differential induction of surface molecules, including those that are involved in cell–cell interactions between NKT cells and other responding cells such as NK cells or DCs.

5. SIGNALING PATHWAYS IN THE DIFFERENTIATION OF CD4 AND CD8 T CELLS

As individual CD4 T cells commit to the Th1 or Th2 effector lineage to the exclusion of the other lineage, they have provided a well-utilized model in which to study cellular and molecular events involved in cell fate choices. Much less is known regarding the Th17 fate choice, although knowledge in this area is rapidly accruing. (For a review of Th17 differentiation, see Harrington *et al.*, 2006.) Th2 differentiation has also been reviewed (Ansel *et al.*, 2003, 2006; Barbulescu *et al.*, 1998; Lee *et al.*, 2006; Szabo *et al.*, 2003). Herein, we review the signaling pathways and cellular events that lead to Th1 differentiation and IFN-γ production, and draw parallels to CD8 T cell effector development and IFN-γ secretion.

5.1. Naive T cells require antigen stimulation for proliferation and effector commitment

While NK and NKT cells are able to secrete IFN-γ within hours of their stimulation, differentiation of naive CD4 and CD8 T cells into efficient IFN-γ secreting cells requires several round of proliferation. While they proliferate, environmental signals influence the expression and activation status of specific receptors, downstream signaling molecules, and transcription factors, which in turn allow these T cells to express IFN-γ, remodel the *Ifng* locus, and commit to the Th1 CD4, or CD8 CTL effector lineage (Murphy and Reiner, 2002; Reiner, 2001; Reiner and Seder, 1999).

For naïve CD8 T cells, signals delivered from the TCR and costimulatory molecules induce differentiation into a fully committed CTL that is capable of IFN-γ secretion and direct killing of infected or transformed cells expressing the cognate MHC class I:peptide complex. The directed commitment of CD8 T cells to become IFN-γ-producing CTLs is thought to result from the constitutive expression of the T-bet paralog Eomesodermin (Eomes) by naive CD8 T cells (Pearce *et al.*, 2003).

By contrast, naive CD4 T cells are more plastic. When stimulated via their TCR and costimulatory molecules, naive CD4 T cells transcribe and produce substantial amount of IL-2 and also produce small amounts of IFN-γ and IL-4 (Grogan and Locksley, 2002), which may poise them to adopt multiple effector phenotypes. The initial coactivation of the *Ifng, Il4*, and Il2 cytokine genes in naive T cells occurs through the concerted activation of constitutively expressed, and rapidly activated transcription factors, including nuclear factor of activated T cells (NFATs), nuclear factor κB (NFκB), and activator protein-1 (AP-1). This coactivation may be facilitated by the juxtaposition of the genes encoding these cytokines in the nuclei of naive T cells, even though these genes are located on different chromosomes (Spilianakis *et al.*, 2005).

Since their differentiation into Th1, Th2, or Th17 effector lineages determines whether the ensuing immune response will be appropriate for specific types of pathogens, it is appropriate that the dominant factor governing these cell fate choices is the cytokine milieu produced by APCs, which use TLRs and other cell surface and cytosolic recognition systems to deduce the nature of the microbial threat and instruct T cell fate choice accordingly (Pulendran and Ahmed, 2006). Other factors involved in lineage choice are the nature of the APC and the magnitude of stimulation via the TCR and costimulatory molecules, with stronger and longer signaling, generally favoring Th1 development. However, the cytokine milieu present during CD4 differentiation remains the best-characterized and most vital influence on naive CD4 T cell priming and early differentiation.

5.2. T cells require cytokine signals for sustained IFN-γ expression

Cytokines play an important role in IFN-γ induction, maintenance, and Th1 differentiation in CD4 T cells. While IFN-γ production appears to be relatively independent of cytokine signals in NKT cells, the importance of IL-12, IL-18, and IFN-γ itself have been well documented in CD4 and CD8 T cells as in NK cells, and many of the downstream signaling pathways are shared among these cell types. The ability of other cytokines, such as type I IFNs, to support IFN-γ expression varies among these cell types, as described below.

Th1 development is heavily influenced by IFN-γ produced by NK cells and by IL-12 and IL-18 produced by DCs and other APCs. Binding of IFN-γ to its receptor on the surface of CD4 T cells leads to STAT1 phosphorylation and nuclear translocation, which in concert with signals from the TCR and CD28 costimulatory molecules, induces T-bet (Lighvani et al., 2001). The induction of IFN-γ transcription and commitment to the Th1 lineage by T-bet is facilitated by its ability to induce two additional transcription factors, Hlx (Mullen et al., 2002) and Runx3 (Djuretic et al., 2007). Runx3 binds cooperatively with T-bet at the *Ifng* promoter to induce its transcription and to the *Il4* silencer to extinguish its expression (Djuretic et al., 2007). Hlx facilitates *Ifng* transcription and chromatin accessibility at the *Ifng* promoter (Mullen et al., 2002) and helps to induce the expression of the IL-12Rβ2 chain (Afkarian et al., 2002). Binding of the IL-12 p35 and p40 heterodimer to its receptor (IL-12Rβ1 and IL-12Rβ2) in turn activates Jak2/Tyk2 and induces signals via phosphorylation and nuclear translocation of STAT4 (Trinchieri et al., 2003), one effect of which is to induce expression of the IL-18 receptor. IL-12 also activates p38 MAPK. Loss of p38α activity in T cells, either by use of a pharmacological inhibitor or p38α-deficient cells, blocks IFN-γ production in response to cytokine stimulation but not TCR stimulation (Berenson et al., 2006b). The combined signals from IFN-γ, IL-12, and IL-18 maximize expansion and optimal activation of Th1 cells (Grogan and Locksley, 2002; Ho and Glimcher, 2002; Murphy and Reiner, 2002), facilitating their permanent commitment to the Th1 lineage.

IFN-γ acts in a positive feedback loop to facilitate its own expression by T cells, as it does in NK cells. In response to activation of naive CD4 T cells via the TCR and costimulatory molecules, the IFN-γ receptor is recruited to the immunologic synapse; this recruitment decreases significantly in the presence of IL-4 and is weaker in naive CD4 T cells from Th2-biased BALB/c mice compared to C57BL/6 mice (Maldonado et al., 2004). The focused recruitment of IFN-γ receptors to the synapse where IFN-γ is being secreted results in an autocrine, positive feedback loop facilitating IFN-γ production and Th1 lineage commitment (Maldonado et al., 2004). While IFN-γ is not required for the induction of Th1 cells, it plays a critical role in suppressing the IL-4-producing potential of Th1 cells (Zhang et al., 2001) and in maintaining the expression of T-bet during CD4 Th1 effector commitment (Afkarian et al., 2002).

Type I IFNs have been implicated in the phosphorylation of STAT4 and Th1 differentiation by human CD4 T cells (Cho et al., 1996; Rogge et al., 1997; Tyler et al., 2007). When combined with IL-18 stimulation, IFN-α/β can drive acute IFN-γ secretion by human CD4 T cells (Athie-Morales et al., 2004; Brinkmann et al., 1993; Matikainen et al., 2001; Parronchi et al., 1992; Sareneva et al., 1998, 2000). In human T and NK cells, IFN-α stimulation upregulates MyD88 mRNA and synergizes with

IL-12 to induce the IL-18 receptor complex, sensitizing these cells to low concentrations of IL-18 (Sareneva et al., 2000). The role of type I IFNs in murine CD4 Th1 development, however, is less clear, as STAT4 phosphorylation in response to IFN-α/β stimulation does not occur as efficiently in mouse T cells as it does in human T cells (Berenson et al., 2004a; Farrar and Murphy, 2000; Farrar et al., 2000a,b; Freudenberg et al., 2002; Persky et al., 2005; Rogge et al., 1997, 1998; Szabo et al., 1997). As a result, mouse CD4 T cells are unable to secrete substantial amounts of IFN-γ or differentiate into Th1 effector cells in response to IFN-α/β stimulation (Berenson et al., 2004a; Persky et al., 2005; Rogge et al., 1998; Wenner et al., 1996). Thus, IL-12R but not type I IFNs can induce the sustained STAT4 activation and IFN-γ production signaling required for Th1 development (Berenson et al., 2006a). In CD8 T cells, which are less dependent on IL-12 for differentiation than CD4 T cells (Carter and Murphy, 1999), IFN-α appears to be sufficient for clonal expansion and gain of cytotoxic effector function on primary stimulation, but is still considerably less effective than IL-12 in facilitating their differentiation into efficient producers of IFN-γ (Curtsinger et al., 2005).

These differences in the ability of IFN-α/β to induce STAT4 phosphorylation likely result from differences between human and murine T cells in the recruitment of STAT4 to the IFNAR. In humans, STAT4 recruitment to IFNAR is STAT2 dependent. However, the murine *Stat2* gene contains a carboxy-terminal minisatellite repeat not found in the human gene (Farrar et al., 2000a; Park et al., 1999; Paulson et al., 1999). Expression of mouse STAT2 in human *STAT2−/−* T cells failed to restore IFNα/β-dependent STAT4 phosphorylation, demonstrating that mouse STAT2 is not functional in human T cells. A chimeric murine N-terminal/human C-terminal *STAT2* gene did restore STAT4 phosphorylation in STAT2-deficient human fibroblasts (Farrar et al., 2000a,b), but was unable to support Th1 development or IFN-γ production in response to type I IFNs in transgenic mice (Persky et al., 2005). The N-terminal domain of STAT4 was shown to interact with the cytoplasmic domain of the human IFNAR2 subunit but not murine IFNAR2 (Tyler et al., 2007), suggesting that differences in both IFNAR and STAT2 contribute to the impaired STAT4 phosphorylation in murine T cells. Together, these data suggest that in murine T cells type I IFNs are unable to induce an effective and sustained STAT4 signal that is necessary for full commitment to IFN-γ production and Th1 differentiation.

While IL-12 is a potent regulator of Th1 immunity, two closely related cytokine family members, IL-23 and IL-27, have been found to play different roles in the development of CD4 effector lineages. Similar to IL-12, IL-23 is produced by activated DCs and its receptor complex is similarly upregulated on NK cells and activated/memory CD4 T cells, DCs, and bone marrow-derived macrophages (Trinchieri et al., 2003).

Early studies implicated IL-23 in later stages of Th1 development or maintenance of IFN-γ production (Oppmann et al., 2000), but IL-23 has recently been shown to be involved in the survival and expansion of the Th17 lineage of CD4 effector T cells (Dong, 2006; Weaver et al., 2006). IL-23 shares the p40 subunit with IL-12, which together with a unique p19 subunit signals through the IL-23 receptor containing the IL-12Rβ1 and IL-23R chains to induce activation of STAT3/STAT4 heterodimers, as compared to the STAT4 homodimer that is induced in response to binding of IL-12 to its receptor.

IL-27 synergizes with IL-12 and IL-18 to induce IFN-γ production, contributes to Th1 differentiation, and can induce the proliferation of Th1 effector cells. IL-27 is a heterodimer expressed by virally infected cells, activated macrophages, and DCs, and is composed of a p28 subunit and an Epstein–Barr virus-induced gene 3 (EBI3) subunit. The receptor for IL-27, composed of GP130 and T cell cytokine receptor (TCCR)/WSX-1 subunits, has strong homology to that of IL-12Rβ2 and is expressed primarily by NK cells and resting CD4 T cells. IL-27 activates both STAT1 and STAT3 proteins; the activation of STAT1 in turn induces T-bet expression, thereby facilitating IFN-γ production and the differentiation of naive CD4 T cells into IL-12-responsive, Th1 effectors (Pflanz et al., 2002). Consequently, mice with deficiencies in IL-27 signaling show diminished CD4 IFN-γ production following primary immunization with keyhole limpet hemocyanin protein (Yoshida et al., 2001) and increased susceptibility to *L. monocytogenes* and *Leishmania major* (Chen et al., 2000; Yoshida et al., 2001). However, while the tempo with which Th1 responses develop *in vivo* is often delayed in IL-27 receptor-deficient mice, given time, these mice ultimately mount Th1 responses in response to chronic infection (Hunter, 2005), likely as a result of alternative pathways to induce T-bet and to facilitate Th1 differentiation (e.g., via IFN-γ and IL-12). By contrast to its redundant role in Th1 development, IL-27 may be essential for repressing the development of Th17 effectors (Hunter, 2005).

5.3. Other factors influencing Th1 lineage commitment

Notch signaling, an evolutionarily conserved pathway involved in cell fate choices, also affects T cell differentiation and Th1 and Th2 effector commitment. In mammals, there are four Notch receptors: Notch1, Notch2, Notch3, and Notch4, and five canonical Notch ligands that can be divided into the Jagged ligand family, containing Jagged1 and Jagged2, and the Delta ligand family, containing Delta-like 1 (DLL1), DLL3, and DLL4 (Baron, 2003). Notch ligands are expressed by APCs and, as described above, TLR receptor recognition of pathogens by APCs likely leads to differences in the expression of cytokines and accessory molecules, such as Notch ligands, that provide early instructional signals to differentiating CD4 T cells. Notch signaling is central to several of the cell lineage choices made during T cell development, including T versus B cell commitment, αβ versus γδ T cell choice, and whether to develop into a CD4 or CD8 functionally mature T cell following selection (reviewed in Osborne and Minter, 2006). A role for Notch receptors in T cell activation, proliferation, and cytokine production has been established (Adler *et al.*, 2003; Palaga *et al.*, 2003). Use of pharmacological inhibitors that block Notch upregulation in response to TCR stimulation in mature CD4 and CD8 T cells significantly impaired T cell proliferation and IFN-γ production (Palaga *et al.*, 2003).

Expression of Delta proteins by APCs during *in vitro* stimulation of naive CD4 T cells favors differentiation into a Th1 cell fate, whereas Th2 development is favored when APCs express the alternate Jagged family of Notch ligands (Amsen *et al.*, 2004). These effects are largely independent of cytokine signaling, suggesting that Notch signaling may have a direct effect on the expression of IL-4 and IFN-γ. Inhibition of Notch function using a γ-secretase inhibitor, which prevents cleavage of the cytoplasmic portion of the Notch receptor, blocked T-bet expression and Th1 commitment in Th1 conditions, but had no effect on developing Th2 cells (Minter *et al.*, 2005). TCR stimulation of naive CD4 T cells in the presence of a DLL1-Fc fusion protein was both necessary and sufficient to significantly increase the expression T-bet and the number of cells secreting IFN-γ (Maekawa *et al.*, 2003). Furthermore, naive CD4 T cells stimulated in Th2 conditions in the presence of DLL1-Fc produced decreased amounts of IL-4. Pretreatment of BALB/c mice with Delta1 prior to *L. major* infection supported the development of a Th1-based immune response, reducing the footpad swelling and promoting IFN-γ secretion by CD4 T cells. In good agreement with these data, pretreatment of mice with the γ-secretase inhibitor delayed the onset of disease in the EAE mouse model of multiple sclerosis, as well as reduced the severity of disease when continuously administered to affected mice (Minter *et al.*, 2005).

5.4. TGF-β and IL-6 negatively regulate IFN-γ production and Th1 development

Similar to NK cells, TGF-β is a potent antagonist of IFN-γ secretion and Th1 development by CD4 T cells. Mice deficient in TGF-β1 or its receptor develop widespread autoimmunity characterized by uncontrolled CD4 T cell activation and excess IFN-γ production (Boivin *et al.*, 1995; Gorelik and Flavell, 2000; Kulkarni *et al.*, 1993; Lucas *et al.*, 2000; Shull *et al.*, 1992). TGF-β blocks Th1 differentiation by inhibiting expression of factors necessary for Th1 development: T-bet, STAT4, IL-12Rβ2, and IFN-γ itself. The SMAD3/4 heterodimers activated by TGF-βR signaling directly inhibit expression of the Th1 master regulator T-bet; conversely, the T-bet paralog, Eomes, which is expressed in NK cells and CD8 T cells but not in CD4 T cells, is not downregulated by SMAD3/4, allowing these cells to maintain some level of IFN-γ production even in the presence of TGF-β. Studies in transgenic mice expressing a dominant-negative form of the TGF-βRII under the control of the CD4 promoter in T cells (Gorelik and Flavell, 2000), indicate that TGF-β signaling in CD4 and NK cells serves to reduce the number and function of Th1 effector cells in mice infected with *L. major*. Thus, while normal BALB/c mice are susceptible to *L. major*, dnTGF-βRII transgenic BALB/c mice generate greater numbers of antigen-specific IFN-γ-producing Th1 effector CD4 T cells and exhibit decreased footpad swelling and parasite numbers (Gorelik and Flavell, 2000; Laouar *et al.*, 2005).

Ablation of TGF-βRII in T cells inhibits NKT cell development and CD8 T cell maturation in the thymus. The remaining peripheral T cells are phenotypically activated Th1 effector CD4 T cells that express T-bet, whereas the frequency of regulatory T cells, which require TGF-β for development, is reduced (Li *et al.*, 2006a). Together these data implicate TGF-β signaling in various aspects of NKT and T cell development, IFN-γ production, and homeostasis.

IL-6 is produced by several cell types including macrophages, DCs, and B cells and is involved in the differentiation of B cells, macrophages, regulatory T cells, and Th17 CD4 effector T cells (Diehl *et al.*, 2000). IL-6 also inhibits Th1 polarization by facilitating NFAT-dependent activation of the *Il4* gene in naive CD4 T cells and by potentiating expression of the suppressor of cytokine signaling-1 (SOCS-1). SOCS-1 blocks signaling from the IFN-γR by preventing the phosphorylation and subsequent activation of STAT1 (Diehl *et al.*, 2000).

5.5. IFN-γ production by memory T cells in response to cytokine stimulation

Formation of a long-lived pool of memory T cells that are capable of a more rapid and robust immune response on reencounter with pathogens forms the basis of vaccination and protective immunity. The ability of memory T cells to mount an effective secondary response is due in part to the increase in memory T cell precursor frequency compared to naive T cell precursor frequencies. In addition, relative to naive T cells, memory T cells have a lower activation threshold, enter the cell cycle more rapidly, and are poised for rapid and potent secretion of IFN-γ (reviewed by Sprent and Surh, 2002). Effector memory T cells, which primarily reside in nonlymphoid tissues, are poised to provide immediate immunity on antigenic reexposure. These cells constitutively express mRNA for *Ifng* and other effector molecules (Bachmann *et al.*, 1999; Cho *et al.*, 1999;

Stetson *et al.*, 2003; Zimmermann *et al.*, 1999), allowing them to secrete IFN-γ within hours of activation (Cho *et al.*, 1999). The efficiency with which memory CD8 T cells protect against reinfection has been demonstrated by studies showing that the transfer of memory CD8 CTL into IFN-γ-deficient hosts that were subsequently infected with wild-type *Listeria monocytogenes* provided greater protection than transferred NK cells (Berg *et al.*, 2005). Generation of memory CD8 T cells capable of responding in a recall response is dependent on the presence of CD4 T cells and IL-2 during the primary response (Sun *et al.*, 2004a), whereas the subsequent maintenance of these cells is dependent, like NK cells, on IL-15 (Nishimura *et al.*, 2000; Schluns *et al.*, 2002). The clonal expansion and long-term survival of memory CD8 T cells is further influenced by IL-12-dependent STAT4 activation during priming, which results in increased expression of Bcl3 and Bcl2-related genes (Li *et al.*, 2006c).

Unlike naive T cells, which require sustained antigen stimulation, costimulation, and cytokine signals to differentiate into effector cells capable of sustained robust expression of IFN-γ, memory T cells can respond to cytokine or antigen stimulation in the absence of costimulation. Stimulation of human memory CD8 T cells with IL-12 and IL-18 is sufficient to induce IFN-γ secretion (Berg *et al.*, 2003; Smeltz, 2007), which can be substantially increased by addition of IL-15 (Smeltz, 2007). The ability of memory T cells to secrete IFN-γ in response to cytokine stimulation in amounts similar to those induced by antigen stimulation is largely due to the fact that effector and memory CD8 T cells express more IL-12Rβ2, IL-18Rα, and IL-18Rβ than naive CD8 T cells (Berg *et al.*, 2003; Raue *et al.*, 2004).

6. TRANSCRIPTION FACTORS DOWNSTREAM OF THE TCR, ACTIVATING NK RECEPTORS, AND CYTOKINE RECEPTORS

Signaling cascades relay information from receptors on the plasma membrane through the cytoplasm and into the nucleus by inducing the expression, posttranslational modification, and/or nuclear translocation of transcription factors. Transcription factors may activate expression by recruiting RNA polymerase-containing complexes to target genes, by recruiting protein complexes that alter chromatin structure such that the binding of other transcriptional activators is facilitated, or by a combination of these mechanisms; the converse is true for transcriptional repressors.

6.1. Factors downstream of the TCR, costimulatory, and activating NK receptors

A number of ubiquitously expressed transcription factors are involved in *Ifng* transcription-induced downstream of the TCR and costimulatory molecules. These include members of the NFAT, NFκB, AP-1, ATF-CREB,

C/EBP, and Ets families (reviewed by Lin and Weiss, 2001; Murphy *et al.*, 2000; Szabo *et al.*, 2003). Nearly all of these factors are also involved in *Ifng* transcription in NK cells (Glimcher *et al.*, 2004; Vivier *et al.*, 2004; Zompi and Colucci, 2005). While T cells require antigen stimulation to express *Ifng* message, resting NK cells express low amounts of this transcript constitutively (Stetson *et al.*, 2003). Thus, the role of these transcription factors in *Ifng* induction in NK cells may be somewhat different than in T cells.

In T cells, cyclic AMP-response element binding protein (CREB) is rapidly activated in response to TCR stimulation (Barton *et al.*, 1996). CREB may also be activated in response to increases in cAMP (reviewed by Kuo and Leiden, 1999). CREB binds as a homodimer or heterodimer in conjunction with activating transcription factor (ATF) proteins. In resting T cells, CREB is maintained in an inactive unphosphorylated state that can bind DNA. Following antigen stimulation, CREB becomes phosphorylated on serine 133, allowing it to interact with an essential coactivator and histone acetyltransferase, CREB binding protein (CBP). The CREB–CBP complex can then recruit and activate the basal transcriptional machinery. By contrast to this positive role, CREB binding to the *IFNG* promoter (Table 2.1) appears to inhibit transcription, apparently by inhibiting the binding of Jun/ATF2 complexes (Penix *et al.*, 1996), perhaps contributing to the known inhibitory effect of cAMP on IFN-γ production.

Activation of T cells through the TCR and CD28 leads to the activation of the MAPKs Erk1 and Erk2, which induces the transcription factor Elk, which in turn upregulates expression of the transcription factor c-Fos (Dong *et al.*, 2002; Lin and Weiss, 2001). In addition, the c-Jun N-terminal kinases 1 and 2 (JNK1, JNK2) and p38 MAPKs are induced resulting in activation of c-Jun and ATF2, respectively, which results in the formation of AP-1, a heterodimer of c-Fos and c-Jun, and of c-Jun/ATF2 heterodimers. Binding of these heterodimers to the *IFNG* promoter helps to activate transcription (Penix *et al.*, 1996; Sica *et al.*, 1997; Sweetser *et al.*, 1998). Consistent with the importance of ATF2, pharmacological inhibitors of the MAPK p38 or dominant-negative forms of this kinase inhibit IFN-γ production by T and NK cells (Berenson *et al.*, 2004b; Rincon *et al.*, 1998).

CCAAT/enhancer binding proteins (C/EBPs) are a family of basic leucine zipper transcription factors (Poli, 1998). Two C/EBP isoforms have been described in lymphocytes: C/EBPγ and C/EBPβ. While most members of the C/EBP family are relatively limited in their tissue distribution, C/EBPγ is unique in that it is ubiquitously expressed and constitutively active (Roman *et al.*, 1990). C/EBPγ can interact with other transcription factors containing a leucine zipper region, and positively or negatively influence their function. C/EBPγ-deficient NK cells and splenocytes produce ~90% less IFN-γ in response to IL-12 or IL-18 than control cells (Kaisho *et al.*, 1999). The reduced IFN-γ secretion is not a

TABLE 2.1 Transcription factors that interact with the promoter and other regulatory elements governing *Ifng* expression and their effects on expression

Transcription factor	Regions bound	Approach used to demonstrate binding[a]	Effects[b]	Comments	References
AP-1	*IFNG*[c] promoter	*in vitro*	A		Barbulescu et al., 1998
ATF2/c-Jun	*Ifng*CNS-6	*in vitro*	A		Penix et al., 1996
	IFNG promoter	*in vitro*	A		Penix et al., 1996
C/EBP	*IFNG* promoter	Inferred based on effects on promoter-driven reporter but not shown directly	A, E	Enhancement requires cooperation with T-bet	Berberich-Siebelt et al., 2000; Tong et al., 2005
Ets-1	*Ifng* promoter	ChIP	A	Enhances in cooperation with T-bet	Grenningloh et al., 2005
NFAT	*IFNG*/*Ifng* promoter	*in vitro*, ChIP	A		Sica et al., 1997
	*Ifng*CNS-6 (*IFNG*CNS-4 kb in human)	ChIP			Lee et al., 2004; Sweetser et al., 1998
NFκB	*Ifng* promoter	p50/50 by ChIP	R		Tato et al., 2006
	IFNG promoter	p50/65 *in vitro*	A		Sica et al., 1997
	IFNG 1st Intron	c-Rel *in vitro*	A		Sica et al., 1992

Runx3	*Ifng* Promoter	ChIP	A		Djuretic *et al.*, 2007
T-bet	IFNG/*Ifng* promoter	ChIP	A, E	Binds cooperatively to the *Ifng* promoter with Runx3	Lee *et al.*, 2004
	*Ifng*CNS-54	ChIP			Shnyreva *et al.*, 2004
	*Ifng*CNS-34	ChIP			Tong *et al.*, 2005
	*Ifng*CNS-22	ChIP			Chang and Aune, 2005
	*Ifng*CNS-6	ChIP			Beima *et al.*, 2006
	*Ifng*CNS+18/20	ChIP			Djuretic *et al.*, 2007; Hatton *et al.*, 2006
STAT 1, 3, 4, 5	IFNG promoter	*in vitro*	A		Xu *et al.*, 1996
	IFNG 1st intron	*in vitro*			Barbulescu *et al.*, 1998
	IFNGCNS-4 kb	*in vitro*, ChIP			Bream *et al.*, 2004; Gonsky *et al.*, 2004; Strengell *et al.*, 2003
STAT6	*Ifng* promoter	ChIP	R		Chang and Aune, 2007
CREB/ATF1	IFNG promoter	*in vitro*	R	Inhibits ATF2 binding	Penix *et al.*, 1996

(*continued*)

TABLE 2.1 (continued)

Transcription factor	Regions bound	Approach used to demonstrate binding[a]	Effects[b]	Comments	References
YY1	Promoter	*in vitro*	A/R		Sweetser *et al.*, 1998; Ye *et al.*, 1996
SMAD3	IFNG promoter	*in vitro*, ChIP	R	Also inhibits T-bet promoter	Yu *et al.*, 2006
GATA-3	IFNG promoter	*in vitro*	R	Repression by GATA-3 may be indirect or be mediated by binding to these regions and not are recruitment of the H3K27 methyltransferase EZH2	Penix *et al.*, 1993
	Ifng promoter	ChIP			Kaminuma *et al.*, 2004
	*Ifng*CNS-54	ChIP			Chang and Aune, 2007

[a] *In vitro* = electromobility shift, *in vitro* footprinting, oligonucleotide DNA precipitation, and/or similar *in vitro* assay; ChIP = chromatin immunoprecipitation.
[b] A = transcriptional activator or coactivator; R = transcriptional repressor or corepressor; E = induces epigenetic modifications, such as histone acetylation or methylation or changes in DNA methylation; many of these studies did not look for effects on epigenetic modifications; thus, factors in addition to those noted might also induce epigenetic modifications.
[c] IFNG denotes the human gene and *Ifng* denotes the mouse gene.

result of defects in the signaling pathways downstream of IL-12 or IL-18 receptors. Although the precise mechanism by which C/EBPγ facilitates IFN-γ production was not demonstrated in these studies, these two transcription factors synergistically activate the *IFNG* promoter and augment IFN-γ production, even in cells in which a critical cytosine in this regulatory element is methylated (Tong *et al.*, 2005). However, in the absence of T-bet, C/EBPβ does not enhance and may inhibit transcription of *Ifng* (Berberich-Siebelt *et al.*, 2000; Tong *et al.*, 2005), suggesting that the function of C/EBP proteins in *Ifng* regulation is context dependent.

Multiple signaling pathways including the TCR, CD28, and IL-18 receptors converge on the NFκB/Rel family of transcription factors. The five NFκB family members include RelA (p65), RelB, and c-Rel, which can transactivate target genes when dimerized, and NFκB1 (p50) and NFκB2 (p52), which do not contain transactivation domains and on their own can repress activation. NFκB proteins are maintained in an inactive form in the cytoplasm by interaction with IκB proteins. T cell activation induces IκB phosphorylation and degradation, which allows NFκB dimers to translocate to the nucleus (Ghosh *et al.*, 1998).

CD4 T cells that are unable to activate NFκB have decreased Th1 responses and IFN-γ production (Aronica *et al.*, 1999; Corn *et al.*, 2003; Tato *et al.*, 2003). Furthermore, the number of CD4 T cells elicited in response to infection by the Th1 pathogen *Toxoplasm gondii* in transgenic mice expressing a degradation-resistant IκB-α is severely decreased, and is associated with the inability of these cells to proliferate and produce IFN-γ. Cytolytic function and IFN-γ secretion of NK cells from these mice is also severely impaired (Tato *et al.*, 2003). RelB-deficient CD4 T cells have impaired T-bet and STAT4 expression resulting in very low IFN-γ secretion (Corn *et al.*, 2005). Consistent with this, RelB-deficient mice are more susceptible to infection with *T. gondii* (Caamano *et al.*, 1999), and have impaired NK cell cytotoxicity and IFN-γ secretion, despite normal IL-12 secretion by APCs (Caamano *et al.*, 1999). c-Rel-deficient NK cells also produce less IFN-γ in response to stimulation with IL-12 and IL-2 (Tato *et al.*, 2006). In contrast, *p50−/−* NK cells proliferate and secrete more IFN-γ in response to IL-12 and IL-2 or *T. gondii* infection, suggesting that it may function as a repressor of IFN-γ. Consistent with this possibility, chromatin immunoprecipitation (ChIP) assays show that p50 is bound to the murine *Ifng* promoter in resting NK cells and binding diminishes in response to IL-12 plus IL-18 (Tato *et al.*, 2006). It is possible that this loss of p50 reflects its replacement by other NFκB proteins that contain activating domains, but if so, it is not c-Rel, because this factor was not detected at the *Ifng* promoter by ChIP in stimulated murine NK cells (Tato *et al.*, 2006). There are other sites in the *IFNG* promoter and first intron to which NFκB proteins can bind, but whether this occurs *in situ* is not known (Sica *et al.*, 1992, 1997). Furthermore, p50, p52, and RelB proteins are required for development of mature NKT cells (Matsuda and Gapin, 2005; Sivakumar *et al.*, 2003).

There are five members of the NFAT family of transcription factors, which are commonly referred to as NFAT1 (NFATp or NFATc2), NFAT2 (NFATc or NFATc1), NFAT3 (NFATc4), NFAT4 (NFATx or NFATc3), and NFAT5, of which NFAT1, NFAT2, and NFAT3 are expressed in lymphocytes (Macian, 2005). NFAT proteins are maintained in an inactive, hyperphosphorylated form within the cytoplasm of resting cells. Following TCR stimulation, sustained elevations in cytosolic calcium activate the phosphatase calcineurin, which dephosphorylates NFATs, resulting in their rapid nuclear import and increased DNA binding affinity. NFAT signaling processes are attenuated by the action of several kinases that phosphorylate NFATs, decreasing their DNA binding, and resulting in their rapid export to the cytoplasm (Rao et al., 1997). NFAT proteins are capable of binding DNA as homodimers or heterodimers and interact with several other transcription factors, including AP-1 (Hogan et al., 2003), making it difficult to assign a simple relationship between the activation of a specific cytokine gene and individual NFAT family members. NFAT proteins can bind to multiple sites within the *IFNG* promoter (Sica et al., 1997; Sweetser et al., 1998) and to the upstream enhancer CNS1/*Ifng*CNS-6 (Table 2.1) (Lee et al., 2004), but the extent to which the effects of NFAT on *Ifng* expression are mediated through these regions of the gene is not clear.

Mice with deficiencies in individual NFAT members have relatively mild immunophenotypes (Macian, 2005). Although NFAT1 and NFAT2 are expressed both by Th1 and Th2 CD4 cells and can induce transcription of either *Ifng* or *Il4* (Monticelli and Rao, 2002; Ranger et al., 1998), NFAT1-deficient CD4 T cells have more sustained IL-4 production following stimulation, resulting in mild Th2-skewing, diminished-Th1 differentiation, and decreased expression of IFN-γ (Hodge et al., 1996; Kiani et al., 1997; Monticelli and Rao, 2002). Enforced expression of a constitutively active form of NFAT2 resulted in increased IFN-γ production, strong Th1 skewing, and incomplete ability to commit to the Th2 effector lineage, even in the presence of IL-4 and blocking antibodies to IL-12 and IFN-γ (Porter and Clipstone, 2002). Conversely, NFAT2-deficient T cells are impaired in their ability to produce IL-4 and other Th2 cytokines (Monticelli and Rao, 2002; Ranger et al., 1998). The immediate phase of human NK cell activation through CD16 (FcγR3) is associated with NFAT1 activation and binding to the *Ifng* promoter (Aramburu et al., 1995). Within 2 h, expression of NFAT2 is increased, though its role in IFN-γ regulation is less clear (Aramburu et al., 1995), as is the role of NFAT in IFN-γ production by NKT cells (Wang et al., 2006).

Multiple signaling pathways, including MAPKs, PI3 kinases, and calcium signaling, are able to activate the winged helix–loop–helix transcription factors of the Ets family. Ets family transcription factors are widely expressed and are important for developmental and cell fate choices (Yordy and Muise-Helmericks, 2000). Ets-1 is the active form in resting T cells, in which it can regulate transcription from Ets-dependent regulatory elements. Following T cell activation, Ets-1 becomes phosphorylated on four serine residues, which abrogates DNA binding, resulting in its rapid degradation (Yordy and Muise-Helmericks, 2000). *Ets1−/−Rag-2−/−* chimeric mice have defects in thymocyte development and numbers of peripheral lymphocytes, implicating Ets-1 in T cell development (Barton *et al.*, 1998; Bories *et al.*, 1995; Muthusamy *et al.*, 1995). Furthermore, while the proportion of peripheral CD4 and CD8 T cells are largely normal in *Ets1−/−Rag-2−/−* chimeric mice, they show proliferative defects and increased apoptosis (Bories *et al.*, 1995; Muthusamy *et al.*, 1995) and greatly impaired cytokine expression (Grenningloh *et al.*, 2005).

Ets-1 contributes to increased STAT4 and IL-12Rβ2 expression in developing Th1 cells, and is recruited to the *Ifng* promoter where it collaborates with T-bet to enhance IFN-γ expression (Grenningloh *et al.*, 2005). Ets-1-deficient CD4 T cells proliferate normally in response to TCR stimulation but have dramatically impaired IFN-γ and IL-2 secretion. By contrast, the effect of Ets-1 deficiency on Th2 cytokine production is more limited, suggesting that Ets-1 is primarily a Th1-specific transcription factor. In good agreement with these data, while wild-type CD4 T ells induced colitis following transfer into severe combined immunodeficiency (SCID) recipients, Ets-1-deficient CD4 T cells did not (Grenningloh *et al.*, 2005).

Unlike CD4 T cells, which require Ets-1 transcription to activate the *Ifng* promoter and upregulate IL-12Rβ2 and STAT4 during Th1 development, NK and NKT cells require Ets-1 for their development, proliferation, and survival (Bories *et al.*, 1995; Muthusamy *et al.*, 1995; Walunas *et al.*, 2000). Ets-1-deficient mice show a drastic reduction of splenic NK cells, and those that do develop are severely impaired in their cytolytic activity and ability to secrete IFN-γ.

In addition to Ets-1, the Ets related molecule (ERM), and the interferon response factor (IRF)-1 have been implicated in Th1 development or IFN-γ production by CD4 T cells. These factors are induced in response to IL-12 in CD4 T cells (Ouyang *et al.*, 1999). However, ERM only modestly enhances *Ifng* expression (Ouyang *et al.*, 1999), whereas IRF-1-deficient mice have impaired Th1 responses *in vivo* (Lohoff *et al.*, 1997). It is unclear if IRF-1 acts downstream of IL-12 to enhance IFN-γ production in T cells in addition to its role in facilitating IL-12 production by APCs.

Tec kinases are a family of tyrosine kinases that are activated downstream of Src family kinases in response to signals from the TCR. TEC kinases in turn phosphorylate PLC-γ, a second messenger necessary for sustained intracellular calcium release and activation of NFAT transcription factors (Schwartzberg *et al.*, 2005). The primary TEC kinases in T lymphocytes and NK cells are Rlk (TXK in humans) and Itk, which contribute to Th1 and Th2 responses, respectively. Itk and Rlk are crucial for the proper expression of Eomes and the subsequent development of conventional CD8 T cells (Atherly et al., 2006; Berg, 2007; Broussard et al., 2006). On stimulation, TXK can translocate to the nucleus, bind to the human *IFNG* promoter (Kashiwakura et al., 1999; Takeba et al., 2002), and enhance the expression of *IFNG* promoter-driven reporters (Kashiwakura et al., 1999; Takeba et al., 2002). In naive mouse CD4 T cells, Rlk expression is extinguished on activation of naive cells but is reexpressed when these cells differentiate into Th1 effectors (Miller et al., 2004). However, Rlk-deficient mice have only minor defects in IFN-γ production or in their response to the Th1-inducing pathogen, *T. gondii* (Schaeffer et al., 1999, 2001), although overexpression of Rlk *in vivo* is associated with Th1 skewing (Takeno et al., 2004). The extent to which the effects of TXK/Rlk are mediated by interaction with the *IFNG* promoter, rather than by the activation of NFAT, are unclear.

6.2. STATs are activated in response to cytokine receptor signaling

STAT1 is activated by the binding of IFN-γ, IL-27, and type I IFNs to their receptors and appears to indirectly influence IFN-γ expression predominantly by potentiating the expression of the transcription factor, T-bet (Afkarian et al., 2002; Lighvani et al., 2001). STAT1 is the only STAT family member activated by IFN-γ, whereas IL-27 and type I IFNs also activate STAT3 (Darnell et al., 1994; Velichko et al., 2002). Despite the ability of multiple cytokine receptors to activate STAT1, only IFN-γR signaling is sufficient to induce sustained expression of T-bet. Doing so creates a positive feedback loop, allowing IFN-γ produced by Th1 cells to provide an autocrine signal back to the cell, thereby facilitating stable Th1 commitment (Lighvani et al., 2001). Given the prominent role of STAT1 in the induction of T-bet, it is surprising that Th1 responses develop in *STAT1−/−* mice infected with lymphocytic choriomeningitis virus (LCMV) or *T. gondii*. This finding may reflect the copious amounts of type I IFNs or IL-12/IL-27 produced, respectively, in response to these infections (Lieberman et al., 2004; Nguyen et al., 2000). In the latter infection, STAT1-deficient CD4 and CD8 T cells expressed T-bet, albeit at reduced levels, which in the presence of high amounts of IL-12 were sufficient to induce functional Th1 immunity. Furthermore, in mice infected with murine cytomegalovirus (MCMV), STAT1 was required for NK cell expansion and cytotoxicity but not for IFN-γ production (Nguyen et al., 2002a).

In contrast to STAT1, STAT4 is critical for Th1 lineage commitment and sustained Th1 responses *in vivo* and for antigen-independent induction of IFN-γ by IL-12 (Carter and Murphy, 1999; Kaplan et al., 1996; Mullen et al., 2001). However, *STAT4−/−* T cells can produce low levels of IFN-γ in response to activation via the TCR (Afkarian et al., 2002), consistent with a role for STAT4 in amplifying rather than initiating IFN-γ production. Consistent with these *in vitro* findings, IFN-γ producing CD4 T cell responses develop, albeit more weakly, in STAT4- and IL-12-deficient mice in response to infection with *L. monocytogenes* and certain viruses (Brombacher et al., 1999; Nguyen et al., 2002b; Oxenius et al., 1999; Schijns et al., 1998). Sustained Th1 immunity to *T. gondii* or *L. major* requires continual expression of IL-12p40; and Th1 immunity can be boosted in IL-12p35- and IL-12p40-deficient mice by provision of IL-12 *in vivo* (Park et al., 2000; Yap et al., 2000). Together, these data suggest that ablation either of IFN-γ signaling via STAT1 or of IL-12 signaling via STAT4 impairs, but does not abolish, IFN-γ production by CD4 T cells, but that both of these transcription factors are necessary for robust and sustained IFN-γ production in response to infection *in vivo*. By contrast to the importance of STAT4 for antigen-dependent and antigen-independent IFN-γ production by CD4 T cells, STAT4 is required in CD8 T cells only for antigen-independent induction of IFN-γ by IL-12 (Carter and Murphy, 1999). And while *STAT4−/−* mice show deficiencies in NK cell cytolysis of target cells (Kaplan et al., 1996; Thierfelder et al., 1996) and are susceptible to infection by *T. gondii* (Cai et al., 2000), their NK cells do secrete IFN-γ, albeit at reduced levels compared to controls, in response to IL-2 and IL-18 *in vitro* (Cai et al., 2000) and infection with MCMV *in vivo* (Nguyen et al., 2002a). These findings suggest that additional STATs are able to compensate in part for lack of IL-12 signals.

Signaling through the IL-2 and IL-15 receptors primarily activates STAT5a and STAT5b, which facilitate NK cell survival, cytolytic activity, and IFN-γ production (Fujii et al., 1998; Ma et al., 2006). The effects of IL-2 and IL-15 on IFN-γ in NK cells may be mediated in part by binding of STAT5a and STAT5b to the upstream regulatory element *Ifng*CNS-6/*IFNG*CNS-4 kb (Table 2.1), respectively, since disruption of this STAT5 binding site abrogated IL-2-dependent stimulation of *IFNG* reporter constructs in human NK-92 cells *in vitro* (Bream et al., 2004). This site is also responsive to STAT5 signaling downstream of CD2, a receptor expressed on the surface of human NK cells and T cells (Gonsky et al., 2004). Exposure to IL-2 has also been reported to activate STAT4 and the MAPKs MKK/ERK in human NK cells (Wang et al., 1999; Yu et al., 2000), consistent with the ability of IL-2 to prime for or induce low-level IFN-γ secretion. By contrast, another group identified a STAT5a-dependent mechanism of Th1 inhibition involving the induction of SOCS-3 (Takatori et al., 2005), which potently inhibited IL-12-induced STAT4 activation in this study. Ablation of STAT5a resulted in increased numbers of Th1 effector cells, particularly in Th2 conditions.

Activation of STAT3 in T and NK cells occurs in response to a variety of cytokines, including IL-6, IL-12, IL-2, and IL-27 (Hibbert et al., 2003;

Jacobson *et al.*, 1995; Lucas *et al.*, 2003; Sun *et al.*, 2004b), of which IL-27 is the most potent (Hibbert *et al.*, 2003). However, *STAT3*−/− mice have no defect in Th1 differentiation, either in IFN-γ production or T-bet expression in CD4 T cells, but do have defects in Th17 development (Mathur *et al.*, 2007; Yang *et al.*, 2007).

6.3. T-box family members are crucial for IFN-γ secretion

T-bet, encoded by the *Tbx21* gene, is the "master regulator" of Th1 development. Enforced expression of T-bet in the EL4 T cell line or in primary naive T cells markedly facilitates activation-induced IFN-γ expression, and in developing Th2 cells impairs the expression of IL-4 and IL-5 while activating IFN-γ expression (Szabo *et al.*, 2000). Conversely, T-bet-deficient CD4 T cells produce little or no IFN-γ even when cultured in strong Th1 conditions. In addition, T-bet-deficient mice on a normally resistant C57BL/6 background are unable to mount an effective Th1 immune response to immunization with protein antigens in the presence of strong, Th1-inducing adjuvants or to infection with *L. major* or LCMV. These mice instead develop Th2 responses characterized by the production of IL-4 and IL-5 (Szabo *et al.*, 2002) These findings suggest that in addition to inducing Th1 development, T-bet is necessary to block Th2 development, which it does so by binding with Runx3 to the *Il4* silencer and by blocking the interaction of GATA-3 with its target DNA sequences in the *Il4* 3′ enhancer (Djuretic *et al.*, 2007; Usui *et al.*, 2006). The interaction of T-bet with GATA-3 requires phosphorylation of Tyr525 on T-bet by the kinase Itk. However, phosphorylation of Tyr525 is not essential for T-bet to inhibit Th2 differentiation or to induce Th1 differentiation (Hwang *et al.*, 2005).

T-bet acts directly on the *Ifng* gene to facilitate its expression and does so by binding to multiple sites within the *Ifng* promoter and within other distal regulatory elements located upstream and downstream of the gene (Table 2.1) (Chang and Aune, 2005; Cho *et al.*, 2003; Hatton *et al.*, 2006; Lee *et al.*, 2004; Shnyreva *et al.*, 2004). T-bet works in concert with at least two other factors to facilitate IFN-γ expression and Th1 immunity: Runx proteins and Hlx. Runx3 and Hlx are direct targets of T-bet, which induce their expression, enabling them to work cooperatively in a feed-forward loop that reinforces Th1 commitment (Singh and Pongubala, 2006). Runx transcription factors are involved in many cellular differentiation processes and have been implicated in CD4 expression and Th1 commitment in T cells. Overexpression of Runx1 in naive CD4 T cells induces Th1 differentiation, whereas expression of a dominant-negative Runx1 favors Th2 differentiation (Komine *et al.*, 2003). Runx3, which is highly expressed in Th1 and CD8 T cells, interacts with T-bet and binds cooperatively with it to the *Ifng* promoter to activate its transcription, and to the *Il4* silencer, to repress its transcription (Djuretic *et al.*, 2007).

The homeobox gene Hlx is expressed by Th1 cells and works synergistically with T-bet to increase the frequency of IFN-γ-producing cells and amount of IFN-γ produced, as well as to enhance the expression of T-bet and IL-12Rβ2 (Mullen et al., 2002). Consistent with this notion, Hlx transgenic mice express less IL-4Rα and generate increased frequencies of IFN-γ-producing cells under Th2-polarizing conditions than controls (Zheng et al., 2004). In contrast, *Hlx* haploinsufficient CD4 T cells have elevated expression of IL-4Rα and a Th2 bias (Mikhalkevich et al., 2006).

While T-bet is sufficient for IFN-γ induction in CD4 T cells, IFN-γ production in CD8 T cells is induced through the concerted action of T-bet and its paralog, Eomes; together they collaborate to assure proper effector differentiation and maintenance of memory CD8 T cells (Pearce et al., 2003). Eomes is expressed in naive and memory CD8 T cells and is important for induction of IFN-γ in these cells. As a consequence, IFN-γ production in response to TCR stimulation is normal in T-bet-deficient CD8 T cells (Szabo et al., 2002). By contrast, IFN-γ expression by CD8 T cells is greatly impaired in the absence of Eomes, and this can be further exacerbated by loss of T-bet, suggesting independent but overlapping roles of these two transcription factors in IFN-γ production (Intlekofer et al., 2005). Similar to CD4 Th1 cells, T-bet expression in CD8 T cells is increased by IL-12 in response to infection, whereas clearance of infection and loss of IL-12 signals increases Eomes expression (Intlekofer et al., 2005; Sullivan et al., 2003). Optimal expansion of effector CD8 T cells requires IL-12-dependent induction of T-bet, as *IL-12Rα−/−* mice maintain high levels of Eomes through primary infection and have reduced numbers of effector cells (Takemoto et al., 2006). Although T-bet deficiency does not affect cytokine production by CD8 T cells in response to *in vitro* stimulation with mitogens, antigen-specific IFN-γ production by T-bet-deficient CTLs following LCMV infection is impaired, suggesting T-bet functions to regulate IFN-γ production or to sustain IFN-γ-producing effector CD8 T cells *in vivo* (Sullivan et al., 2003). Perhaps consistent with the latter possibility, Eomes, which is highly expressed by memory CD8 T cell populations in both humans and mice, induces expression of the high-affinity IL-15Rα chain (Intlekofer et al., 2005). Together, these data indicate that in CD8 T cells, expression of genes associated with effector function and migration are initiated by Eomes and augmented by T-bet, whereas Eomes functions to induce genes associated with self-renewal and homeostasis.

NK cells are poised to rapidly produce IFN-γ immediately after activation, which is enabled by their constitutive expression of *Ifng* mRNA (Stetson et al., 2003; Tato et al., 2004). Consistent with this finding, T-bet and Eomes are constitutively expressed in NK cells (Szabo et al., 2002), and the *Ifng* promoter in these cells displays epigenetic marks indicative of transcriptionally permissive chromatin (Tato et al., 2004). In the absence of T-bet, NK cells have dramatically decreased IFN-γ production and cytolytic activity (Szabo *et al.*, 2002). Surprisingly, T-bet and Eomes are also required to maintain the expression of receptors necessary for NK cell development (Intlekofer *et al.*, 2005; Townsend *et al.*, 2004). Eomes is required for expression of the IL-15Rα, Eomes heterozygous mice show substantially reduced frequencies of NK cells compared to wild-type mice, and this defect can be further exacerbated by the loss of one T-b*et al*lele (Intlekofer *et al.*, 2005).

While CD8 and NK cells express both Eomes and T-bet, only T-bet is expressed by NKT cells and is crucial to their development. T-bet-deficient mice are nearly devoid of NKT cells (Intlekofer *et al.*, 2005; Townsend *et al.*, 2004), likely due to the lack of IL-2R/IL-15Rβ (CD122), which is required for IL-15-induced proliferation and survival of developing NKT cells (Matsuda and Gapin, 2005). Ectopic expression of T-bet in developing NKT cells results in increased expression of select genes associated with their maturation, but expression of IFN-γ, granzyme B and perforin are only modestly enhanced (Matsuda *et al.*, 2006). It is unknown how T-bet is induced during NKT cell differentiation, as *IFN-γ–/–*, *IFN-γR–/–*, and *STAT1–/–* mice show no defects in NKT cell development or survival (Townsend *et al.*, 2004).

7. EPIGENETIC PROCESSES GOVERN PLASTICITY OF CELL FATE CHOICES AND HELP TO IDENTIFY DISTAL REGULATORY ELEMENTS

In principle, the transcription factors that function as "master regulators" of T and NK cell effector function could be both necessary and sufficient for the initiation and faithful propagation of their respective effector lineages. In practice, transcription factors must bind to their recognition sites within regulatory elements and then recruit general transcriptional activators or repressors to regulate gene transcription.

The ability of NK and NKT cells to rapidly produce substantial amounts of IFN-γ on stimulation derives from their constitutive expression of Eomes and T-bet (NK cells) or of T-b*et al*one (NKT cells), and the fact that the regulatory elements to which these and other transcription factors must bind in the *Ifng* locus are contained within accessible chromatin, thereby facilitating activation of *Ifng* transcription (Bream *et al.*, 2004; Chang and Aune, 2005; Stetson *et al.*, 2003). This is not the case for naive CD4 T cells, which can produce only small amounts of IFN-γ immediately following TCR stimulation and require multiple rounds of cell division under appropriate conditions to gain the capacity to produce IFN-γ efficiently (Mullen *et al.*, 2001). This lag reflects the time required for these cells to induce the expression of T-bet, for T-bet to induce the expression of its partners Runx3 and Hlx, and for these factors together to induce *Ifng* expression, to modify the chromatin of the *Ifng* locus to make it more accessible, and to induce IL-12Rβ2 and IL-18R. Naive CD8 T cells are more efficient in gaining the capacity to produce substantial amounts of IFN-γ than naive CD4 T cells, reflecting, at least in part, their constitutive expression of Eomes. Thus, the ability of T-bet and Eomes to induce IFN-γ expression and enforce the Th1 and CTL effector fates appears to derive not only from the ability of these transcription factors to initiate the transcription of *Ifng* and other target genes, but also from the ability of these transcription factors to induce epigenetic modifications within the *Ifng* gene to assure heritability of expression thereafter.

Eukaryotic cells face the challenge of fitting a genome consisting of several billion nucleotides into a nucleus only microns in diameter, while maintaining spatial organization and accessibility to factors that govern the transcription, repair, and replication processes. This is achieved by the association of DNA with proteins in chromatin, the basic unit of which is the nucleosome, consisting of an octamer of histone proteins with two copies each of H2A, H2B, H3, and H4, around which ~150 bp of DNA is wrapped (Felsenfeld and Groudine, 2003). Differences in nucleosome composition, relative position and interactions with DNA, posttranslational modifications to histone tails, and methylation of cytosines within CpG dinucleotides of the DNA itself encode information without affecting the underlying DNA sequence, and constitute the epigenetic code of that cell. The epigenetic code influences transcription factor binding, transcription initiation, and progression, and thus, along with transcription factor abundance and activity, determines in a given cell whether a gene is or can be expressed and to what degree. Unlike the fixed information encoded by DNA sequence, epigenetic information is plastic but potentially heritable, and thus can be propagated from parental cells to daughter cells in a manner conducive to maintaining the overall program of gene expression that characterizes a specific cell type.

These epigenetic modifications can be detected by several techniques, and because they are often targeted to regulatory elements, these techniques are one means by which to search for *cis*-regulatory elements (Nardone *et al.*, 2004; Wilson and Merkenschlager, 2006). Hypersensitivity of specific sequences to digestion by low DNase I concentrations indicates specific sites where nucleosomes have been displaced or their conformation altered, either reflecting or facilitating the binding of transcription factors to important regulatory elements. Modification of histone tails can provide a signal for the binding of protein complexes associated with transcriptional activation or silencing. Among the histone modifications associated with permissiveness to gene transcription are acetylation of lysines of histone H3 (^{Ac}H3) or H4 (^{Ac}H4) and di- or trimethylation of histone H3 lysine 4 (H3-K4^{me2} or H3-K4^{me3}). In contrast, di- or trimethylation of H3 lysine 27 (H3-K27$^{me2/me3}$) or lysine 9 (H3-K9$^{me2/me3}$) are repressive marks characteristic of Polycomb-mediated repression and heterochromatic silencing, respectively (Kouzarides, 2007). DNA methylation is mediated by DNA methyltransferases and is linked to the formation of repressive chromatin through the recruitment of proteins that mediate transcriptionally silent histone modifications and ATP-dependent nucleosomal remodeling (Vire *et al.*, 2006). These epigenetic modifications often are found together in varying degrees; the extent of gene expression or potential for gene expression is influenced by the combined contribution of multiple epigenetic processes working in concert with transcription factors and regulatory elements to activate or inhibit gene expression.

Changes in cytokine expression during Th1 and Th2 commitment by CD4 T cells are associated with changes in the epigenetic modifications at the Th2 cytokine locus and at the *Ifng* gene (Ansel *et al.*, 2003, 2006; Lee *et al.*, 2006; Wilson and Merkenschlager, 2006). In general, both the *Il4/Il13/Il5* and *Ifng* loci are poised in naive CD4 cells containing epigenetic modifications indicative both of transcriptional repression and permissiveness. During Th1 development, the *Ifng* locus gains transcriptionally favorable histone modifications, loses CpG methylation, and thereby gains accessibility to binding of transcriptional activators, while the *Il4/Il13/Il5* locus gains repressive modifications including H3-K27^{me3}. Selected areas of the *Ifng* locus maintain accessibility in naive and Th2 cells, suggesting that these regions may play a role in preserving locus plasticity in these cell types.

8. TRANSCRIPTIONAL REGULATORY ELEMENTS WITHIN THE *IFNG* GENE

We will begin by reviewing the promoter and intronic regulatory elements, the transcription factors binding to them, their functions, and the epigenetic modifications to these regions that are typical of naive and effector T cells and NK cells. Then in the following sections, we will discuss newer information on the structure of the extended *Ifng* locus, distal regulatory elements within the locus, and what is known regarding their functions.

8.1. Regulatory elements within the *Ifng* promoter and gene

The *Ifng* gene is composed of four exons spanning ∼5.5 kb in humans and rodents, upstream of which is its core ∼600 bp promoter. The binding of transcription factors and the subsequent epigenetic and chromatin changes within the *Ifng* gene have been studied in some detail. These studies together with a number of functional assays have defined the *Ifng* promoter and two intronic enhancers (Fig. 2.1). The *Ifng* promoter directs Th1-specific expression of transgenic reporter constructs *in vivo* (Young *et al.*, 1989; Zhu *et al.*, 2001), which is consistent with the presence of multiple binding sites for T-bet in the promoter (Cho *et al.*, 2003) and

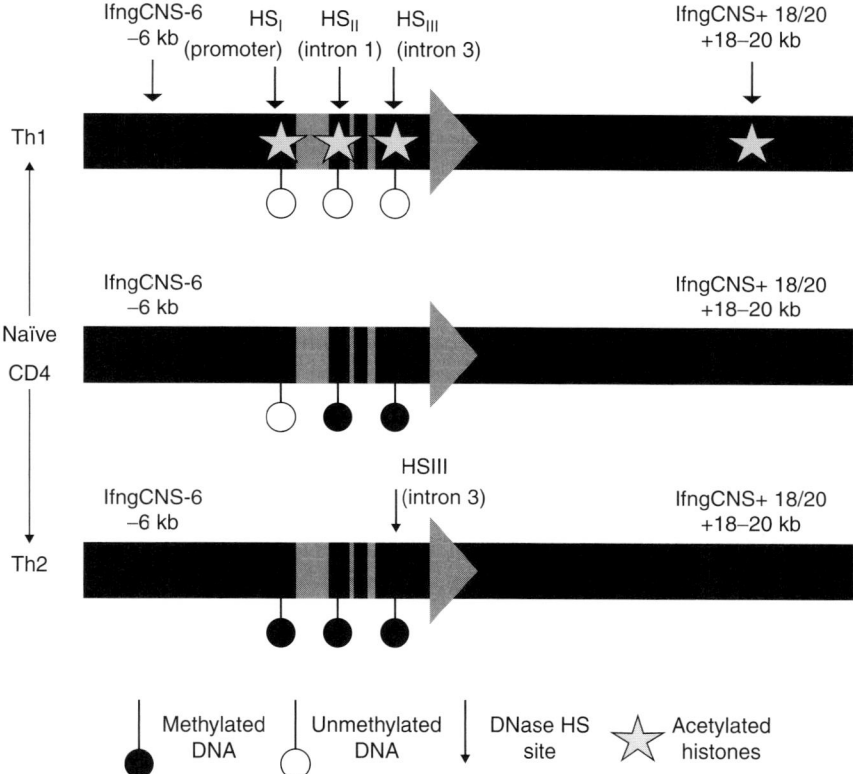

FIGURE 2.1 The *Ifng* promoter, gene, and nearby regulatory regions in naïve, Th1 and Th2 CD4 T cells. In naive CD4 T cells, the *Ifng* promoter and gene lack histone H3 and H4 lysine acetylation (^{Ac}H3/^{Ac}H4) or methylation (H3K4^{me2}), CpGs in the proximal *Ifng* promoter are demethylated, but CpGs in the introns and nearby regulatory regions are methylated. Activation initiates epigenetic remodeling, the initial 12–24 h of which are dependent on signals downstream of the TCR and similar in Th1 or Th2 conditions. Thereafter, Th1 differentiation results in progressive increases in ^{Ac}H3, ^{Ac}H4, and H3K4^{me2} at the promoter. These histone modifications along with CpG demethylation occur within the *Ifng* gene and at enhancers located (in the mouse) 6 kb upstream (*Ifng*CNS-6) and 18–20 kb downstream (*Ifng*CNS+ 18/20), DNase HS sites appear at the promoter (HS$_I$), introns 1 (HS$_{II}$) and 3 (HS$_{III}$) and *Ifng*CNS+ 18/20, and an activation-inducible HS site appears at *Ifng*CNS-6. By contrast, Th2 differentiation results in the acquisition of the DNase HS site at intron 3, promoter CpG methylation, and gain of the repressive H3K27me2/3 modification throughout these regions. (See Plate 3 in Color Plate Section.)

the demonstration of T-bet binding to the promoter in Th1 cells by ChIP (Beima *et al.*, 2006; Chang and Aune, 2005; Hatton *et al.*, 2006; Shnyreva *et al.*, 2004). In transgenic reporter assays, addition of introns 1 and 3 enhances expression from the *Ifng* promoter but abolishes Th1-specificity by as yet unknown mechanisms (Young *et al.*, 1989; Zhu *et al.*, 2001).

In addition to T-bet, a number of other transcription factors can bind to the *Ifng* promoter and/or introns 1 and 3 and transactivate reporter constructs containing these elements, including AP-1, ATF-2/c-Jun, Ets-1, NFκB, NFAT, STATs, and T-bet (reviewed in Murphy *et al.*, 2000) (Barbulescu *et al.*, 1998; Glimcher *et al.*, 2004; Persky *et al.*, 2005; Szabo *et al.*, 2003). Conversely, CREB can bind to the promoter and inhibit expression (Penix *et al.*, 1996), as can GATA-3; repression by GATA-3 may be dependent on its ability to interact with and inhibit the function of T-bet and STAT4 (Hwang *et al.*, 2005; Kaminuma *et al.*, 2004) and/or to recruit the H3K27-methyltransferase and Polycomb protein EZH2 to *Ifng* (Chang and Aune, 2007). YY1, a ubiquitously expressed transcription factor has been reported to inhibit or facilitate expression driven by the *Ifng* promoter (Sweetser *et al.*, 1998; Ye *et al.*, 1996). YY1 and other transcription factors are also likely to mediate chromatin remodeling at the *Ifng* promoter.

As noted above, naive T cells are poised to express low levels of *Ifng* mRNA shortly after activation (Grogan and Locksley, 2002; Mullen *et al.*, 2002). In these cells, the *Ifng* promoter and introns lack transcriptionally favorable or repressive histone marks, and CpG dinucleotides in the introns are heavily methylated (Agarwal and Rao, 1998; Avni *et al.*, 2002; Fields *et al.*, 2002; Mullen *et al.*, 2002). Naive T cells also lack the DNase I hypersensitive sites in the *Ifng* promoter (HS1) and introns 1 (HSII) and 3 (HSIII), which are characteristic of Th1 cells (Agarwal and Rao, 1998). However, the *Ifng* promoter is demethylated in naive T cells and is juxtaposed to the Th2 cytokine locus, perhaps creating a hub that allows *Ifng* and *Il4* to compete for limiting amounts of transcription activators present in naive T cells prior to lineage specification (Spilianakis *et al.*, 2005).

Initial activation of naive CD4 T cells results in the acetylation of histones H3 and H4 in the *Ifng* promoter under Th0, Th1, and Th2 conditions, but sustained acetylation and extension of acetylation further downstream into the gene is Th1 specific and requires STAT4 and T-bet (Avni *et al.*, 2002; Chang and Aune, 2005; Fields *et al.*, 2002; Mullen *et al.*, 2001). Similarly, T-bet in concert with Hlx induces Th1-specific DNA demethylation that extends from the promoter into the intronic regions of the gene and also induces DNase HSII and HSIII in introns 1 and 3 (Agarwal and Rao, 1998). These actions are likely mediated in part by the binding of T-bet to a T-box half-site immediately upstream of a C/EBP-AP1-ATF site; together these transcription factors cause the dissociation of the mSin3a repressor, which is associated with histone deacetylase activity (Tong *et al.*, 2005). As noted above, Runx3 also binds cooperatively with T-bet to the *Ifng* promoter (Djuretic *et al.*, 2007). Th1 differentiation also results in nucleosome remodeling at the *Ifng* promoter, as demonstrated by the formation of DNase HSI. This remodeling utilizes the ATP-dependent nucleosome remodeling complex Swi-SNF, including the subunit Brg-1, which is recruited to the *Ifng* promoter in a STAT-4 dependent, Th1-dependent manner and likely facilitates transcription factor binding and/or the transit of RNA polymerase containing complexes required for efficient transcription (Zhang and Boothby, 2006).

Th2 cells, as well as cells that are incapable of producing IFN-γ, such as fibroblasts and hepatocytes, exhibit extensive CpG methylation within 200 bp of the *Ifng* promoter (Schoenborn *et al.*, 2007; Winders *et al.*, 2004). However, enforced expression of T-bet into terminally differentiated Th2 cells induces expression of IFN-γ (Szabo *et al.*, 2000). This may be due, in part, to the ability of T-bet to bind to the *Ifng* promoter even when it is methylated (Tong *et al.*, 2005). These data demonstrate a mechanism by which forced expression of T-bet in Th2 cells can override repressive epigenetic modification to induce *Ifng* expression.

Consistent with the constitutive transcription of *Ifng* by NK cells and their ability to rapidly produce large amounts of this cytokine, the *Ifng* gene in NK cells displays permissive epigenetic marks similar to those found in Th1 cells; DNase hypersensitive sites are present, histones are acetylated, and intronic as well as promoter CpGs are demethylated (Chang and Aune, 2007; Tato *et al.*, 2004).

8.2. Identification of candidate distal regulatory elements in the *Ifng* locus

In addition to the promoter, proper expression of mammalian genes is dependent on other regulatory elements that may be located in the introns or in upstream and downstream flanking sequences. These regulatory elements may include enhancers, silencers, boundary elements, and locus control regions, which can influence locus accessibility and the initiation and maintenance of gene expression. Such elements are most commonly located within 50–75 kb on either side of the gene they regulate, but may be located up to hundreds of kb away. Thus, to fully understand the regulation of a gene such as *Ifng* one must identify each of its regulatory elements, characterize the function of these elements, and identify and understand the actions of the transcription factors that bind to them.

As described above, *Ifng* is regulated in part through its promoter and intronic regulatory elements, but these regions are not sufficient for proper control (Young *et al.*, 1989; Zhu *et al.*, 2001) (Soutto *et al.*, 2002; Young *et al.*, 1989). By contrast, a 191 kb bacterial artificial chromosome (BAC) containing the human *IFNG* gene and ~90 kb of upstream and downstream flanking sequences resulted in high level, CD8- and Th1-specific IFN-γ production (Soutto *et al.*, 2002), suggesting that distal transcriptional regulatory elements are required for proper expression, the essential elements are present within this extended region, and elements from the human *IFNG* locus function properly in mice.

To identify additional regulatory elements, our lab and other laboratories have utilized evolutionary conservation to distinguish two conserved noncoding sequences (CNS) 6 kb upstream (known as CNS1 or *Ifng*CNS-6) and 18–20 kb downstream (referred to as CNS2 or *Ifng*CNS+ 18–20) from the mouse *Ifng* promoter (Fig. 2.1); the region analogous to *Ifng*CNS-6 in mice is located ~4 kb upstream in humans (*IFNG*CNS-4 kb) (Bream *et al.*, 2004). These CNS elements were subsequently shown to contain Th1-restricted permissive histone marks and DNase hypersensitive sites and to enhance IFN-γ expression in reporter assays (Bream *et al.*, 2004; Lee *et al.*, 2004; Shnyreva *et al.*, 2004). Another group subsequently identified additional CNSs, some of which were enriched in acetylated histones in Th1 effectors and/or NK cells, extending >50 kb upstream and downstream of *Ifng* (Chang and Aune, 2005), suggesting that additional regulatory elements were present within this region.

Unlike the Th2 cytokine locus, which consists of three coordinately expressed cytokine genes and a housekeeping gene, the extended *Ifng* locus contains up to two alternately expressed cytokine genes, a housekeeping gene, and, in rodents, a series of genomic aberrations. In humans and rodents, the *Il22* and *Mdm1* genes are located upstream of *Ifng* in the same transcriptional orientation (Fig. 2.2). *Mdm1*, a housekeeping gene, is expressed ubiquitously. *Il22* encodes a proinflammatory member of the IL-10 cytokine family (Wolk *et al.*, 2004), which is most highly expressed by activated Th17 CD4 T cells (Liang *et al.*, 2006; Zheng *et al.*, 2007). The distance between *Il22* and *Ifng* is much greater in the mouse than human genome due to the presence of a complex set of structural rearrangements and segmental duplications located between these genes in mice. This region includes the *Iltifβ* paralog, an inverted duplication of the *Il22* gene in which a portion of the promoter has been lost precluding its expression (Dumoutier *et al.*, 2000), and is flanked on either side by six highly conserved short tandem sequence duplications. *Iltifβ* is present in C57BL/6 and 129 strain mice but not in the BALB/c or DBA/2 strains nor in the rat. The gene encoding *IL26*, also a member of the IL-10 cytokine family, is located between *IL22* and *IFNG* in humans. Orthologs of *IL26* are found in all vertebrates for which information is available but is not present in rodents (Igawa *et al.*, 2006), suggesting that the locus order found in humans (*MDM1* → *IL22* → *IL26* → *IFNG*) is ancestral. Downstream of *Ifng*, no known coding genes are found for >500 kb, but a

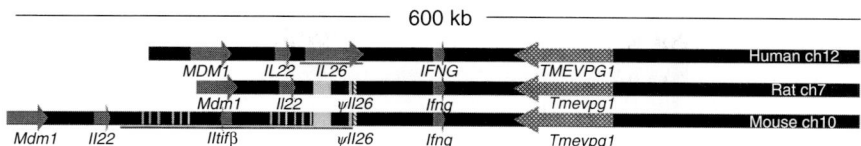

FIGURE 2.2 Structure of the *Ifng* locus in human, rat, and mouse. Alignment of 600 kb flanking the *Ifng* gene on mouse chromosome 10 with sytenic regions of rat chromosome 7 and human chromosome 12. Genes are denoted by blue arrows, indicating the direction of transcription. The stippled arrow denotes *Tmevpg1*, an antisense, noncoding transcript. Red horizontal lines below the human and mouse chromosomes indicate the location of a complex segmental duplication in the C57BL/6 mouse genome. The blue hatched bar denoted as $\Psi Il26$ represents sequences homologous to exon 5 of the human *IL26* gene; orange bars indicate LINE and LTR-LINE-LTR insertions. Modified from Schoenborn et al. (2007). (See Plate 4 in Color Plate Section.)

noncoding antisense transcript, *Tmevpg1*, extends to within 100 kb of the *Ifng* start site. *Tmevpg1* is expressed by naïve CD4, CD8, and NK cells and is downregulated on activation, suggesting a potential role in *Ifng* gene regulation (Vigneau et al., 2001, 2003). It is unknown if there are differences in *Tmevpg1* expression among mouse strains or Th1, Th2, or Th17 cells.

Despite the striking differences in genomic structure in the *Ifng* locus between rodents and humans, the pattern of *Ifng* expression is substantially similar in these species, suggesting that most, if not all, regulatory elements needed for proper expression are proximal to the region where synteny between rodents and other vertebrate species is lost. Based in part on this notion, several groups including ours have utilized bioinformatics and various experimental approaches to identify candidate regulatory elements extending over ~120 kb surrounding the *Ifng* locus (Chang and Aune, 2005; Hatton et al., 2006; Schoenborn et al., 2007). Eight CNS elements were identified based on their exhibiting >70% homology between human and C57BL/6 mouse for 100 or more base pairs. Seven of these eight CNSs exhibit differential epigenetic modifications in naive, Th1, Th2, and NK cells that suggested they might participate in the *cis*-regulation of *Ifng* (Chang and Aune, 2005, 2007; Hatton et al., 2006; Schoenborn et al., 2007).

In naive CD4 and CD8 T cells, these CNSs (like the *Ifng* promoter and introns) lack the transcriptionally favorable histone modifications H3-K4^{me2} and ^{Ac}H4 with two exceptions—low levels of H3-K4^{me2} are present at two conserved regions termed *Ifng*CNS-34 and *Ifng*CNS-22 based on their distance in kilobases from the *Ifng* promoter (Fig. 2.3) (Hatton et al., 2006; Schoenborn et al., 2007). Conversely, naive CD4 T cells have moderate levels of the repressive histone modification H3-K27^{me3} in the region

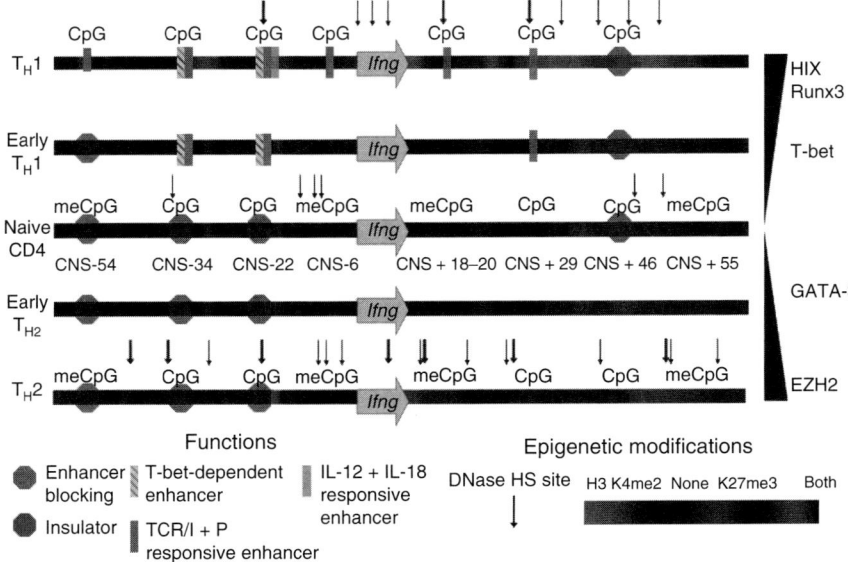

FIGURE 2.3 Epigenetic evolution of the extended *Ifng* locus and functions of distal regulatory regions in naïve, Th1 and Th2 CD4 T cells. In naive CD4 T cells, the extended *Ifng* locus is in a poised state in which CpGs in the *Ifng* promoter and two upstream elements, *Ifng*CNS-34 and *Ifng*CNS-22, and two downstream elements, *Ifng*CNS+ 18/20 and *Ifng*CNS+ 29 are demethylated, there are low levels of permissive H3K4^{me2} at *Ifng*CNS-34 and *Ifng*CNS-22; and small amounts of repressive H3K27^{me2} downstream of *Ifng*CNS+ 29. In this context, the insulator activity of *Ifng*CNS+ 46 and enhancer-blocking activity of *Ifng*CNS-54, *Ifng*CNS-34, and *Ifng*CNS-22 are proposed to be in effect. Th1 differentiation results in the induction of the transcription factors T-bet, then Runx3 and Hlx; the acquisition of DNase HS sites in the *Ifng* promoter, introns I and III, and within several *Ifng*CNSs; gain of H3K4^{me2} and complete CpG demethylation at all *Ifng*CNSs except *Ifng*CNS+ 55; and the loss of H3K27^{me2} throughout the locus. In this context, the enhancer activity of each of the upstream *Ifng*CNSs (with *Ifng*CNS-55 the weakest), *Ifng*CNS+ 18/20 and *Ifng*CNS+ 29 and insulator activity of *Ifng*CNS+ 46 are proposed to be in effect. Th2 differentiation results in increased expression of the transcription factor Gata3, then recruitment of the Polycomb protein/H3K27 methyltransferase EZH2; the acquisition of multiple DNase HS sites, almost all of which, unlike those seen in Th1 cells, are adjacent to but not within *Ifng*CNSs; spreading of repressive H3K27^{me2} throughout the locus; and CpG methylation in the *Ifng* promoter. In this context, the insulator activity of *Ifng*CNS+ 46 is proposed to be lost and the promoters and enhancers silenced. The symbols along the bottom indicate the symbols denoting specific functions and epigenetic modifications. For DNase HS sites, thicker arrows denote strong HS sites and thinner arrows weak HS sites; a heat map is used to depict posttranslational modifications of H3. (See Plate 5 in Color Plate Section.)

between *Ifng*CNS+ 29 and *Ifng*CNS+ 46 (Schoenborn *et al.*, 2007). Interestingly, in naive CD4 T cells these four CNSs lack CpG methylation, and two of them, *Ifng*CNS-34 and *Ifng*CNS+ 46, have weak HS sites (hs-35 and hs+ 49/53) nearby but not directly associated with them, as mapped by a new high-resolution method (Schoenborn *et al.*, 2007). Together, the modifications found at these CNSs in naive T cells suggest that they may play an early role in the initiation of *Ifng* activation or repression during CD4 T cell functional differentiation.

Th1 differentiation is associated with the acquisition of DNase HS sites at *Ifng*CNS-22, *Ifng*CNS-6, *Ifng*CNS+ 18, and *Ifng*CNS+ 29, and discrete peaks of enrichment for H3-K4^{me2}, ^{Ac}H3, and ^{Ac}H4 at these CNSs and at *Ifng*CNS-54 (Fig. 2.3) (Hatton *et al.*, 2006; Lee *et al.*, 2004; Schoenborn *et al.*, 2007; Shnyreva *et al.*, 2004). *Ifng*CNS-54, *Ifng*CNS-6, and *Ifng*CNS+ 18–20, which have methylated CpGs in naive T cells, become demethylated during Th1 differentiation in parallel with a complete loss of H3-K27^{me3} in the *Ifng* locus (Schoenborn *et al.*, 2007). Collectively, these changes result in heightened transcriptional accessibility within the *Ifng* locus. Perhaps counter intuitively, the *Ifng* promoter and regions upstream and downstream of the gene, with the exception of the region from *Ifng*CNS-34 and *Ifng*CNS-22, appear to gain and retain, at least for several days, the typically repressive H3K9^{me2} mark (Chang and Aune, 2007), which here is associated with active transcription as it sometimes is (Vakoc *et al.*, 2005). The gain in histone acetylation is largely STAT4- and T-bet-dependent (Chang and Aune, 2005), and the gain of H3-K4^{me2} and loss of H3-K27^{me3} are largely T-bet-dependent (our unpublished observations). Similarly, as naive CD8 T cells differentiate into CTL effectors they gain histone acetylation and H3-K4^{me2} at conserved elements (Zhou *et al.*, PNAS 2004 and our unpublished observations). Consistent with their constitutive expression of *Ifng* mRNA, freshly isolated NK cells have modest amounts of ^{Ac}H4 not only within the *Ifng* gene but also near *Ifng*CNS-22 and between *Ifng*CNS+ 29 and *Ifng*CNS+ 46 (Chang and Aune, 2005). ^{Ac}H4 increases at these sites and ^{Ac}H4 is acquired at *Ifng*CNS-6 and at 50 kb downstream of the gene when NK cells are cultured in IL-2 (Chang and Aune, 2005). In these studies, histone modifications were not examined at high resolution throughout the *Ifng* locus; thus these modifications may be present at additional sites as well. Together, these data suggest that in IFN-γ-producing Th1 and CD8 effector T cells and in NK cells, multiple regulatory elements are made more accessible as a means to facilitate *Ifng* expression. This is likely to be true in NKT cells as well, but, to our knowledge, studies to test this possibility have not been done.

Conversely, Th2 effector CD4 T cells must silence the expression of *Ifng*, which they do both by altering the expression of key transcription factors and the epigenetic modifications in the *Ifng* locus. In Th2 CD4

T cells, the *Ifng* locus is marked by CpG methylation of the *Ifng* promoter and absence of H3-K4^{me2}, with the exception of modest enrichment at *Ifng*CNS-34 and *Ifng*CNS-22 similar to that seen in naive CD4 T cells (Fig. 2.3) (Hatton *et al.*, 2006; Schoenborn *et al.*, 2007). Furthermore, the *Ifng* locus in Th2 CD4 T cells is extensively marked by the repressive histone modifications H3-K27$^{me2/me3}$ (Chang and Aune, 2007; Schoenborn *et al.*, 2007). While T-bet is crucial for inducing IFN-γ expression, GATA-3 acts to silence *Ifng*. In developing Th2 effector T cells, GATA-3 can bind to the promoter and intron 1 of *Ifng*, as well as *Ifng*CNS-54, and forced expression of GATA-3 in developing Th1 cells inhibits *Ifng* transcription and results in the recruitment of the H3K27 methyltransferase EZH2 to the *Ifng* locus and the acquisition of H3-K27^{me2} and loss of the H3-K9^{me2}, such that H3-K9^{me2} is only transiently present in the *Ifng* locus in Th2 conditions (Chang and Aune, 2007). Since the expression of *Ifng* is markedly repressed by day 7–8 in Th2 cells, the finding that four distal elements (*Ifng*CNS-34, *Ifng*CNS-22, *Ifng*CNS+ 29, *Ifng*CNS+ 46) remain completely demethylated during Th2 differentiation suggests that CpG methylation of these elements is not required for the initial repression of *Ifng*, though it cannot be excluded that these elements may gain CpG methylation during extended culture. Remarkably, Th2 cells acquire a number of HS sites, including HSIII in intron 3 and HS-22 within *Ifng*CNS-22 that are also found in Th1 but not in naive CD4 T cells, as well as four strong Th2-specific HS sites located at HS-40, HS-35, HS+ 8, and HS+ 26 (Fig. 2.3). In addition, HS+ 49 is strongest in Th2 cells, intermediate in naive CD4 T cells and just above background in Th1 cells, and a number of weaker sites located throughout the *Ifng* locus are found exclusively in Th2 effector T cells. While the HS sites found in Th1 cells are located within conserved sequences, nearly all of the Th2-specific HS sites are adjacent to conserved sequences. The location of these Th2-specific HS sites adjacent to CNSs may reflect nucleosome sliding or displacement from nearby clustered regulatory elements that are employed differentially in Th1 cells versus Th2 and naive CD4 T cells.

9. FUNCTIONAL ANALYSIS OF CANDIDATE DISTAL REGULATORY ELEMENTS IN THE *IFNG* LOCUS

The ability of these CNSs to function as enhancers has been evaluated primarily by *in vitro* transfection studies. In these studies, CNSs have been linked to the *Ifng* gene or reporter genes driven by the *Ifng* promoter, and their ability to enhance expression is determined following transfection of these constructs into NK and T cell lines or into primary CD4 and CD8 T cells. *Ifng*CNS-6 and *Ifng*CNS-22 enhance *Ifng* expression in each of these cell types, whereas other IfngCNSs are more limited in their ability to enhance expression (Fig. 2.3). *Ifng*CNS-6, which contains dense CpG methylation in naive T cells, becomes completely demethylated and gains accessibility in Th1 and CD8 effector T cells (Lee *et al.*, 2004; Schoenborn *et al.*, 2007; Shnyreva *et al.*, 2004; Zhou *et al.*, 2004b). *Ifng*CNS-6 appears to be responsive to T-bet and to signals downstream of the TCR, which is consistent with the binding of T-bet and NFAT1 to this element (Lee *et al.*, 2004; Schoenborn *et al.*, 2007; Shnyreva *et al.*, 2004). *Ifng*CNS-22 has predicted binding sites for a number of transcription factors involved in T cell development and effector function and is demethylated and packaged in accessible chromatin in naive, Th1 and Th2 CD4 T lineage cells. *Ifng*CNS-22 is a T-bet-dependent enhancer that is responsive to signals downstream of the TCR and to IL-12 plus IL-18 *in vitro*, with the latter function demonstrable in NK cells and Th0/Th1 cells but not in CD8 effectors (Hatton *et al.*, 2006; Schoenborn *et al.*, 2007). The importance of *Ifng*CNS-22 in IFN-γ expression has also been demonstrated *in vivo* (Hatton *et al.*, 2006). Deletion of this element from a 160 kb BAC in which a Thy1.1 reporter was inserted into exon 1 of murine *Ifng* resulted in greatly reduced expression of this reporter in NK, CD8, and CD4 Th1 T cells following activation via the TCR, ionomycin plus phorbol myristate acetate (PMA), or IL-12 plus IL-18. These data demonstrate that *Ifng*CNS-22 is essential to assure proper, high-level expression of IFN-γ *in vivo*, and appears to be responsive to T-bet, STAT4, and NFκB.

Five other CNSs show less clear and consistent enhancer activity. *Ifng*CNS-34 enhances IFN-γ production in stimulated, T-bet-transfected EL4 cells, requiring the presence of two dimeric Brachyury *cis*-regulatory elements in the distal *Ifng* promoter for its activity (Hatton *et al.*, 2006; Schoenborn *et al.*, 2007). *Ifng*CNS-34 also enhances *Ifng* expression in NK cells and primary CD8 T cells in response to signals from the TCR or ionomycin plus PMA, respectively. *Ifng*CNS-54 exhibits enhancer activity only in NK cells in which it has a modest ability to enhance IFN-γ expression on its own that is increased in the presence of ionomycin plus PMA. *Ifng*CNS+ 29 enhances expression in EL-4 cells and primary Th0 CD4 T cells in response to ionomycin plus PMA. *Ifng*CNS+ 46 enhances basal expression in NK cells, but this effect was not influenced by ionomycin plus PMA or IL-12 plus IL-18, suggesting that this element may be responsive to transcription factors constitutively expressed and active in NK cells. *Ifng*CNS+ 18–20 has very weak enhancer activity detected only in EL-4 cells stimulated with ionomycin alone (Shnyreva *et al.*, 2004).

In addition to enhancers, proper expression of genes requires that they be protected by boundary elements from the unwanted effects of regulatory elements associated with nearby genes and surrounding chromatin domains (Valenzuela and Kamakaka, 2006; West and Fraser, 2005). Insulators protect transcriptionally active genes from the negative effects of neighboring repressive chromatin or maintain a silenced locus despite its location near transcriptionally permissive genes. Enhancer-blocking elements are a second type of boundary element that function to shield a promoter from the actions of distal enhancers, without preventing more proximal enhancers from influencing the promoter (Gaszner and Felsenfeld, 2006).

To date, four *Ifng* CNS elements have demonstrated boundary element function in *in vitro* assays (Fig. 2.3). In these studies, *Ifng*CNS+ 46 acts as an insulator, suggesting that it may form a functional 3' boundary of the *Ifng* locus, keeping *Ifng* poised by limiting the intrusion of repressive chromatin and/or encroachment by the downstream noncoding *Tmevpg1* transcript (Vigneau *et al.*, 2001, 2003). Naive CD4 T cells lack CpG methylation at *Ifng*CNS+ 46 and contain two nearby HS sites (hs+ 49 and hs+ 53); however, the surrounding histones are modestly enriched for H3-K27^{me3}. These two features are more prominent in Th2 cells and markedly diminished in Th1 cells. The presence of H3-K27^{me3} has been suggested to serve as a mark for regulatory regions poised for silencing on differentiation (Bernstein *et al.*, 2006). Thus, the region surrounding *Ifng*CNS+ 46 and HS sites (hs+ 49/hs+ 53) may also serve as a developmental switch that protects *Ifng* locus accessibility in naive and Th1 CD4 T cells, but facilitates silencing in Th2 cells. By contrast, *Ifng*CNS-54, *Ifng*CNS-34, and *Ifng*CNS-22 exhibit enhancer-blocking activity, suggesting that these three elements may serve as sequential barriers to segregate *Ifng* from upstream transposable elements and segmental duplications and from the regulatory elements associated with *Il22* and *Mdm1*, or vice versa. Thus, in addition to their ability to act as T-bet-dependent enhancers in Th1 cells, *Ifng*CNS-34 and *Ifng*CNS-22 may serve a basal function in protecting the locus from ectopic activation by surrounding elements. However, with the exception of the demonstration that *Ifng*CNS-22 is an important enhancer *in vivo*, the actual contribution of these other distal elements to proper *Ifng* expression will require further evaluation, using transgenic and knockout approaches *in vivo*.

10. CONCLUSIONS AND FUTURE DIRECTIONS

IFN-γ is crucial for immunity against viral and intracellular bacterial infections and tumor control; however, aberrant IFN-γ expression has been associated with a number of autoinflammatory and autoimmune diseases. During infection, the innate recognition of pathogens leads to the production of IFN-γ by NK and/or NKT cells, which in turn influences the generation of IFN-γ-producing CD4 and CD8 T cells. In NK and NKT cells, the *Ifng* locus is open and accessible, allowing them to produce IFN-γ rapidly, in response to signals that activate STAT4, NFκB, and AP-1.

In contrast, differentiation of CD8 and CD4 T cells into high IFN-γ-producing effector T cells takes longer and is more complex, requiring the induction of transcription factors that facilitate IFN-γ production, remodel the *Ifng* locus and imprint on these cells an "epigenetic memory" of the context in which they first encountered antigen. This allows the resulting memory T cells to retain the ability to faithfully reiterate the correct effector program such that, on subsequent antigen encounter, that effector program, and not an alternate program, is rapidly executed.

The past few years have brought much insight regarding the molecular regulation of IFN-γ, particularly with the identification and *in vitro* characterization of distal regulatory elements. When compared to the Th2 locus, the epigenetic modifications and three-dimensional structure of the *Ifng* locus in naive CD4 T cells suggest a model whereby the chromatin architecture of *Ifng* is poised to facilitate either rapid opening or silencing during Th1 or Th2 differentiation, respectively. A number of recently described enhancers, boundary elements, or dual-purpose elements in the *Ifng* locus are likely to be involved in these changes and may function by binding transcription factors, such as T-bet, Eomes, and STAT4 to enable *Ifng* expression, or by binding other factors, like GATA-3, to silence *Ifng*. Future studies should provide more details and, through deleting these regulatory elements from the endogenous locus, reveal their functions in the most physiological context.

ACKNOWLEDGMENTS

The authors' work described herein was supported by NIH grants AI071272 and HD18184 and a grant from the March of Dimes. JRS was supported by predoctoral training grants from the NIH (CA009537) and Cancer Research Institute.

REFERENCES

Adler, S. H., Chiffoleau, E., Xu, L., Dalton, N. M., Burg, J. M., Wells, A. D., Wolfe, M. S., Turka, L. A., and Pear, W. S. (2003). Notch signaling augments T cell responsiveness by enhancing CD25 expression. *J. Immunol.* **171**, 2896–2903.

Afkarian, M., Sedy, J. R., Yang, J., Jacobson, N. G., Cereb, N., Yang, S. Y., Murphy, T. L., and Murphy, K. M. (2002). T-bet is a STAT1-induced regulator of IL-12R expression in naive CD4+ T cells. *Nat. Immunol.* **3**, 549–557.

Agarwal, S., and Rao, A. (1998). Modulation of chromatin structure regulates cytokine gene expression during T cell differentiation. *Immunity* **9**, 765–775.

Amprey, J. L., Im, J. S., Turco, S. J., Murray, H. W., Illarionov, P. A., Besra, G. S., Porcelli, S. A., and Spath, G. F. (2004). A subset of liver NK T cells is activated during *Leishmania donovani* infection by CD1d-bound lipophosphoglycan. *J. Exp. Med.* **200**, 895–904.

Amsen, D., Blander, J. M., Lee, G. R., Tanigaki, K., Honjo, T., and Flavell, R. A. (2004). Instruction of distinct CD4 T helper cell fates by different notch ligands on antigen-presenting cells. *Cell* **117**, 515–526.

Ansel, K. M., Lee, D. U., and Rao, A. (2003). An epigenetic view of helper T cell differentiation. *Nat. Immunol.* **4,** 616–623.

Ansel, K. M., Djuretic, I., Tanasa, B., and Rao, A. (2006). Regulation of Th2 differentiation and *Il4* locus accessibility. *Annu. Rev. Immunol.* **24,** 607–656.

Aramburu, J., Azzoni, L., Rao, A., and Perussia, B. (1995). Activation and expression of the nuclear factors of activated T cells, NFATp and NFATc, in human natural killer cells: Regulation upon CD16 ligand binding. *J. Exp. Med.* **182,** 801–810.

Aronica, M. A., Mora, A. L., Mitchell, D. B., Finn, P. W., Johnson, J. E., Sheller, J. R., and Boothby, M. R. (1999). Preferential role for NF-kappa B/Rel signaling in the type 1 but not type 2 T cell-dependent immune response *in vivo*. *J. Immunol.* **163,** 5116–5124.

Atherly, L. O., Lucas, J. A., Felices, M., Yin, C. C., Reiner, S. L., and Berg, L. J. (2006). The Tec family tyrosine kinases Itk and Rlk regulate the development of conventional CD8+ T cells. *Immunity* **25,** 79–91.

Athie-Morales, V., Smits, H. H., Cantrell, D. A., and Hilkens, C. M. (2004). Sustained IL-12 signaling is required for Th1 development. *J. Immunol.* **172,** 61–69.

Avni, O., Lee, D., Macian, F., Szabo, S. J., Glimcher, L. H., and Rao, A. (2002). T(H) cell differentiation is accompanied by dynamic changes in histone acetylation of cytokine genes. *Nat. Immunol.* **3,** 643–651.

Bachmann, M. F., Barner, M., Viola, A., and Kopf, M. (1999). Distinct kinetics of cytokine production and cytolysis in effector and memory T cells after viral infection. *Eur. J. Immunol.* **29,** 291–299.

Barbulescu, K., Becker, C., Schlaak, J. F., Schmitt, E., Meyer zum Buschenfelde, K. H., and Neurath, M. F. (1998). IL-12 and IL-18 differentially regulate the transcriptional activity of the human IFN-gamma promoter in primary CD4+ T lymphocytes. *J. Immunol.* **160,** 3642–3647.

Baron, M. (2003). An overview of the Notch signalling pathway. *Semin. Cell Dev. Biol.* **14,** 113–119.

Barton, K., Muthusamy, N., Chanyangam, M., Fischer, C., Clendenin, C., and Leiden, J. M. (1996). Defective thymocyte proliferation and IL-2 production in transgenic mice expressing a dominant-negative form of CREB. *Nature* **379,** 81–85.

Barton, K., Muthusamy, N., Fischer, C., Ting, C. N., Walunas, T. L., Lanier, L. L., and Leiden, J. M. (1998). The Ets-1 transcription factor is required for the development of natural killer cells in mice. *Immunity* **9,** 555–563.

Beima, K. M., Miazgowicz, M. M., Lewis, M. D., Yan, P. S., Huang, T. H., and Weinmann, A. S. (2006). T-bet binding to newly identified target gene promoters is cell type-independent but results in variable context-dependent functional effects. *J. Biol. Chem.* **281,** 11992–12000.

Berberich-Siebelt, F., Klein-Hessling, S., Hepping, N., Santner-Nanan, B., Lindemann, D., Schimpl, A., Berberich, I., and Serfling, E. (2000). C/EBPbeta enhances IL-4 but impairs IL-2 and IFN-gamma induction in T cells. *Eur. J. Immunol.* **30,** 2576–2585.

Berenson, L. S., Farrar, J. D., Murphy, T. L., and Murphy, K. M. (2004a). Frontline: Absence of functional STAT4 activation despite detectable tyrosine phosphorylation induced by murine IFN-alpha. *Eur. J. Immunol.* **34,** 2365–2374.

Berenson, L. S., Ota, N., and Murphy, K. M. (2004b). Issues in T-helper 1 development–resolved and unresolved. *Immunol. Rev.* **202,** 157–174.

Berenson, L. S., Gavrieli, M., Farrar, J. D., Murphy, T. L., and Murphy, K. M. (2006a). Distinct characteristics of murine STAT4 activation in response to IL-12 and IFN-alpha. *J. Immunol.* **177,** 5195–5203.

Berenson, L. S., Yang, J., Sleckman, B. P., Murphy, T. L., and Murphy, K. M. (2006b). Selective requirement of p38alpha MAPK in cytokine-dependent, but not antigen receptor-dependent, Th1 responses. *J. Immunol.* **176,** 4616–4621.

Berg, L. J. (2007). Signalling through TEC kinases regulates conventional versus innate CD8(+) T-cell development. *Nat. Rev. Immunol.* **7,** 479–485.

Berg, R. E., Crossley, E., Murray, S., and Forman, J. (2003). Memory CD8+ T cells provide innate immune protection against Listeria monocytogenes in the absence of cognate antigen. *J. Exp. Med.* **198,** 1583–1593.

Berg, R. E., Crossley, E., Murray, S., and Forman, J. (2005). Relative contributions of NK and CD8 T cells to IFN-gamma mediated innate immune protection against *Listeria monocytogenes*. *J. Immunol.* **175,** 1751–1757.

Bernstein, B. E., Mikkelsen, T. S., Xie, X., Kamal, M., Huebert, D. J., Cuff, J., Fry, B., Meissner, A., Wernig, M., Plath, K., Jaenisch, R., Wagschal, R., *et al.* (2006). A bivalent chromatin structure marks key developmental genes in embryonic stem cells. *Cell* **125,** 315–326.

Bischoff, S. C. (2007). Role of mast cells in allergic and non-allergic immune responses: Comparison of human and murine data. *Nat. Rev. Immunol.* **7,** 93–104.

Boehm, U., Klamp, T., Groot, M., and Howard, J. C. (1997). Cellular responses to interferon-gamma. *Annu. Rev. Immunol.* **15,** 749–795.

Boivin, G. P., O'Toole, B. A., Orsmby, I. E., Diebold, R. J., Eis, M. J., Doetschman, T., and Kier, A. B. (1995). Onset and progression of pathological lesions in transforming growth factor-beta 1-deficient mice. *Am. J. Pathol.* **146,** 276–288.

Bories, J. C., Willerford, D. M., Grevin, D., Davidson, L., Camus, A., Martin, P., Stehelin, D., and Alt, F. W. (1995). Increased T-cell apoptosis and terminal B-cell differentiation induced by inactivation of the Ets-1 proto-oncogene. *Nature* **377,** 635–638.

Bouma, G., and Strober, W. (2003). The immunological and genetic basis of inflammatory bowel disease. *Nat. Rev. Immunol.* **3,** 521–533.

Bream, J. H., Hodge, D. L., Gonsky, R., Spolski, R., Leonard, W. J., Krebs, S., Targan, S., Morinobu, A., O'Shea, J. J., and Young, H. A. (2004). A distal region in the interferon-gamma gene is a site of epigenetic remodeling and transcriptional regulation by interleukin-2. *J. Biol. Chem.* **279,** 41249–41257.

Brinkmann, V., Geiger, T., Alkan, S., and Heusser, C. H. (1993). Interferon alpha increases the frequency of interferon gamma-producing human CD4+ T cells. *J. Exp. Med.* **178,** 1655–1663.

Brombacher, F., Dorfmuller, A., Magram, J., Dai, W. J., Kohler, G., Wunderlin, A., Palmer-Lehmann, K., Gately, M. K., and Alber, G. (1999). IL-12 is dispensable for innate and adaptive immunity against low doses of *Listeria monocytogenes*. *Int. Immunol.* **11,** 325–332.

Broussard, C., Fleischacker, C., Horai, R., Chetana, M., Venegas, A. M., Sharp, L. L., Hedrick, S. M., Fowlkes, B. J., and Schwartzberg, P. L. (2006). Altered development of CD8+ T cell lineages in mice deficient for the Tec kinases Itk and Rlk. *Immunity* **25,** 93–104.

Brutkiewicz, R. R. (2006). CD1d ligands: The good, the bad, and the ugly. *J. Immunol.* **177,** 769–775.

Caamano, J., Alexander, J., Craig, L., Bravo, R., and Hunter, C. A. (1999). The NF-kappa B family member RelB is required for innate and adaptive immunity to *Toxoplasma gondii*. *J. Immunol.* **163,** 4453–4461.

Cai, G., Radzanowski, T., Villegas, E. N., Kastelein, R., and Hunter, C. A. (2000). Identification of STAT4-dependent and independent mechanisms of resistance to *Toxoplasma gondii*. *J. Immunol.* **165,** 2619–2627.

Carson, W. E., Fehniger, T. A., Haldar, S., Eckhert, K., Lindemann, M. J., Lai, C. F., Croce, C. M., Baumann, H., and Caligiuri, M. A. (1997). A potential role for interleukin-15 in the regulation of human natural killer cell survival. *J. Clin. Invest.* **99,** 937–943.

Carter, L. L., and Murphy, K. M. (1999). Lineage-specific requirement for signal transducer and activator of transcription (Stat)4 in interferon gamma production from CD4(+) versus CD8(+) T cells. *J. Exp. Med.* **189,** 1355–1360.

Casanova, J. L., and Abel, L. (2002). Genetic dissection of immunity to mycobacteria: The human model. *Annu. Rev. Immunol.* **20**, 581–620.

Chang, S., and Aune, T. M. (2005). Histone hyperacetylated domains across the Ifng gene region in natural killer cells and T cells. *Proc. Natl. Acad. Sci. USA* **102**, 17095–17100.

Chang, S., and Aune, T. M. (2007). Dynamic changes in histone-methylation 'marks' across the locus encoding interferon-gamma during the differentiation of T helper type 2 cells. *Nat. Immunol.* **8**(7), 723–731.

Chen, Q., Ghilardi, N., Wang, H., Baker, T., Xie, M. H., Gurney, A., Grewal, I. S., and de Sauvage, F. J. (2000). Development of Th1-type immune responses requires the type I cytokine receptor TCCR. *Nature* **407**, 916–920.

Cho, S. S., Bacon, C. M., Sudarshan, C., Rees, R. C., Finbloom, D., Pine, R., and O'Shea, J. J. (1996). Activation of STAT4 by IL-12 and IFN-alpha: Evidence for the involvement of ligand-induced tyrosine and serine phosphorylation. *J. Immunol.* **157**, 4781–4789.

Cho, B. K., Wang, C., Sugawa, S., Eisen, H. N., and Chen, J. (1999). Functional differences between memory and naive CD8 T cells. *Proc. Natl. Acad. Sci. USA* **96**, 2976–2981.

Cho, J. Y., Grigura, V., Murphy, T. L., and Murphy, K. (2003). Identification of cooperative monomeric Brachyury sites conferring T-bet responsiveness to the proximal IFN-gamma promoter. *Int. Immunol.* **15**, 1149–1160.

Cooper, M. A., Bush, J. E., Fehniger, T. A., VanDeusen, J. B., Waite, R. E., Liu, Y., Aguila, H. L., and Caligiuri, M. A. (2002). In vivo evidence for a dependence on interleukin 15 for survival of natural killer cells. *Blood* **100**, 3633–3638.

Corn, R. A., Aronica, M. A., Zhang, F., Tong, Y., Stanley, S. A., Kim, S. R., Stephenson, L., Enerson, B., McCarthy, S., Mora, A., and Boothby, M. (2003). T cell-intrinsic requirement for NF-kappa B induction in postdifferentiation IFN-gamma production and clonal expansion in a Th1 response. *J. Immunol.* **171**, 1816–1824.

Corn, R. A., Hunter, C., Liou, H. C., Siebenlist, U., and Boothby, M. R. (2005). Opposing roles for RelB and Bcl-3 in regulation of T-box expressed in T cells, GATA-3, and Th effector differentiation. *J. Immunol.* **175**, 2102–2110.

Curtsinger, J. M., Valenzuela, J. O., Agarwal, P., Lins, D., and Mescher, M. F. (2005). Type I IFNs provide a third signal to CD8 T cells to stimulate clonal expansion and differentiation. *J. Immunol.* **174**, 4465–4469.

Dalton, D. K., Pitts-Meek, S., Keshav, S., Figari, I. S., Bradley, A., and Stewart, T. A. (1993). Multiple defects of immune cell function in mice with disrupted interferon-gamma genes. *Science* **259**, 1739–1742.

Darnell, J. E., Jr., Kerr, I. M., and Stark, G. R. (1994). Jak-STAT pathways and transcriptional activation in response to IFNs and other extracellular signaling proteins. *Science* **264**, 1415–1421.

de Jong, R., Altare, F., Haagen, I. A., Elferink, D. G., Boer, T., van Breda Vriesman, P. J., Kabel, P. J., Draaisma, J. M., van Dissel, J. T., Kroon, F. P., Casanova, J.-L., and Ottenhoff, T. H. M. (1998). Severe mycobacterial and *Salmonella* infections in interleukin-12 receptor-deficient patients. *Science* **280**, 1435–1438.

Decker, T., Muller, M., and Stockinger, S. (2005). The yin and yang of type I interferon activity in bacterial infection. *Nat. Rev. Immunol.* **5**, 675–687.

Diehl, S., Anguita, J., Hoffmeyer, A., Zapton, T., Ihle, J. N., Fikrig, E., and Rincon, M. (2000). Inhibition of Th1 differentiation by IL-6 is mediated by SOCS1. *Immunity* **13**, 805–815.

Djuretic, I. M., Levanon, D., Negreanu, V., Groner, Y., Rao, A., and Ansel, K. M. (2007). Transcription factors T-bet and Runx3 cooperate to activate Ifng and silence Il4 in T helper type 1 cells. *Nat. Immunol.* **8**, 145–153.

Dong, C. (2006). Diversification of T-helper-cell lineages: Finding the family root of IL-17-producing cells. *Nat. Rev. Immunol.* **6**, 329–333.

Dong, C., Davis, R. J., and Flavell, R. A. (2002). MAP kinases in the immune response. *Annu. Rev. Immunol.* **20,** 55–72.

Dumoutier, L., Van Roost, E., Ameye, G., Michaux, L., and Renauld, J. C. (2000). IL-TIF/IL-22: Genomic organization and mapping of the human and mouse genes. *Genes Immun.* **1,** 488–494.

Egawa, T., Eberl, G., Taniuchi, I., Benlagha, K., Geissmann, F., Hennighausen, L., Bendelac, A., and Littman, D. R. (2005). Genetic evidence supporting selection of the Valpha14i NKT cell lineage from double-positive thymocyte precursors. *Immunity* **22,** 705–716.

Farrar, J. D., and Murphy, K. M. (2000). Type I interferons and T helper development. *Immunol. Today* **21,** 484–489.

Farrar, J. D., Smith, J. D., Murphy, T. L., Leung, S., Stark, G. R., and Murphy, K. M. (2000a). Selective loss of type I interferon-induced STAT4 activation caused by a minisatellite insertion in mouse Stat2. *Nat. Immunol.* **1,** 65–69.

Farrar, J. D., Smith, J. D., Murphy, T. L., and Murphy, K. M. (2000b). Recruitment of Stat4 to the human interferon-alpha/beta receptor requires activated Stat2. *J. Biol. Chem.* **275,** 2693–2697.

Felsenfeld, G., and Groudine, M. (2003). Controlling the double helix. *Nature* **421,** 448–453.

Fields, P. E., Kim, S. T., and Flavell, R. A. (2002). Cutting edge: Changes in histone acetylation at the IL-4 and IFN-gamma loci accompany Th1/Th2 differentiation. *J. Immunol.* **169,** 647–650.

Filipe-Santos, O., Bustamante, J., Chapgier, A., Vogt, G., de Beaucoudrey, L., Feinberg, J., Jouanguy, E., Boisson-Dupuis, S., Fieschi, C., Picard, C., and Casanova, J. L. (2006). Inborn errors of IL-12/23- and IFN-gamma-mediated immunity: Molecular, cellular, and clinical features. *Semin. Immunol.* **18,** 347–361.

Fischer, K., Scotet, E., Niemeyer, M., Koebernick, H., Zerrahn, J., Maillet, S., Hurwitz, R., Kursar, M., Bonneville, M., Kaufmann, S. H., and Schaible, U. E. (2004). Mycobacterial phosphatidylinositol mannoside is a natural antigen for CD1d-restricted T cells. *Proc. Natl. Acad. Sci. USA* **101,** 10685–10690.

Fitzgerald, K. A., Bowie, A. G., Skeffington, B. S., and O'Neill, L. A. (2000). Ras, protein kinase C zeta, and I kappa B kinases 1 and 2 are downstream effectors of CD44 during the activation of NF-kappa B by hyaluronic acid fragments in T-24 carcinoma cells. *J. Immunol.* **164,** 2053–2063.

Fontenot, J. D., Rasmussen, J. P., Williams, L. M., Dooley, J. L., Farr, A. G., and Rudensky, A. Y. (2005). Regulatory T cell lineage specification by the forkhead transcription factor foxp3. *Immunity* **22,** 329–341.

Freudenberg, M. A., Merlin, T., Kalis, C., Chvatchko, Y., Stubig, H., and Galanos, C. (2002). Cutting edge: A murine, IL-12-independent pathway of IFN-gamma induction by gram-negative bacteria based on STAT4 activation by Type I IFN and IL-18 signaling. *J. Immunol.* **169,** 1665–1668.

Fujii, H., Ogasawara, K., Otsuka, H., Suzuki, M., Yamamura, K., Yokochi, T., Miyazaki, T., Suzuki, H., Mak, T. W., Taki, S., and Taniguchi, T. (1998). Functional dissection of the cytoplasmic subregions of the IL-2 receptor betac chain in primary lymphocyte populations. *EMBO J.* **17,** 6551–6557.

Gansert, J. L., Kiessler, V., Engele, M., Wittke, F., Rollinghoff, M., Krensky, A. M., Porcelli, S. A., Modlin, R. L., and Stenger, S. (2003). Human NKT cells express granulysin and exhibit antimycobacterial activity. *J. Immunol.* **170,** 3154–3161.

Gapin, L., Matsuda, J. L., Surh, C. D., and Kronenberg, M. (2001). NKT cells derive from double-positive thymocytes that are positively selected by CD1d. *Nat. Immunol.* **2,** 971–978.

Gaszner, M., and Felsenfeld, G. (2006). Insulators: Exploiting transcriptional and epigenetic mechanisms. *Nat. Rev. Genet.* **7,** 703–713.

Gerosa, F., Gobbi, A., Zorzi, P., Burg, S., Briere, F., Carra, G., and Trinchieri, G. (2005). The reciprocal interaction of NK cells with plasmacytoid or myeloid dendritic cells profoundly affects innate resistance functions. *J. Immunol.* **174,** 727–734.

Ghosh, P., Tan, T. H., Rice, N. R., Sica, A., and Young, H. A. (1993). The interleukin 2 CD28-responsive complex contains at least three members of the NF kappa B family: c-Rel, p50, and p65. *Proc. Natl. Acad. Sci. USA* **90,** 1696–1700.

Ghosh, S., May, M. J., and Kopp, E. B. (1998). NF-kappa B and Rel proteins: Evolutionarily conserved mediators of immune responses. *Annu. Rev. Immunol.* **16,** 225–260.

Glimcher, L. H., Townsend, M. J., Sullivan, B. M., and Lord, G. M. (2004). Recent developments in the transcriptional regulation of cytolytic effector cells. *Nat. Rev. Immunol.* **4,** 900–911.

Gonsky, R., Deem, R. L., Bream, J., Young, H. A., and Targan, S. R. (2004). Enhancer role of STAT5 in CD2 activation of IFN-gamma gene expression. *J. Immunol.* **173,** 6241–6247.

Gorelik, L., and Flavell, R. A. (2000). Abrogation of TGFbeta signaling in T cells leads to spontaneous T cell differentiation and autoimmune disease. *Immunity* **12,** 171–181.

Grenningloh, R., Kang, B. Y., and Ho, I. C. (2005). Ets-1, a functional cofactor of T-bet, is essential for Th1 inflammatory responses. *J. Exp. Med.* **201,** 615–626.

Grimbaldeston, M. A., Metz, M., Yu, M., Tsai, M., and Galli, S. J. (2006). Effector and potential immunoregulatory roles of mast cells in IgE-associated acquired immune responses. *Curr. Opin. Immunol.* **18,** 751–760.

Grogan, J. L., and Locksley, R. M. (2002). T helper cell differentiation: On again, off again. *Curr. Opin. Immunol.* **14,** 366–372.

Hanabuchi, S., Watanabe, N., Wang, Y. H., Wang, Y. H., Ito, T., Shaw, J., Cao, W., Qin, F. X., and Liu, Y. J. (2006). Human plasmacytoid predendritic cells activate NK cells through glucocorticoid-induced tumor necrosis factor receptor-ligand (GITRL). *Blood* **107,** 3617–3623.

Harrington, L. E., Mangan, P. R., and Weaver, C. T. (2006). Expanding the effector CD4 T-cell repertoire: The Th17 lineage. *Curr. Opin. Immunol.* **18,** 349–356.

Harty, J. T., and Bevan, M. J. (1995). Specific immunity to *Listeria monocytogenes* in the absence of IFN gamma. *Immunity* **3,** 109–117.

Harty, J. T., Tvinnereim, A. R., and White, D. W. (2000). CD8+ T cell effector mechanisms in resistance to infection. *Annu. Rev. Immunol.* **18,** 275–308.

Hatton, R. D., Harrington, L. E., Luther, R. J., Wakefield, T., Janowski, K. M., Oliver, J. R., Lallone, R. L., Murphy, K. M., and Weaver, C. T. (2006). A distal conserved sequence element controls *Ifng* gene expression by T cells and NK cells. *Immunity* **25,** 717–729.

Hibbert, L., Pflanz, S., De Waal Malefyt, R., and Kastelein, R. A. (2003). IL-27 and IFN-alpha signal via Stat1 and Stat3 and induce T-Bet and IL-12Rbeta2 in naive T cells. *J. Interferon Cytokine Res.* **23,** 513–522.

Hill, N., and Sarvetnick, N. (2002). Cytokines: Promoters and dampeners of autoimmunity. *Curr. Opin. Immunol.* **14,** 791–797.

Ho, I. C., and Glimcher, L. H. (2002). Transcription: Tantalizing times for T cells. *Cell* **109** (Suppl.), S109–S120.

Hodge, M. R., Ranger, A. M., Charles de la Brousse, F., Hoey, T., Grusby, M. J., and Glimcher, L. H. (1996). Hyperproliferation and dysregulation of IL-4 expression in NF-ATp-deficient mice. *Immunity* **4,** 397–405.

Hogan, P. G., Chen, L., Nardone, J., and Rao, A. (2003). Transcriptional regulation by calcium, calcineurin, and NFAT. *Genes Dev.* **17,** 2205–2232.

Hoshino, K., Tsutsui, H., Kawai, T., Takeda, K., Nakanishi, K., Takeda, Y., and Akira, S. (1999). Cutting edge: Generation of IL-18 receptor-deficient mice: Evidence for IL-1 receptor-related protein as an essential IL-18 binding receptor. *J. Immunol.* **162,** 5041–5044.

Huang, S., Hendriks, W., Althage, A., Hemmi, S., Bluethmann, H., Kamijo, R., Vilcek, J., Zinkernagel, R. M., and Aguet, M. (1993). Immune response in mice that lack the interferon-gamma receptor. *Science* **259**, 1742–1745.

Hunter, C. A. (2005). New IL-12-family members: IL-23 and IL-27, cytokines with divergent functions. *Nat. Rev. Immunol.* **5**, 521–531.

Hunter, C. A., Gabriel, K. E., Radzanowski, T., Neyer, L. E., and Remington, J. S. (1997). Type I interferons enhance production of IFN-gamma by NK cells. *Immunol. Lett.* **59**, 1–5.

Hwang, E. S., Szabo, S. J., Schwartzberg, P. L., and Glimcher, L. H. (2005). T helper cell fate specified by kinase-mediated interaction of T-bet with GATA-3. *Science* **307**, 430–433.

Igawa, D., Sakai, M., and Savan, R. (2006). An unexpected discovery of two interferon gamma-like genes along with interleukin (IL)-22 and -26 from teleost: IL-22 and -26 genes have been described for the first time outside mammals. *Mol. Immunol.* **43**, 999–1009.

Ikeda, H., Old, L. J., and Schreiber, R. D. (2002). The roles of IFN gamma in protection against tumor development and cancer immunoediting. *Cytokine Growth Factor Rev.* **13**, 95–109.

Intlekofer, A. M., Takemoto, N., Wherry, E. J., Longworth, S. A., Northrup, J. T., Palanivel, V. R., Mullen, A. C., Gasink, C. R., Kaech, S. M., Miller, J. D., Gapin, L., Ryan, L., et al. (2005). Effector and memory CD8+ T cell fate coupled by T-bet and eomesodermin. *Nat. Immunol.* **6**, 1236–1244.

Jacobson, N. G., Szabo, S. J., Weber-Nordt, R. M., Zhong, Z., Schreiber, R. D., Darnell, J. E., Jr., and Murphy, K. M. (1995). Interleukin 12 signaling in T helper type 1 (Th1) cells involves tyrosine phosphorylation of signal transducer and activator of transcription (Stat)3 and Stat4. *J. Exp. Med.* **181**, 1755–1762.

John, B., Rajagopal, D., Pashine, A., Rath, S., George, A., and Bal, V. (2002). Role of IL-12-independent and IL-12-dependent pathways in regulating generation of the IFN-gamma component of T cell responses to Salmonella typhimurium. *J. Immunol.* **169**, 2545–2552.

Jouanguy, E., Doffinger, R., Dupuis, S., Pallier, A., Altare, F., and Casanova, J. L. (1999). IL-12 and IFN-gamma in host defense against mycobacteria and Salmonella in mice and men. *Curr. Opin. Immunol.* **11**, 346–351.

Kaisho, T., Tsutsui, H., Tanaka, T., Tsujimura, T., Takeda, K., Kawai, T., Yoshida, N., Nakanishi, K., and Akira, S. (1999). Impairment of natural killer cytotoxic activity and interferon gamma production in CCAAT/enhancer binding protein gamma-deficient mice. *J. Exp. Med.* **190**, 1573–1582.

Kaminuma, O., Kitamura, F., Kitamura, N., Miyagishi, M., Taira, K., Yamamoto, K., Miura, O., and Miyatake, S. (2004). GATA-3 suppresses IFN-gamma promoter activity independently of binding to cis-regulatory elements. *FEBS Lett.* **570**, 63–68.

Kaplan, D. H., Shankaran, V., Dighe, A. S., Stockert, E., Aguet, M., Old, L. J., and Schreiber, R. D. (1998). Demonstration of an interferon gamma-dependent tumor surveillance system in immunocompetent mice. *Proc. Natl. Acad. Sci. USA* **95**, 7556–7561.

Kaplan, M. H., Sun, Y. L., Hoey, T., and Grusby, M. J. (1996). Impaired IL-12 responses and enhanced development of Th2 cells in Stat4-deficient mice. *Nature* **382**, 174–177.

Kashiwakura, J., Suzuki, N., Nagafuchi, H., Takeno, M., Takeba, Y., Shimoyama, Y., and Sakane, T. (1999). Txk, a nonreceptor tyrosine kinase of the Tec family, is expressed in T helper type 1 cells and regulates interferon gamma production in human T lymphocytes. *J. Exp. Med.* **190**, 1147–1154.

Kawano, T., Cui, J., Koezuka, Y., Toura, I., Kaneko, Y., Motoki, K., Ueno, H., Nakagawa, R., Sato, H., Kondo, E., Koseki, H., and Taniguchi, M. (1997). CD1d-restricted and TCR-mediated activation of valpha14 NKT cells by glycosylceramides. *Science* **278**, 1626–1629.

Kiani, A., Viola, J. P., Lichtman, A. H., and Rao, A. (1997). Down-regulation of IL-4 gene transcription and control of Th2 cell differentiation by a mechanism involving NFAT1. *Immunity* **7,** 849–860.

Kobayashi, M., Fitz, L., Ryan, M., Hewick, R. M., Clark, S. C., Chan, S., Loudon, R., Sherman, F., Perussia, B., and Trinchieri, G. (1989). Identification and purification of natural killer cell stimulatory factor (NKSF), a cytokine with multiple biologic effects on human lymphocytes. *J. Exp. Med.* **170,** 827–845.

Koka, R., Burkett, P. R., Chien, M., Chai, S., Chan, F., Lodolce, J. P., Boone, D. L., and Ma, A. (2003). Interleukin (IL)-15R[alpha]-deficient natural killer cells survive in normal but not IL-15R[alpha]-deficient mice. *J. Exp. Med.* **197,** 977–984.

Komine, O., Hayashi, K., Natsume, W., Watanabe, T., Seki, Y., Seki, N., Yagi, R., Sukzuki, W., Tamauchi, H., Hozumi, K., Habu, S., Kubo, S., *et al*. (2003). The Runx1 transcription factor inhibits the differentiation of naive CD4+ T cells into the Th2 lineage by repressing GATA3 expression. *J. Exp. Med.* **198,** 51–61.

Kouzarides, T. (2007). Chromatin modifications and their function. *Cell* **128,** 693–705.

Kronenberg, M. (2005). Toward an understanding of NKT cell biology: Progress and paradoxes. *Annu. Rev. Immunol.* **23,** 877–900.

Kronenberg, M., and Gapin, L. (2002). The unconventional lifestyle of NKT cells. *Nat. Rev. Immunol.* **2,** 557–568.

Kulkarni, A. B., Huh, C. G., Becker, D., Geiser, A., Lyght, M., Flanders, K. C., Roberts, A. B., Sporn, M. B., Ward, J. M., and Karlsson, S. (1993). Transforming growth factor beta 1 null mutation in mice causes excessive inflammatory response and early death. *Proc. Natl. Acad. Sci. USA* **90,** 770–774.

Kuo, C. T., and Leiden, J. M. (1999). Transcriptional regulation of T lymphocyte development and function. *Annu. Rev. Immunol.* **17,** 149–187.

Lanier, L. L. (2005). NK cell recognition. *Annu. Rev. Immunol.* **23,** 225–274.

Laouar, Y., Sutterwala, F. S., Gorelik, L., and Flavell, R. A. (2005). Transforming growth factor-beta controls T helper type 1 cell development through regulation of natural killer cell interferon-gamma. *Nat. Immunol.* **6,** 600–607.

Lee, C. K., Rao, D. T., Gertner, R., Gimeno, R., Frey, A. B., and Levy, D. E. (2000). Distinct requirements for IFNs and STAT1 in NK cell function. *J. Immunol.* **165,** 3571–3577.

Lee, D. U., Avni, O., Chen, L., and Rao, A. (2004). A distal enhancer in the interferon-gamma (IFN-gamma) locus revealed by genome sequence comparison. *J. Biol. Chem.* **279,** 4802–4810.

Lee, G. R., Kim, S. T., Spilianakis, C. G., Fields, P. E., and Flavell, R. A. (2006). T helper cell differentiation: Regulation by cis elements and epigenetics. *Immunity* **24,** 369–379.

Li, M. O., Sanjabi, S., and Flavell, R. A. (2006a). Transforming growth factor-beta controls development, homeostasis, and tolerance of T cells by regulatory T cell-dependent and -independent mechanisms. *Immunity* **25,** 455–471.

Li, M. O., Wan, Y. Y., Sanjabi, S., Robertson, A. K., and Flavell, R. A. (2006b). Transforming growth factor-beta regulation of immune responses. *Annu. Rev. Immunol.* **24,** 99–146.

Li, Q., Eppolito, C., Odunsi, K., and Shrikant, P. A. (2006c). IL-12-programmed long-term CD8+ T cell responses require STAT4. *J. Immunol.* **177,** 7618–7625.

Liang, S. C., Tan, X. Y., Luxenberg, D. P., Karim, R., Dunussi-Joannopoulos, K., Collins, M., and Fouser, L. A. (2006). Interleukin (IL)-22 and IL-17 are coexpressed by Th17 cells and cooperatively enhance expression of antimicrobial peptides. *J. Exp. Med.* **203,** 2271–2279.

Lieberman, L. A., and Hunter, C. A. (2002). The role of cytokines and their signaling pathways in the regulation of immunity to *Toxoplasma gondii*. *Int. Rev. Immunol.* **21,** 373–403.

Lieberman, L. A., Banica, M., Reiner, S. L., and Hunter, C. A. (2004). STAT1 plays a critical role in the regulation of antimicrobial effector mechanisms, but not in the development of Th1-type responses during toxoplasmosis. *J. Immunol.* **172,** 457–463.

Lighvani, A. A., Frucht, D. M., Jankovic, D., Yamane, H., Aliberti, J., Hissong, B. D., Nguyen, B. V., Gadina, M., Sher, A., Paul, W. E., and O'Shea, J. J. (2001). T-bet is rapidly induced by interferon-gamma in lymphoid and myeloid cells. *Proc. Natl. Acad. Sci. USA* **98,** 15137–15142.

Lin, J., and Weiss, A. (2001). T cell receptor signalling. *J. Cell Sci.* **114,** 243–244.

Lin, J. T., Martin, S. L., Xia, L., and Gorham, J. D. (2005). TGF-beta 1 uses distinct mechanisms to inhibit IFN-gamma expression in CD4+ T cells at priming and at recall: Differential involvement of Stat4 and T-bet. *J. Immunol.* **174,** 5950–5958.

Lohoff, M., Ferrick, D., Mittrucker, H. W., Duncan, G. S., Bischof, S., Rollinghoff, M., and Mak, T. W. (1997). Interferon regulatory factor-1 is required for a T helper 1 immune response *in vivo*. *Immunity* **6,** 681–689.

Lohr, J., Knoechel, B., and Abbas, A. K. (2006). Regulatory T cells in the periphery. *Immunol. Rev.* **212,** 149–162.

Lucas, M., Schachterle, W., Oberle, K., Aichele, P., and Diefenbach, A. (2007). Dendritic cells prime natural killer cells by trans-presenting interleukin 15. *Immunity* **26,** 503–517.

Lucas, P. J., Kim, S. J., Melby, S. J., and Gress, R. E. (2000). Disruption of T cell homeostasis in mice expressing a T cell-specific dominant negative transforming growth factor beta II receptor. *J. Exp. Med.* **191,** 1187–1196.

Lucas, S., Ghilardi, N., Li, J., and de Sauvage, F. J. (2003). IL-27 regulates IL-12 responsiveness of naive CD4+ T cells through Stat1-dependent and -independent mechanisms. *Proc. Natl. Acad. Sci. USA* **100,** 15047–15052.

Ma, A., Koka, R., and Burkett, P. (2006). Diverse functions of IL-2, IL-15, and IL-7 in lymphoid homeostasis. *Annu. Rev. Immunol.* **24,** 657–679.

Macian, F. (2005). NFAT proteins: Key regulators of T-cell development and function. *Nat. Rev. Immunol.* **5,** 472–484.

MacLennan, C., Fieschi, C., Lammas, D. A., Picard, C., Dorman, S. E., Sanal, O., MacLennan, J. M., Holland, S. M., Ottenhoff, T. H., Casanova, J. L., and Kumararatne, D. S. (2004). Interleukin (IL)-12 and IL-23 are key cytokines for immunity against Salmonella in humans. *J. Infect. Dis.* **190,** 1755–1757.

Maekawa, Y., Tsukumo, S., Chiba, S., Hirai, H., Hayashi, Y., Okada, H., Kishihara, K., and Yasutomo, K. (2003). Delta1-Notch3 interactions bias the functional differentiation of activated CD4+ T cells. *Immunity* **19,** 549–559.

Maldonado, R. A., Irvine, D. J., Schreiber, R., and Glimcher, L. H. (2004). A role for the immunological synapse in lineage commitment of CD4 lymphocytes. *Nature* **431,** 527–532.

Marshall, J. D., Heeke, D. S., Abbate, C., Yee, P., and Van Nest, G. (2006). Induction of interferon-gamma from natural killer cells by immunostimulatory CpG DNA is mediated through plasmacytoid-dendritic-cell-produced interferon-alpha and tumour necrosis factor-alpha. *Immunology* **117,** 38–46.

Massague, J. (1998). TGF-beta signal transduction. *Annu. Rev. Biochem.* **67,** 753–791.

Mathur, A. N., Chang, H. C., Zisoulis, D. G., Stritesky, G. L., Yu, Q., O'Malley, J. T., Kapur, R., Levy, D. E., Kansas, G. S., and Kaplan, M. H. (2007). Stat3 and Stat4 direct development of IL-17-secreting Th cells. *J. Immunol.* **178,** 4901–4907.

Matikainen, S., Paananen, A., Miettinen, M., Kurimoto, M., Timonen, T., Julkunen, I., and Sareneva, T. (2001). IFN-alpha and IL-18 synergistically enhance IFN-gamma production in human NK cells: Differential regulation of Stat4 activation and IFN-gamma gene expression by IFN-alpha and IL-12. *Eur. J. Immunol.* **31,** 2236–2245.

Matsuda, J. L., and Gapin, L. (2005). Developmental program of mouse Valpha14i NKT cells. *Curr. Opin. Immunol.* **17,** 122–130.

Matsuda, J. L., Gapin, L., Baron, J. L., Sidobre, S., Stetson, D. B., Mohrs, M., Locksley, R. M., and Kronenberg, M. (2003). Mouse Valpha 14i natural killer T cells are resistant to cytokine polarization *in vivo*. *Proc. Natl. Acad. Sci. USA* **100,** 8395–8400.

Matsuda, J. L., Zhang, Q., Ndonye, R., Richardson, S. K., Howell, A. R., and Gapin, L. (2006). T-bet concomitantly controls migration, survival, and effector functions during the development of Valpha14i NKT cells. *Blood* **107**, 2797–2805.

Mikhalkevich, N., Becknell, B., Caligiuri, M. A., Bates, M. D., Harvey, R., and Zheng, W. P. (2006). Responsiveness of naive CD4 T cells to polarizing cytokine determines the ratio of Th1 and Th2 cell differentiation. *J. Immunol.* **176**, 1553–1560.

Miller, A. T., Wilcox, H. M., Lai, Z., and Berg, L. J. (2004). Signaling through Itk promotes T helper 2 differentiation via negative regulation of T-bet. *Immunity* **21**, 67–80.

Minter, L. M., Turley, D. M., Das, P., Shin, H. M., Joshi, I., Lawlor, R. G., Cho, O. H., Palaga, T., Gottipati, S., Telfer, J. C., Kostura, L., Fauq, L., et al. (2005). Inhibitors of gamma-secretase block *in vivo* and *in vitro* T helper type 1 polarization by preventing Notch upregulation of Tbx21. *Nat. Immunol.* **6**, 680–688.

Monticelli, S., and Rao, A. (2002). NFAT1 and NFAT2 are positive regulators of IL-4 gene transcription. *Eur. J. Immunol.* **32**, 2971–2978.

Mullen, A. C., High, F. A., Hutchins, A. S., Lee, H. W., Villarino, A. V., Livingston, D. M., Kung, A. L., Cereb, N., Yao, T. P., Yang, S. Y., and Reiner, S. L. (2001). Role of T-bet in commitment of TH1 cells before IL-12-dependent selection. *Science* **292**, 1907–1910.

Mullen, A. C., Hutchins, A. S., High, F. A., Lee, H. W., Sykes, K. J., Chodosh, L. A., and Reiner, S. L. (2002). Hlx is induced by and genetically interacts with T-bet to promote heritable T(H)1 gene induction. *Nat. Immunol.* **3**, 652–658.

Murphy, K. M., Ouyang, W., Farrar, J. D., Yang, J., Ranganath, S., Asnagli, H., Afkarian, M., and Murphy, T. L. (2000). Signaling and transcription in T helper development. *Annu. Rev. Immunol.* **18**, 451–494.

Murphy, K. M., and Reiner, S. L. (2002). The lineage decisions of helper T cells. *Nat. Rev. Immunol.* **2**, 933–944.

Muthusamy, N., Barton, K., and Leiden, J. M. (1995). Defective activation and survival of T cells lacking the Ets-1 transcription factor. *Nature* **377**, 639–642.

Nagarajan, N. A., and Kronenberg, M. (2007). Invariant NKT cells amplify the innate immune response to lipopolysaccharide. *J. Immunol.* **178**, 2706–2713.

Nardone, J., Lee, D. U., Ansel, K. M., and Rao, A. (2004). Bioinformatics for the 'bench biologist': How to find regulatory regions in genomic DNA. *Nat. Immunol.* **5**, 768–774.

Neurath, M. F., Weigmann, B., Finotto, S., Glickman, J., Nieuwenhuis, E., Iijima, H., Mizoguchi, A., Mizoguchi, E., Mudter, J., Galle, P. R., Bhan, A., Autschbach, A., et al. (2002). The transcription factor T-bet regulates mucosal T cell activation in experimental colitis and Crohn's disease. *J. Exp. Med.* **195**, 1129–1143.

Newman, K. C., and Riley, E. M. (2007). Whatever turns you on: Accessory-cell-dependent activation of NK cells by pathogens. *Nat. Rev. Immunol.* **7**, 279–291.

Nguyen, K. B., Cousens, L. P., Doughty, L. A., Pien, G. C., Durbin, J. E., and Biron, C. A. (2000). Interferon alpha/beta-mediated inhibition and promotion of interferon gamma: STAT1 resolves a paradox. *Nat. Immunol.* **1**, 70–76.

Nguyen, K. B., Salazar-Mather, T. P., Dalod, M. Y., Van Deusen, J. B., Wei, X. Q., Liew, F. Y., Caligiuri, M. A., Durbin, J. E., and Biron, C. A. (2002a). Coordinated and distinct roles for IFN-alpha beta, IL-12, and IL-15 regulation of NK cell responses to viral infection. *J. Immunol.* **169**, 4279–4287.

Nguyen, K. B., Watford, W. T., Salomon, R., Hofmann, S. R., Pien, G. C., Morinobu, A., Gadina, M., O'Shea, J. J., and Biron, C. A. (2002b). Critical role for STAT4 activation by type 1 interferons in the interferon-gamma response to viral infection. *Science* **297**, 2063–2066.

Nishimura, H., Yajima, T., Naiki, Y., Tsunobuchi, H., Umemura, M., Itano, K., Matsuguchi, T., Suzuki, M., Ohashi, P. S., and Yoshikai, Y. (2000). Differential roles of interleukin 15 mRNA isoforms generated by alternative splicing in immune responses *in vivo*. *J. Exp. Med.* **191**, 157–170.

Oki, S., Chiba, A., Yamamura, T., and Miyake, S. (2004). The clinical implication and molecular mechanism of preferential IL-4 production by modified glycolipid-stimulated NKT cells. *J. Clin. Invest.* **113**, 1631–1640.

Oppmann, B., Lesley, R., Blom, B., Timans, J. C., Xu, Y., Hunte, B., Vega, F., Yu, N., Wang, J., Singh, K., Zonin, F., Vaisberg, F., *et al.* (2000). Novel p19 protein engages IL-12p40 to form a cytokine, IL-23, with biological activities similar as well as distinct from IL-12. *Immunity* **13**, 715–725.

Osborne, B. A., and Minter, L. M. (2007). Notch signalling during peripheral T-cell activation and differentiation. *Nat. Rev. Immunol.* **7**, 64–75.

Ouyang, W., Jacobson, N. G., Bhattacharya, D., Gorham, J. D., Fenoglio, D., Sha, W. C., Murphy, T. L., and Murphy, K. M. (1999). The Ets transcription factor ERM is Th1-specific and induced by IL-12 through a Stat4-dependent pathway. *Proc. Natl. Acad. Sci. USA* **96**, 3888–3893.

Oxenius, A., Karrer, U., Zinkernagel, R. M., and Hengartner, H. (1999). IL-12 is not required for induction of type 1 cytokine responses in viral infections. *J. Immunol.* **162**, 965–973.

Palaga, T., Miele, L., Golde, T. E., and Osborne, B. A. (2003). TCR-mediated Notch signaling regulates proliferation and IFN-gamma production in peripheral T cells. *J. Immunol.* **171**, 3019–3024.

Park, A. Y., Hondowicz, B. D., and Scott, P. (2000). IL-12 is required to maintain a Th1 response during *Leishmania major* infection. *J. Immunol.* **165**, 896–902.

Park, C., Lecomte, M. J., and Schindler, C. (1999). Murine Stat2 is uncharacteristically divergent. *Nucleic Acids Res.* **27**, 4191–4199.

Parronchi, P., De Carli, M., Manetti, R., Simonelli, C., Sampognaro, S., Piccinni, M. P., Macchia, D., Maggi, E., Del Prete, G., and Romagnani, S. (1992). IL-4 and IFN (alpha and gamma) exert opposite regulatory effects on the development of cytolytic potential by Th1 or Th2 human T cell clones. *J. Immunol.* **149**, 2977–2983.

Paulson, M., Pisharody, S., Pan, L., Guadagno, S., Mui, A. L., and Levy, D. E. (1999). Stat protein transactivation domains recruit p300/CBP through widely divergent sequences. *J. Biol. Chem.* **274**, 25343–25349.

Pearce, E. L., Mullen, A. C., Martins, G. A., Krawczyk, C. M., Hutchins, A. S., Zediak, V. P., Banica, M., DiCioccio, C. B., Gross, D. A., Mao, C. A., Shen, H., Cereb, H., *et al.* (2003). Control of effector CD8+ T cell function by the transcription factor Eomesodermin. *Science* **302**, 1041–1043.

Penix, L. A., Sweetser, M. T., Weaver, W. M., Hoeffler, J. P., Kerppola, T. K., and Wilson, C. B. (1996). The proximal regulatory element of the interferon-gamma promoter mediates selective expression in T cells. *J. Biol. Chem.* **271**, 31964–31972.

Penix, L., Weaver, W. M., Pang, Y., Young, H. A., and Wilson, C. B. (1993). Two essential regulatory elements in the human interferon gamma promoter confer activation specific expression in T cells. *J. Exp. Med.* **178**, 1483–1496.

Persky, M. E., Murphy, K. M., and Farrar, J. D. (2005). IL-12, but not IFN-alpha, promotes STAT4 activation and Th1 development in murine CD4+ T cells expressing a chimeric murine/human Stat2 gene. *J. Immunol.* **174**, 294–301.

Pflanz, S., Timans, J. C., Cheung, J., Rosales, R., Kanzler, H., Gilbert, J., Hibbert, L., Churakova, T., Travis, M., Vaisberg, E., Blumenschein, W. M., Mattson, W. M., *et al.* (2002). IL-27, a heterodimeric cytokine composed of EBI3 and p28 protein, induces proliferation of naive CD4(+) T cells. *Immunity* **16**, 779–790.

Poli, V. (1998). The role of C/EBP isoforms in the control of inflammatory and native immunity functions. *J. Biol. Chem.* **273**, 29279–29282.

Porter, C. M., and Clipstone, N. A. (2002). Sustained NFAT signaling promotes a Th1-like pattern of gene expression in primary murine CD4+ T cells. *J. Immunol.* **168**, 4936–4945.

Pulendran, B., and Ahmed, R. (2006). Translating innate immunity into immunological memory: Implications for vaccine development. *Cell* **124**, 849–863.

Puzanov, I. J., Bennett, M., and Kumar, V. (1996). IL-15 can substitute for the marrow microenvironment in the differentiation of natural killer cells. *J. Immunol.* **157**, 4282–4285.

Ranger, A. M., Oukka, M., Rengarajan, J., and Glimcher, L. H. (1998). Inhibitory function of two NFAT family members in lymphoid homeostasis and Th2 development. *Immunity* **9**, 627–635.

Rao, A., Luo, C., and Hogan, P. G. (1997). Transcription factors of the NFAT family: Regulation and function. *Annu. Rev. Immunol.* **15**, 707–747.

Raue, H. P., Brien, J. D., Hammarlund, E., and Slifka, M. K. (2004). Activation of virus-specific CD8+ T cells by lipopolysaccharide-induced IL-12 and IL-18. *J. Immunol.* **173**, 6873–6881.

Reiner, S. L. (2001). Helper T cell differentiation, inside and out. *Curr. Opin. Immunol.* **13**, 351–355.

Reiner, S. L., and Seder, R. A. (1999). Dealing from the evolutionary pawnshop: How lymphocytes make decisions. *Immunity* **11**, 1–10.

Rincon, M., Enslen, H., Raingeaud, J., Recht, M., Zapton, T., Su, M. S., Penix, L. A., Davis, R. J., and Flavell, R. A. (1998). Interferon-gamma expression by Th1 effector T cells mediated by the p38 MAP kinase signaling pathway. *EMBO J.* **17**, 2817–2829.

Rogge, L., Barberis-Maino, L., Biffi, M., Passini, N., Presky, D. H., Gubler, U., and Sinigaglia, F. (1997). Selective expression of an interleukin-12 receptor component by human T helper 1 cells. *J. Exp. Med.* **185**, 825–831.

Rogge, L., D'Ambrosio, D., Biffi, M., Penna, G., Minetti, L. J., Presky, D. H., Adorini, L., and Sinigaglia, F. (1998). The role of Stat4 in species-specific regulation of Th cell development by type I IFNs. *J. Immunol.* **161**, 6567–6574.

Roman, C., Platero, J. S., Shuman, J., and Calame, K. (1990). Ig/EBP-1: A ubiquitously expressed immunoglobulin enhancer binding protein that is similar to C/EBP and heterodimerizes with C/EBP. *Genes Dev.* **4**, 1404–1415.

Rosenzweig, S. D., and Holland, S. M. (2005). Defects in the interferon-gamma and interleukin-12 pathways. *Immunol. Rev.* **203**, 38–47.

Sakaguchi, S. (2005). Naturally arising Foxp3-expressing CD25+ CD4+ regulatory T cells in immunological tolerance to self and non-self. *Nat. Immunol.* **6**, 345–352.

Sareneva, T., Matikainen, S., Kurimoto, M., and Julkunen, I. (1998). Influenza A virus-induced IFN-alpha/beta and IL-18 synergistically enhance IFN-gamma gene expression in human T cells. *J. Immunol.* **160**, 6032–6038.

Sareneva, T., Julkunen, I., and Matikainen, S. (2000). IFN-alpha and IL-12 induce IL-18 receptor gene expression in human NK and T cells. *J. Immunol.* **165**, 1933–1938.

Savinov, A. Y., Wong, F. S., and Chervonsky, A. V. (2001). IFN-gamma affects homing of diabetogenic T cells. *J. Immunol.* **167**, 6637–6643.

Schaeffer, E. M., Debnath, J., Yap, G., McVicar, D., Liao, X. C., Littman, D. R., Sher, A., Varmus, H. E., Lenardo, M. J., and Schwartzberg, P. L. (1999). Requirement for Tec kinases Rlk and Itk in T cell receptor signaling and immunity. *Science* **284**, 638–641.

Schaeffer, E. M., Yap, G. S., Lewis, C. M., Czar, M. J., McVicar, D. W., Cheever, A. W., Sher, A., and Schwartzberg, P. L. (2001). Mutation of Tec family kinases alters T helper cell differentiation. *Nat. Immunol.* **2**, 1183–1188.

Schijns, V. E., Haagmans, B. L., Wierda, C. M., Kruithof, B., Heijnen, I. A., Alber, G., and Horzinek, M. C. (1998). Mice lacking IL-12 develop polarized Th1 cells during viral infection. *J. Immunol.* **160**, 3958–3964.

Schluns, K. S., Williams, K., Ma, A., Zheng, X. X., and Lefrancois, L. (2002). Cutting edge: Requirement for IL-15 in the generation of primary and memory antigen-specific CD8 T cells. *J. Immunol.* **168**, 4827–4831.

Schmieg, J., Yang, G., Franck, R. W., and Tsuji, M. (2003). Superior protection against malaria and melanoma metastases by a C-glycoside analogue of the natural killer T cell ligand alpha-Galactosylceramide. *J. Exp. Med.* **198,** 1631–1641.

Schoenborn, J. R., Dorschner, M. O., Sekimata, M., Santer, D. M., Shnyreva, M., Fitzpatrick, D. R., Stamatoyannopoulos, J. A., and Wilson, C. B. (2007). Comprehensive epigenetic profiling identifies multiple distal regulatory elements directing transcription of the gene encoding interferon-gamma. *Nat. Immunol.* **8,** 732–742.

Schwartzberg, P. L., Finkelstein, L. D., and Readinger, J. A. (2005). TEC-family kinases: Regulators of T-helper-cell differentiation. *Nat. Rev. Immunol.* **5,** 284–295.

Shnyreva, M., Weaver, W. M., Blanchette, M., Taylor, S. L., Tompa, M., Fitzpatrick, D. R., and Wilson, C. B. (2004). Evolutionarily conserved sequence elements that positively regulate IFN-gamma expression in T cells. *Proc. Natl. Acad. Sci. USA* **101,** 12622–12627.

Shull, M. M., Ormsby, I., Kier, A. B., Pawlowski, S., Diebold, R. J., Yin, M., Allen, R., Sidman, C., Proetzel, G., Calvin, D., Annunziata, N., and Doetschman, T. (1992). Targeted disruption of the mouse transforming growth factor-beta 1 gene results in multifocal inflammatory disease. *Nature* **359,** 693–699.

Sica, A., Tan, T. H., Rice, N., Kretzschmar, M., Ghosh, P., and Young, H. A. (1992). The c-rel protooncogene product c-Rel but not NF-kappa B binds to the intronic region of the human interferon-gamma gene at a site related to an interferon-stimulable response element. *Proc. Natl. Acad. Sci. USA* **89,** 1740–1744.

Sica, A., Dorman, L., Viggiano, V., Cippitelli, M., Ghosh, P., Rice, N., and Young, H. A. (1997). Interaction of NF-kappaB and NFAT with the interferon-gamma promoter. *J. Biol. Chem.* **272,** 30412–30420.

Singh, H., and Pongubala, J. M. (2006). Gene regulatory networks and the determination of lymphoid cell fates. *Curr. Opin. Immunol.* **18,** 116–120.

Sivakumar, V., Hammond, K. J., Howells, N., Pfeffer, K., and Weih, F. (2003). Differential requirement for Rel/nuclear factor kappa B family members in natural killer T cell development. *J. Exp. Med.* **197,** 1613–1621.

Skurkovich, B., and Skurkovich, S. (2003). Anti-interferon-gamma antibodies in the treatment of autoimmune diseases. *Curr. Opin. Mol. Ther.* **5,** 52–57.

Smeltz, R. B. (2007). Profound enhancement of the IL-12/IL-18 pathway of IFN-gamma secretion in human CD8+ memory T cell subsets via IL-15. *J. Immunol.* **178,** 4786–4792.

Soutto, M., Zhou, W., and Aune, T. M. (2002). Cutting edge: Distal regulatory elements are required to achieve selective expression of IFN-gamma in Th1/Tc1 effector cells. *J. Immunol.* **169,** 6664–6667.

Spilianakis, C. G., Lalioti, M. D., Town, T., Lee, G. R., and Flavell, R. A. (2005). Interchromosomal associations between alternatively expressed loci. *Nature* **435,** 637–645.

Sprent, J., and Surh, C. D. (2002). T cell memory. *Annu. Rev. Immunol.* **20,** 551–579.

Stetson, D. B., Mohrs, M., Reinhardt, R. L., Baron, J. L., Wang, Z.-E., Gapin, L., Kronenberg, M., and Locksley, R. M. (2003). Constitutive cytokine mRNAs mark natural killer (NK) and NK T cells poised for rapid effector function. *J. Exp. Med.* **198,** 1069–1076.

Strengell, M., Matikainen, S., Siren, J., Lehtonen, A., Foster, D., Julkunen, I., and Sareneva, T. (2003). IL-21 in synergy with IL-15 or IL-18 enhances IFN-gamma production in human NK and T cells. *J. Immunol.* **170,** 5464–5469.

Sullivan, B. M., Juedes, A., Szabo, S. J., von Herrath, M., and Glimcher, L. H. (2003). Antigen-driven effector CD8 T cell function regulated by T-bet. *Proc. Natl. Acad. Sci. USA* **100,** 15818–15823.

Sun, J. C., Williams, M. A., and Bevan, M. J. (2004a). CD4+ T cells are required for the maintenance, not programming, of memory CD8+ T cells after acute infection. *Nat. Immunol.* **5,** 927–933.

Sun, R., Tian, Z., Kulkarni, S., and Gao, B. (2004b). IL-6 prevents T cell-mediated hepatitis via inhibition of NKT cells in CD4+ T cell- and STAT3-dependent manners. *J. Immunol.* **172,** 5648–5655.

Sweetser, M. T., Hoey, T., Sun, Y. L., Weaver, W. M., Price, G. A., and Wilson, C. B. (1998). The roles of nuclear factor of activated T cells and ying-yang 1 in activation-induced expression of the interferon-gamma promoter in T cells. *J. Biol. Chem.* **273,** 34775–34783.

Szabo, S. J., Dighe, A. S., Gubler, U., and Murphy, K. M. (1997). Regulation of the interleukin (IL)-12R beta 2 subunit expression in developing T helper 1 (Th1) and Th2 cells. *J. Exp. Med.* **185,** 817–824.

Szabo, S. J., Kim, S. T., Costa, G. L., Zhang, X., Fathman, C. G., and Glimcher, L. H. (2000). A novel transcription factor, T-bet, directs Th1 lineage commitment. *Cell* **100,** 655–669.

Szabo, S. J., Sullivan, B. M., Stemmann, C., Satoskar, A. R., Sleckman, B. P., and Glimcher, L. H. (2002). Distinct effects of T-bet in TH1 lineage commitment and IFN-gamma production in CD4 and CD8 T cells. *Science* **295,** 338–342.

Szabo, S. J., Sullivan, B. M., Peng, S. L., and Glimcher, L. H. (2003). Molecular mechanisms regulating Th1 immune responses. *Annu. Rev. Immunol.* **21,** 713–758.

Takatori, H., Nakajima, H., Kagami, S., Hirose, K., Suto, A., Suzuki, K., Kubo, M., Yoshimura, A., Saito, Y., and Iwamoto, I. (2005). Stat5a inhibits IL-12-induced Th1 cell differentiation through the induction of suppressor of cytokine signaling 3 expression. *J. Immunol.* **174,** 4105–4112.

Takeba, Y., Nagafuchi, H., Takeno, M., Kashiwakura, J., and Suzuki, N. (2002). Txk, a member of nonreceptor tyrosine kinase of Tec family, acts as a Th1 cell-specific transcription factor and regulates IFN-gamma gene transcription. *J. Immunol.* **168,** 2365–2370.

Takemoto, N., Intlekofer, A. M., Northrup, J. T., Wherry, E. J., and Reiner, S. L. (2006). Cutting edge: IL-12 inversely regulates T-bet and eomesodermin expression during pathogen-induced CD8+ T cell differentiation. *J. Immunol.* **177,** 7515–7519.

Takeno, M., Yoshikawa, H., Kurokawa, M., Takeba, Y., Kashiwakura, J. I., Sakaguchi, M., Yasueda, H., and Suzuki, N. (2004). Th1-dominant shift of T cell cytokine production, and subsequent reduction of serum immunoglobulin E response by administration in vivo of plasmid expressing Txk/Rlk, a member of Tec family tyrosine kinases, in a mouse model. *Clin. Exp. Allergy* **34,** 965–970.

Tassi, I., and Colonna, M. (2005). The cytotoxicity receptor CRACC (CS-1) recruits EAT-2 and activates the PI3K and phospholipase Cgamma signaling pathways in human NK cells. *J. Immunol.* **175,** 7996–8002.

Tassi, I., Presti, R., Kim, S., Yokoyama, W. M., Gilfillan, S., and Colonna, M. (2005). Phospholipase C-gamma 2 is a critical signaling mediator for murine NK cell activating receptors. *J. Immunol.* **175,** 749–754.

Tato, C. M., Villarino, A., Caamano, J. H., Boothby, M., and Hunter, C. A. (2003). Inhibition of NF-kappa B activity in T and NK cells results in defective effector cell expansion and production of IFN-gamma required for resistance to *Toxoplasma gondii*. *J. Immunol.* **170,** 3139–3146.

Tato, C. M., Martins, G. A., High, F. A., DiCioccio, C. B., Reiner, S. L., and Hunter, C. A. (2004). Cutting edge: Innate production of IFN-gamma by NK cells is independent of epigenetic modification of the IFN-gamma promoter. *J. Immunol.* **173,** 1514–1517.

Tato, C. M., Mason, N., Artis, D., Shapira, S., Caamano, J. C., Bream, J. H., Liou, H. C., and Hunter, C. A. (2006). Opposing roles of NF-kappaB family members in the regulation of NK cell proliferation and production of IFN-gamma. *Int. Immunol.* **18,** 505–513.

Thierfelder, W. E., van Deursen, J. M., Yamamoto, K., Tripp, R. A., Sarawar, S. R., Carson, R. T., Sangster, M. Y., Vignali, D. A., Doherty, P. C., Grosveld, G. C., and Ihle, J. N. (1996). Requirement for Stat4 in interleukin-12-mediated responses of natural killer and T cells. *Nature* **382,** 171–174.

Tong, Y., Aune, T., and Boothby, M. (2005). T-bet antagonizes mSin3a recruitment and transactivates a fully methylated IFN-gamma promoter via a conserved T-box half-site. *Proc. Natl. Acad. Sci. USA* **102**, 2034–2039.

Townsend, M. J., Weinmann, A. S., Matsuda, J. L., Salomon, R., Farnham, P. J., Biron, C. A., Gapin, L., and Glimcher, L. H. (2004). T-bet regulates the terminal maturation and homeostasis of NK and Valpha14i NKT cells. *Immunity* **20**, 477–494.

Trinchieri, G., Pflanz, S., and Kastelein, R. A. (2003). The IL-12 family of heterodimeric cytokines: New players in the regulation of T cell responses. *Immunity* **19**, 641–644.

Tyler, D. R., Persky, M. E., Matthews, L. A., Chan, S., and Farrar, J. D. (2007). Pre-assembly of STAT4 with the human IFN-alpha/beta receptor-2 subunit is mediated by the STAT4 N-domain. *Mol. Immunol.* **44**, 1864–1872.

Usui, T., Preiss, J. C., Kanno, Y., Yao, Z. J., Bream, J. H., O'Shea, J. J., and Strober, W. (2006). T-bet regulates Th1 responses through essential effects on GATA-3 function rather than on IFNG gene acetylation and transcription. *J. Exp. Med.* **203**, 755–766.

Vakoc, C. R., Mandat, S. A., Olenchock, B. A., and Blobel, G. A. (2005). Histone H3 lysine 9 methylation and HP1gamma are associated with transcription elongation through mammalian chromatin. *Mol. Cell* **19**, 381–391.

Valenzuela, L., and Kamakaka, R. T. (2006). Chromatin insulators. *Annu. Rev. Genet.* **40**, 107–138.

Velichko, S., Wagner, T. C., Turkson, J., Jove, R., and Croze, E. (2002). STAT3 activation by type I interferons is dependent on specific tyrosines located in the cytoplasmic domain of interferon receptor chain 2c. Activation of multiple STATS proceeds through the redundant usage of two tyrosine residues. *J. Biol. Chem.* **277**, 35635–35641.

Vigneau, S., Levillayer, F., Crespeau, H., Cattolico, L., Caudron, B., Bihl, F., Robert, C., Brahic, M., Weissenbach, J., and Bureau, J. F. (2001). Homology between a 173-kb region from mouse chromosome 10, telomeric to the Ifng locus, and human chromosome 12q15. *Genomics* **78**, 206–213.

Vigneau, S., Rohrlich, P. S., Brahic, M., and Bureau, J. F. (2003). Tmevpg1, a candidate gene for the control of Theiler's virus persistence, could be implicated in the regulation of gamma interferon. *J. Virol.* **77**, 5632–5638.

Vire, E., Brenner, C., Deplus, R., Blanchon, L., Fraga, M., Didelot, C., Morey, L., Van Eynde, A., Bernard, D., Vanderwinden, J. M., Bollen, M., Esteller, M., et al. (2006). The Polycomb group protein EZH2 directly controls DNA methylation. *Nature* **439**, 871–874.

Vivier, E., Nunes, J. A., and Vely, F. (2004). Natural killer cell signaling pathways. *Science* **306**, 1517–1519.

Vosshenrich, C. A., Ranson, T., Samson, S. I., Corcuff, E., Colucci, F., Rosmaraki, E. E., and Di Santo, J. P. (2005). Roles for common cytokine receptor gamma-chain-dependent cytokines in the generation, differentiation, and maturation of NK cell precursors and peripheral NK cells in vivo. *J. Immunol.* **174**, 1213–1221.

Walunas, T. L., Wang, B., Wang, C. R., and Leiden, J. M. (2000). Cutting edge: The Ets1 transcription factor is required for the development of NK T cells in mice. *J. Immunol.* **164**, 2857–2860.

Wang, K. S., Ritz, J., and Frank, D. A. (1999). IL-2 induces STAT4 activation in primary NK cells and NK cell lines, but not in T cells. *J. Immunol.* **162**, 299–304.

Wang, K. S., Frank, D. A., and Ritz, J. (2000). Interleukin-2 enhances the response of natural killer cells to interleukin-12 through up-regulation of the interleukin-12 receptor and STAT4. *Blood* **95**, 3183–3190.

Wang, Z. Y., Kusam, S., Munugalavadla, V., Kapur, R., Brutkiewicz, R. R., and Dent, A. L. (2006). Regulation of Th2 cytokine expression in NKT cells: Unconventional use of Stat6, GATA-3, and NFAT2. *J. Immunol.* **176**, 880–888.

Weaver, C. T., Harrington, L. E., Mangan, P. R., Gavrieli, M., and Murphy, K. M. (2006). Th17: An effector CD4 T cell lineage with regulatory T cell ties. *Immunity* **24**, 677–688.

Wenner, C. A., Guler, M. L., Macatonia, S. E., O'Garra, A., and Murphy, K. M. (1996). Roles of IFN-gamma and IFN-alpha in IL-12-induced T helper cell-1 development. *J. Immunol.* **156,** 1442–1447.

West, A. G., and Fraser, P. (2005). Remote control of gene transcription. *Hum. Mol. Genet.* **14**(Spec. No. 1), R101–R111.

Williams, M. A., Holmes, B. J., Sun, J. C., and Bevan, M. J. (2006). Developing and maintaining protective CD8+ memory T cells. *Immunol. Rev.* **211,** 146–153.

Williams, N. S., Klem, J., Puzanov, I. J., Sivakumar, P. V., Schatzle, J. D., Bennett, M., and Kumar, V. (1998). Natural killer cell differentiation: Insights from knockout and transgenic mouse models and *in vitro* systems. *Immunol. Rev.* **165,** 47–61.

Wilson, C. B., and Merkenschlager, M. (2006). Chromatin structure and gene regulation in T cell development and function. *Curr. Opin. Immunol.* **18,** 143–151.

Winders, B. R., Schwartz, R. H., and Bruniquel, D. (2004). A distinct region of the murine IFN-gamma promoter is hypomethylated from early T cell development through mature naive and Th1 cell differentiation, but is hypermethylated in Th2 cells. *J. Immunol.* **173,** 7377–7384.

Wolk, K., Kunz, S., Witte, E., Friedrich, M., Asadullah, K., and Sabat, R. (2004). IL-22 increases the innate immunity of tissues. *Immunity* **21,** 241–254.

Xu, X., Sun, Y. L., and Hoey, T. (1996). Cooperative DNA binding and sequence-selective recognition conferred by the STAT amino-terminal domain. *Science* **273,** 794–797.

Yang, X. O., Panopoulos, A. D., Nurieva, R., Chang, S. H., Wang, D., Watowich, S. S., and Dong, C. (2007). STAT3 regulates cytokine-mediated generation of inflammatory helper T cells. *J. Biol. Chem.* **282,** 9358–9363.

Yap, G., Pesin, M., and Sher, A. (2000). Cutting edge: IL-12 is required for the maintenance of IFN-gamma production in T cells mediating chronic resistance to the intracellular pathogen, *Toxoplasma gondii*. *J. Immunol.* **165,** 628–631.

Ye, J., Cippitelli, M., Dorman, L., Ortaldo, J. R., and Young, H. A. (1996). The nuclear factor YY1 suppresses the human gamma interferon promoter through two mechanisms: Inhibition of AP1 binding and activation of a silencer element. *Mol. Cell. Biol.* **16,** 4744–4753.

Yordy, J. S., and Muise-Helmericks, R. C. (2000). Signal transduction and the Ets family of transcription factors. *Oncogene* **19,** 6503–6513.

Yoshida, H., Hamano, S., Senaldi, G., Covey, T., Faggioni, R., Mu, S., Xia, M., Wakeham, A. C., Nishina, H., Potter, J., Saris, C. J., and Mak, T. W. (2001). WSX-1 is required for the initiation of Th1 responses and resistance to *L. major* infection. *Immunity* **15,** 569–578.

Young, H. A., Komschlies, K. L., Ciccarone, V., Beckwith, M., Rosenberg, M., Jenkins, N. A., Copeland, N. G., and Durum, S. K. (1989). Expression of human IFN-gamma genomic DNA in transgenic mice. *J. Immunol.* **143,** 2389–2394.

Yu, J., Wei, M., Becknell, B., Trotta, R., Liu, S., Boyd, Z., Jaung, M. S., Blaser, B. W., Sun, J., Benson, D. M., Jr., Mao, H., Yokohama, H., *et al.* (2006). Pro- and antiinflammatory cytokine signaling: Reciprocal antagonism regulates interferon-gamma production by human natural killer cells. *Immunity* **24,** 575–590.

Yu, T. K., Caudell, E. G., Smid, C., and Grimm, E. A. (2000). IL-2 activation of NK cells: Involvement of MKK1/2/ERK but not p38 kinase pathway. *J. Immunol.* **164,** 6244–6251.

Zhang, F., and Boothby, M. (2006). T helper type 1-specific Brg1 recruitment and remodeling of nucleosomes positioned at the IFN-gamma promoter are Stat4 dependent. *J. Exp. Med.* **203,** 1493–1505.

Zhang, Y., Apilado, R., Coleman, J., Ben-Sasson, S., Tsang, S., Hu-Li, J., Paul, W. E., and Huang, H. (2001). Interferon gamma stabilizes the T helper cell type 1 phenotype. *J. Exp. Med.* **194,** 165–172.

Zheng, W. P., Zhao, Q., Zhao, X., Li, B., Hubank, M., Schatz, D. G., and Flavell, R. A. (2004). Up-regulation of Hlx in immature Th cells induces IFN-gamma expression. *J. Immunol.* **172**, 114–122.

Zheng, Y., Danilenko, D. M., Valdez, P., Kasman, I., Eastham-Anderson, J., Wu, J., and Ouyang, W. (2007). Interleukin-22, a T(H)17 cytokine, mediates IL-23-induced dermal inflammation and acanthosis. *Nature* **445**, 648–651.

Zhou, D., Mattner, J., Cantu, C., 3rd, Schrantz, N., Yin, N., Gao, Y., Sagiv, Y., Hudspeth, K., Wu, Y. P., Yamashita, T., Teneberg, S., Wang, S., *et al.* (2004a). Lysosomal glycosphingolipid recognition by NKT cells. *Science* **306**, 1786–1789.

Zhou, W., Chang, S., and Aune, T. M. (2004b). Long-range histone acetylation of the Ifng gene is an essential feature of T cell differentiation. *Proc. Natl. Acad. Sci. USA* **101**, 2440–2445.

Zhu, H., Yang, J., Murphy, T. L., Ouyang, W., Wagner, F., Saparov, A., Weaver, C. T., and Murphy, K. M. (2001). Unexpected characteristics of the IFN-gamma reporters in non-transformed T cells. *J. Immunol.* **167**, 855–865.

Zimmermann, C., Prevost-Blondel, A., Blaser, C., and Pircher, H. (1999). Kinetics of the response of naive and memory CD8 T cells to antigen: Similarities and differences. *Eur. J. Immunol.* **29**, 284–290.

Zompi, S., and Colucci, F. (2005). Anatomy of a murder–signal transduction pathways leading to activation of natural killer cells. *Immunol. Lett.* **97**, 31–39.

CHAPTER 3

The Expansion and Maintenance of Antigen-Selected CD8$^+$ T Cell Clones

Douglas T. Fearon*

Contents		
	1. Background	105
	1.1. Introduction	105
	1.2. Subsets of antigen-selected CD8$^+$ T cells: Central memory, effector memory, and effector cells	106
	1.3. Central memory CD8$^+$ T cells generate new effector CD8$^+$ T cells	107
	1.4. Current models for development of antigen-stimulated CD8$^+$ T cells	108
	2. The Behavior of the CD8$^+$ T Cell in Persistent Viral Infections	111
	2.1. Persistent CD8$^+$ T cell stimulation and expansion: "Inflationary" epitopes	111
	2.2. Cellular senescence despite continued clonal expansion	113
	2.3. Clonal persistence versus clonal succession	114
	2.4. Molecular requirements for clonal persistence	114
	3. Clarifying the Role of IL-2 in the Clonal Expansion and Effector Differentiation of the CD8$^+$ T Cell	115
	3.1. Is IL-2 "the" or "a" mediator of CD8$^+$ T cell clonal expansion?	115
	3.2. CD8$^+$ T cell clonal expansion without IL-2R signaling	116
	4. Coreceptors Mediating IL-2-Independent CD8$^+$ T Cell Clonal Expansion	118

* Wellcome Trust Immunology Unit, University of Cambridge, Medical Research Council Centre, Cambridge CB2 2QH, United Kingdom

Advances in Immunology, Volume 96
ISSN 0065-2776, DOI: 10.1016/S0065-2776(07)96003-4

© 2007 Elsevier Inc.
All rights reserved.

		4.1. CD27	118
		4.2. Other coreceptors	120
	5.	Modifying the Antiproliferative Effects of Types I and II IFN	120
		5.1. The effects of types I and II IFN on CD8$^+$ T cells	121
		5.2. Regulating IFN-γR expression	121
		5.3. Stat1 as a "Switch" determining the effects of types I and II IFN on proliferation	122
	6.	Transcriptional Control of Replicative Senescence: Bmi-1, Blimp-1, and BCL6/BCL6b	123
	7.	A Refined Model for CD8$^+$ T Cell Clonal Expansion: Sequential Phases of CD27-Dependent Self-Renewal and IL-2-Dependent Differentiation	126
	8.	Clinical Extensions of the TCR/CD27 Pathway: Adoptive CD8$^+$ T Cell Therapy	127
Acknowledgments			129
References			129

Abstract

The biological purpose of the mature, postthymic CD8$^+$ T cell is to respond to microbial antigens with a developmental program of clonal expansion and concomitant differentiation leading to effector cells (T$_{EFF}$) that provide antimicrobial defense. Because many microbial infections persist into a chronic phase, this antigen-stimulated developmental program must be capable of continually generating T$_{EFF}$, perhaps for the lifetime of the individual. This chapter proposes that the ability of a CD8$^+$ T cell clone to maintain the continual production of T$_{EFF}$ during periods of persistent antigenic stimulation is based on a program that has two sequential phases of clonal expansion: an initial stage that occurs mainly in the secondary lymphoid tissues and is mediated by ligation of the T cell receptor (TCR) and CD27, and a subsequent, IL-2-dependent phase that occurs predominantly in peripheral, nonlymphoid tissues. The TCR/CD27-dependent phase establishes a nondifferentiating, self-renewing pool of clonally expanding cells, and the IL-2-dependent phase mediates continued clonal expansion that is coupled to the development of T$_{EFF}$. The two pools are linked by the process of asymmetrical division within the self-renewing subset so that, at steady state of cellular replication in this TCR/CD27-dependent subset, one daughter cell remains undifferentiated and the other initiates its commitment to IL-2-dependent terminal differentiation. Superimposed on this basic scheme are a shift in the CD8$^+$ T cell response to type I and II interferon (IFN) from anti- to proproliferative and transcriptional control of replicative senescence by Bmi-1, Blimp-1, and BCL6/BCL6b. This developmental program ensures that despite the occurrence of cellular senescence antiviral CD8$^+$ T cell clones are maintained for the duration of persistent viral infections.

1. BACKGROUND

1.1. Introduction

Of the two biological challenges for the CD8$^+$ T cell, maintaining the continuous production of effector cells (T$_{EFF}$) during persistent infections and preserving clones that have been selected during primary infections for enhanced responses to subsequent infections, immunologists have tended to focus their efforts on explaining the basis for the latter capability, perhaps because it holds the promise of leading to improved vaccines. However, when considering whether it might be more detrimental to the individual if the CD8$^+$ T cell system could not continually produce T$_{EFF}$ during persistent infections or could not maintain antigen-selected clones after a primary infection, it is clear that the former circumstance would be disastrous, whereas the latter may or may not be, depending on whether a secondary infection occurs. Some have even argued that surviving the primary infection indicates that a primary CD8$^+$ T cell response for that microorganism is sufficient. Furthermore, as immunological control of persistent viral infections is often mainly mediated by CD8$^+$ T cells while control of secondary viral infections that lack a latent or persistent phase is often based on the humoral response, the ability of the CD8$^+$ T cell system to produce continually T$_{EFF}$ may be more biologically important than is its ability to persist in an apparently quiescent, yet alert state between acute infections. Understanding the signals that mediate continual replication of CD8$^+$ T cells *in vivo* without clonal senescence, if this is the mechanism for long-term CD8$^+$ T cell control of persistent viral infections, is not only inherently interesting but also, by defining the rules for maintaining the replicative function of the antigen-experienced CD8$^+$ T cell, it may lead to strategies for improved adoptive CD8$^+$ T cell immunotherapy. Therefore, the purpose of this chapter is to present an interpretation of the literature in relation to the developmental program that enables the CD8$^+$ T cell system to cope with persistent microbial infections.

With this being the intention of the chapter, it is necessary to discuss briefly the nomenclature used by investigators in the field of CD8$^+$ T cell biology. Despite the less compelling role for the CD8$^+$ T cell system in classical immunological memory, which is the ability to respond to a second infection more effectively than to the first, the nomenclature of the antigen-dependent phase of CD8$^+$ T cell development is dominated by memory terminology. All CD8$^+$ T cells generated during antigen-dependent development, between the naive, antigen-inexperienced cell, and the terminally differentiated cell, are referred to as memory cells, even in the context of persistent infections where such terminology is probably inappropriate. This usage is too common to change now, but by defining the functional characteristics of each "memory" subset, the potential role of that subset in persistent, as well as repetitive, infections will become evident.

1.2. Subsets of antigen-selected CD8$^+$ T cells: Central memory, effector memory, and effector cells

In 1999, Lanzavecchia and his colleagues (Sallusto et al., 1999) reported that four subsets of human peripheral blood CD8$^+$ T cells could be distinguished when assessed for the expression of CD45RA and the chemokine receptor, CCR7. These were

1. A CD45RA$^+$, CCR7$^+$ subset, which was considered to be composed of naive cells because they had the naive CD45 isoform and lacked potential effector functions, such as the capacity for rapid, T cell receptor (TCR)-induced interferon (IFN)-γ production and preformed perforin.
2. A CD45RA$^-$, CCR7$^+$ subset, which was considered to represent antigen-experienced cells because they lacked the naive CD45 isoform, but, interestingly, resembled naive cells in lacking potential effector functions. The expression of CCR7 would enable these cells to respond to CCL21 in secondary lymphoid organs. Hence, these cells were termed "central memory" T cells (T$_{CM}$).
3. A CD45RA$^-$, CCR7$^-$ subset that was considered also to include memory cells, but which differed from the T$_{CM}$ subset not only by lacking expression of CCR7 but also in having a capacity for rapid, TCR-induced IFN-γ and preformed cytoplasmic granules containing perforin. The absence of CCR7 and expression of CCR5 implied impaired homing to secondary lymphoid organs and, instead, more likely localization to peripheral tissue sites where effector function would be appropriate. Therefore, this population was termed the effector memory (T$_{EM}$) subset.
4. A fourth subset of CD45RA$^+$, CCR7$^-$ cells that also were capable of rapid production of IFN-γ and stained even more intensely for perforin. These were termed effector cells (T$_{EFF}$) that had reverted to the CD45RA$^+$ isoform of naive cells, which is a characteristic of highly differentiated, antigen-experienced cells (Michie et al., 1992).

This report also indicated that *in vitro* stimulation of CD4$^+$ T$_{CM}$ caused these cells to acquire a T$_{EM}$ phenotype, but that stimulation of CD4$^+$ T$_{EM}$

did not induce a T_{CM} phenotype. *In vitro* stimulation of naive CD4$^+$ T cells generated cells with a T_{CM} phenotype. These findings suggested a pathway of antigen-dependent development that was, naïve → T_{CM} → T_{EM}, proceeding from least to most differentiated with respect to effector function. However, the relevance of *in vitro* to *in vivo* differentiation is unclear, and this analysis was performed with CD4$^+$ T cells rather than CD8$^+$ T cells.

Finding that homing potential correlated with effector function suggested that antigen-dependent development of CD8$^+$ T cells was regulated rather than stochastic, and by defining distinct subsets of antigen-experienced cells (Sallusto et al., 1999) implicitly raised important questions of what the functions of the subsets are, and how these subsets develop during cellular responses to acute and persistent infections. Left unresolved by this study was the question of whether these subsets of cells identified only by CD45 isoforms and CCR7 were themselves developmentally homogenous or even could be further subdivided, and whether analysis of resting cells in peripheral blood could be extended, without modification, to actively proliferating cells in lymphoid and extralymphoid tissue. Nevertheless, in many ways this study has set the agenda for research in the development of antigen-experienced CD8$^+$ T cells, which is remarkable in that it was a noninterventional study of human lymphocytes rather than an interventional study of the murine CD8$^+$ T cell system.

1.3. Central memory CD8$^+$ T cells generate new effector CD8$^+$ T cells

To establish the relevance of Sallusto et al. (1999) to the problem of how CD8$^+$ T cells generate new T_{EFF} in persistent or recurrent infections, subsets of murine memory CD8$^+$ T cell subsets corresponding to the human subsets needed to be identified and assessed for differences in their replicative function when challenged by viral infection *in vivo*. In 2003, Ahmed and his colleagues (Wherry et al., 2003) reported that in mice that had been infected with *Lymphocytic choriomeningitis virus* (LCMV) 2–3 months previously, two populations of memory CD8$^+$ T cells could be distinguished based on their relative expression of the homing receptor, CD62L. The CD62Lhigh cells were CCR7$^+$ and the CD62Llow cells were CCR7$^-$, so these two murine memory CD8$^+$ T cell populations appeared to correspond to the human T_{CM} and T_{EM} subsets, a correlation that seemed reasonable since both CCR7 and CD62L promote migration of cells from blood into secondary lymphoid organs. However, the correspondence was not exact because the murine T_{CM} and T_{EM} both were capable of rapid production of IFN-γ while only human T_{EM} but not T_{CM} had this function. Despite these differences, especially with respect to an important marker of effector differentiation, the CD62Lhigh memory $CD8^+$ T cell was considered to be equivalent to the human T_{CM} and to be developmentally distinct from the CD62Llow memory $CD8^+$ T cell, which were termed T_{EM}, even though both had the effector function of potential IFN-γ production.

Nevertheless, the distinction of LCMV-specific memory subsets by relative CD62L expression was useful because after adoptive transfer and challenge of recipient mice with relevant viruses, the two subsets were functionally distinct. T_{CM} provided more effective immunity to the viral challenge than T_{EM} and demonstrated greater clonal expansion and generation of new T_{EFF}, even though only a modest enhancement in replicative function relative to T_{EM} was observed *in vitro*. This observation that CD62Lhigh memory $CD8^+$ T_{CM} were more effective in secondary clonal expansion than CD62Llow T_{EM} has been confirmed in almost all studies using a similar protocol of assessing adoptively transferred memory cells obtained from mice that had resolved acute, primary viral infections (Bachmann *et al.*, 2005; Bouneaud *et al.*, 2005; Marzo *et al.*, 2005). Perhaps even more importantly for the intent of this chapter, this functional distinction even holds for $CD8^+$ T cells specific for the "inflationary" immediate-early 1 (IE1) epitope taken from mice with persistent infection with murine cytomegalovirus (mCMV) (Pahl-Seibert *et al.*, 2005) (see Section 2). Therefore, T_{CM} are responsible for generating new T_{EFF}, whether taken from mice during antigen-free "memory" periods or during persistent viral infections in which there has been continual antigenic stimulation.

1.4. Current models for development of antigen-stimulated $CD8^+$ T cells

With the discovery of heterogeneity in memory $CD8^+$ T cells that, in humans at least, was reflected by cells differing in their extent of effector differentiation and in the mouse by cells with differing *in vivo* proliferative potential, two general proposals have been advanced to account for the development of antigen-stimulated $CD8^+$ T cells. The first was suggested by analogy with the developmental pathways of other organ systems, and postulated that, "A stem cell-like capacity for self-renewal could be the basis for the continual generation of effector lymphocytes from the memory pool" (Fearon *et al.*, 2001). This stem cell stage would provide the pool of replicating precursor cells from which could emerge $CD8^+$ T cells that commit to effector differentiation. These differentiating cells would have limited replicative potential, but asymmetrical division in the stem cell pool would insure maintenance of undifferentiated cells with relatively unlimited replicative potential. Since the cytokine, IL-2, was known to drive effector differentiation, clonal expansion mediated by

IL-2 was considered to be responsible for generating the T_{EM} and T_{EFF} subsets of antigen-experienced $CD8^+$ T cells. Therefore, the existence of an IL-2-independent pathway for $CD8^+$ T cell clonal expansion was proposed for the development of the self-renewing, nondifferentiating pool of cells that would maintain replicative function (Fearon et al., 2006). Being undifferentiated, the self-renewing cells would be contained within the T_{CM} subset, so the proposal was consistent with the apparent precursor relationship of human $CD4^+$ T_{CM} to $T_{EM/EFF}$ (Sallusto et al., 1999).

This stem cell model has been supported by recent findings of an IL-2-independent pathway of clonal expansion for the $CD8^+$ T cell that generates cells having a T_{CM} phenotype and lacking effector functions (Carr et al., 2006), which will be discussed in more detail in Section 4, and by the demonstration of asymmetrical division of $CD8^+$ T cells in vivo (Chang et al., 2007). Asymmetrical division, which is central to the concept of a self-renewing cell, was supported in this study by the finding that at the first cellular division of $CD8^+$ T cells in response to microbial antigenic challenge, one daughter cell contained granzyme B, was capable of IFN-γ production, exhibited short-term host defense, but had diminished long-term protective capability, while the other daughter cell lacked immediate effector functions but had better long-term antimicrobial activity, possibly reflecting better replicative function, although this was not shown.

The second, and perhaps more generally accepted model is termed "linear differentiation," and was presented as a consequence of finding that memory $CD8^+$ T cells with a T_{CM} phenotype had better secondary proliferative function than did T_{EM} cells and that T_{CM} apparently could not be detected during acute, primary $CD8^+$ T cell clonal expansion (Wherry et al., 2003). Adoptive transfer to naive recipients of LCMV-specific $CD8^+$ T cells taken from mice at various times during and after acute LCMV infection followed by viral challenge showed that $CD8^+$ T cells with secondary proliferative function were not detectable at the peak of the acute primary response, when many T_{EFF} were present, but gradually appeared during the weeks following resolution of the primary infection. As this coincided with a change in phenotype of the LCMV-specific $CD8^+$ T cells from $T_{EFF/EM}$ (CD62Llow) to T_{CM} (CD62Lhigh), the authors concluded, "Thus, the findings of our study and the proposed model of linear differentiation (Naïve → Effector → T_{EM} → T_{CM}) are likely to provide the paradigm for acute infections. We propose that this will be the natural course of memory T cell differentiation in the absence of antigen. It is possible, however, that under certain conditions, especially chronic infections where antigen persists at high amounts, one may see a different pattern of memory T cell differentiation." The latter comment refers to the phenomenon of "exhaustion" of $CD8^+$ T cells in mice infected with clone 13 LCMV. High levels of virus persist beyond the acute phase of infection, and antigen-specific $CD8^+$ T cells demonstrate impaired TCR signaling and an inability to replicate, perhaps because of the expression of the inhibitory receptor PD-1 (Barber et al., 2006) or excessive IL-10 production (Brooks et al., 2006). However, these interesting observations may not be relevant to the response of the $CD8^+$ T cell system to other persistent viral infections in which $CD8^+$ T cells do maintain an ability to generate new T_{EFF}, as will be discussed in Section 2.

Perhaps the most controversial aspect of the linear differentiation pathway is its requirement that antigen-experienced $CD8^+$ T cells be capable of dedifferentiation in the absence of antigenic stimulation, with a gradual loss of effector function and regaining of a level of proliferative function that is at least equivalent to that of the naive $CD8^+$ T cell. Although the findings in this study (Wherry et al., 2003) seemed to indicate that dedifferentiation occurred, and the complexity of dealing with diverse and changing microbial targets has often selected for unique biological capabilities in the adaptive immune system, such a capability would be unusual in a general developmental biology context. Furthermore, a subsequent study reported that cells that had converted their CD62L phenotype during the memory phase did not have replicative function (Bouneaud et al., 2005), another found that antigen-specific CD62Lhigh, $CD8^+$ T cells with high proliferative function could be found during the peak of the primary response (Bachmann et al., 2005), and a third concluded that a capacity for conversion from T_{EM} to T_{CM} may reflect incomplete differentiation (Marzo et al., 2005). Moreover, in addition to differentiation-associated changes in transcription of the CD62L gene, low expression of CD62L may be induced by TCR-induced, metalloproteinase-mediated cleavage of the ectodomain of CD62L, which is reversible if further changes in the developmental status of the cell have not occurred (Chao et al., 1997; Jung et al., 1988). Finally, cells with a T_{CM} phenotype survive better than do those with T_{EM} and T_{EFF} phenotypes, perhaps because they maintain expression of CD127, the IL-7Rα chain (Kaech et al., 2003; Schluns et al., 2000), and the gradual increase in the T_{CM}/T_{EM} ratio during the memory phase could also have been caused by the homeostatic expansion of T_{CM} in response to IL-15 (Becker et al., 2002; Goldrath et al., 2002; Tan et al., 2002). Therefore, although dedifferentiation might occur with a "transitional" type of antigen-stimulated $CD8^+$ T cell, perhaps accounting for the secondary replicative function of memory $CD8^+$ T cells that had received IL-2R signals during the primary response (Williams et al., 2006), the essential prediction of the linear differentiation model is that the first step in development of the antigen-stimulated $CD8^+$ T cell is differentiation to T_{EFF}, and that cells with long-term replicative function are derived from these T_{EFF}.

This prediction of the linear differentiation pathway implies that whenever an acute infection is not cleared, the $CD8^+$ T cell system will necessarily fail. There would be an inability of antigen-specific CD8$^+$ T cells to maintain the production of new T$_{EFF}$ because there would be no antigen-free "rest period" to allow dedifferentiation of T$_{EFF}$ to T$_{CM}$ with acquisition of replicative function. Although this circumstance is compatible with the CD8$^+$ T cell response to an acute viral infection, such as LCMV, it cannot explain the success of the CD8$^+$ T cell system in controlling persistent viral infections, such as those caused by the herpes viruses. However, as these viruses cease replicating after the acute phase and enter into the latent phase of infection, one could suggest that latency is equivalent to an antigen-free period. Section 2 will discuss whether antigenic stimulation ever fully ceases during the latent phase of herpes virus infection, at least for all antigens, but even more persuasive evidence for the ability of the CD8$^+$ T cell system to maintain a continuous production of T$_{EFF}$ during long-term, active viral infections is its ability to control infection with human immunodeficiency virus (HIV) until the CD4$^+$ T cell response is lost. The occurrence of HIV escape mutants must indicate the continued generation of functional T$_{EFF}$ to account for the selection of the mutants. Therefore, better insight into the developmental program of antigen-experienced CD8$^+$ T cells may be gained by an analysis of persistent rather than acute viral infections.

2. THE BEHAVIOR OF THE CD8$^+$ T CELL IN PERSISTENT VIRAL INFECTIONS

This section will review the evidence that persistent viral infections cause continuous antigenic stimulation of the CD8$^+$ T cell, that cellular replicative senescence occurs with highly differentiated CD8$^+$ T cells, that despite the occurrence of senescence, an antigen-specific CD8$^+$ T cell response is maintained that reflects clonal maintenance rather than clonal succession, all of which strongly suggests the existence of a self-renewing stage of antigen-dependent development.

2.1. Persistent CD8$^+$ T cell stimulation and expansion: "Inflationary" epitopes

Reddehase and colleagues have proposed a "silencing/desilencing and immune sensing" hypothesis by which CD8$^+$ T cells control CMV latency by epitope-specific sensing of transcriptional reactivation of the virus. CD8$^+$ T cells specific for an IE1 epitope recognize and terminate virus reactivation *in vivo* at the first opportunity in the reactivated gene expression program (Reddehase *et al.*, 1989; Simon *et al.*, 2006). This CD8$^+$ T cell response is caused by viral reactivation rather than being a unique attribute of the IE1 peptide epitope because similar CD8$^+$ T cell responses occur with other epitopes when their expression is regulated by the IE1 promoter in recombinant viruses (Karrer *et al.*, 2004). The continual presentation of the IE1 epitope in a small proportion of latently infected cells causes an "inflation," or continuous expansion, over time of $CD8^+$ T cells specific for this epitope (Karrer *et al.*, 2003). Most $CD8^+$ T cells having an inflationary response had a highly differentiated phenotype of CD28low, CD27low, CD122low, and CD62Llow, whereas $CD8^+$ T cells specific for epitopes that induced expansion during the acute phase but not during the latent phase of mCMV infection showed "a slow reversion" to the T_{CM} phenotype (Sierro *et al.*, 2005), as had been observed with memory $CD8^+$ T cells specific for LCMV (Wherry *et al.*, 2003). From this and other similar observations (Holtappels *et al.*, 2000; Munks *et al.*, 2006), it was concluded that a particular memory phenotype is determined by the frequency of TCR stimulation, with continually presented epitopes causing a $T_{EM/EFF}$ phenotype, and epitopes that are not presented during latency being associated with a T_{CM} phenotype. The prediction of the linear differentiation model would be that clones composed of $CD8^+$ T cells with highly differentiated $T_{EM/EFF}$ phenotype would no longer generate new effector cells, but since the continued $CD8^+$ T cell response to the IE1 epitope is necessary to control reactivation of mCMV (Simon *et al.*, 2006), it is likely that there is continued generation of new T_{EFF}, although cellular turnover studies are needed to confirm this conclusion. Either $CD8^+$ T cells with a $T_{EFF/EM}$ phenotype can replicate in mCMV-infected mice, in contrast to the response of T_{EFF} from LCMV-infected mice (Wherry *et al.*, 2003), or there is a source in mCMV-infected mice of IE1-specific $CD8^+$ T cells having the T_{CM} phenotypic characteristic of replicative competence. The latter possibility was experimentally supported when a population of CD62Lhigh, IE1-specific $CD8^+$ T cells in the lymph nodes of infected mice was found to have remarkable proliferative function after adoptive transfer to naive mice and challenge with mCMV (Pahl-Seibert *et al.*, 2005).

Although mCMV is perhaps the best experimental system for addressing the question of how the $CD8^+$ T cell responds to continual stimulation, analyses of other murine herpes viruses are consistent with the conclusions drawn from the mCMV studies. For example, latent infection with herpes simplex virus (HSV) infection is also associated with continual $CD8^+$ T cell stimulation (Khanna *et al.*, 2003; van Lint *et al.*, 2005), and these $CD8^+$ T cells may be required to prevent reactivation in sensory ganglions (Liu *et al.*, 2000). Also, HSV-specific $CD8^+$ T cells from latently infected mice having a CD62Lhigh, T_{CM} phenotype proliferated as well as naive cells after adoptive transfer and viral challenge while cells with the same specificity but a T_{EFF} phenotype did not (Stock *et al.*, 2006). Therefore, the $CD8^+$ T cell system copes with persistent antigenic stimulation and maintains a capacity for generating new T_{EFF} apparently by maintaining a pool of less differentiated cells within the phenotypic T_{CM} subset.

2.2. Cellular senescence despite continued clonal expansion

The studies with mCMV and HSV suggest that the continued production of T_{EFF} is mediated by replication of less differentiated CD8$^+$ T cells, but additional evidence for replicative senescence in T_{EFF} is necessary to exclude the possibility that T_{EFF} numbers are sustained by replication of these cells. Indeed, a reasonable objection to a proposal for a self-renewing, less differentiated subset of antigen-experienced CD8$^+$ T cells has been based on the well-established ability of immunologists to maintain clones of murine CD8$^+$ T cells *in vitro* by periodic restimulation with antigen and IL-2. However, the occurrence of replicative senescence has been demonstrated in several circumstances.

Senescence was induced *in vivo* by repetitive cycles of adoptive transfer of LCMV-specific TCR transgenic CD8$^+$ T cells and infection with LCMV. After each cycle of infection, a higher proportion of the CD8$^+$ T cells expressed the killer cell lectin-like receptor G1 (KLRG1), and the expansion of the restimulated CD8$^+$ T cells correspondingly diminished (Voehringer et al., 2001). The KLRG1$^+$ CD8$^+$ T cells also demonstrated diminished proliferation *in vitro* in response to antigenic stimulation. It is not obvious why in these experiments a self-renewing subset was not maintained, but this may have been caused by the use of splenocytes for recovery of the antigen-experienced, LCMV-specific CD8$^+$ T cells rather than lymph node cells, which may select more effectively for CD62L high cells.

The association between expression of KLRG1 and persistent antigenic stimulation has also been demonstrated for human CD8$^+$ T cells specific for CMV, Epstein-Barr virus (EBV), and HIV (Thimme et al., 2005). The KLRG1$^+$ human CD8$^+$ T cells replicated poorly in response to stimulation with phytohemagglutinin and IL-2 (Voehringer et al., 2002). Similar findings of replicative senescence in association with the expression of CD57 on antigen-experienced human CD8$^+$ T cells have been reported (Brenchley et al., 2003) and are possibly an important consequence of depletion of CD4$^+$ T cells in HIV-infected patients (Papagno et al., 2004). CD8$^+$ T cells from "nonprogressor" patients maintain *in vitro* replicative function while CD8$^+$ T cells from "progressor" patients do not (Migueles et al., 2002). Of course, it is not possible to determine from these studies of CD8$^+$ T cells from HIV patients whether the replication-incompetent state contributed to loss of control of HIV replication, or whether uncontrolled replication caused senescence of the HIV-specific CD8$^+$ T cells. However, in persistent viral infections that are controlled, senescent, antigen-experienced CD8$^+$ T cells can be identified, so that their presence does not necessarily indicate overwhelming viral infection. Rather, it may be a normal developmental outcome of continued antigenic stimulation, as it appears to be in mice and humans with CMV infections.

2.3. Clonal persistence versus clonal succession

Two general processes could maintain the long-term generation of $CD8^+$ T_{EFF} in persistent viral infections: the maintenance of clones that are selected by antigen early in the antiviral response or the replacement of depleted clones by recruitment of naive $CD8^+$ T cells, a process that is termed clonal succession. A recent report has suggested that clonal succession contributes to the murine response to persistent infection with polyomavirus, although a requirement for clonal succession was not demonstrated (Vezys et al., 2006). The relevance of this finding to the response of human $CD8^+$ T cells to CMV (Khan et al., 2002; Weekes et al., 1999), EBV, and HIV (Cohen et al., 2002) is unclear as long-term persistence of virus-specific clones was demonstrated in each of these analyses. Given these studies tracking $CD8^+$ T cell clones by the use of CDR3-specific probes, and the continued control of persistent viral infections in aging adults experiencing normal thymic involution, it seems likely that the essential means for maintaining an antiviral response in persistent infections is clonal maintenance rather than clonal succession.

2.4. Molecular requirements for clonal persistence

Relative to the many studies of different genetically modified mice in classical memory protocols, there are relatively few reports of the signaling pathways that mediate the maintenance of repetitively stimulated $CD8^+$ T cells in persistent viral infections. In mice infected with γ-herpesvirus or HSV there is a requirement for $CD4^+$ T cells (Cardin et al., 1996), at least in part for their role in "licensing" of dendritic cells (Smith et al., 2004) by stimulation through CD40 (Sarawar et al., 2001), and for CD27 on either $CD4^+$ or $CD8^+$ T cells (Kemball et al., 2006). The need for $CD4^+$ T cell-dependent activation of dendritic cells through CD40L–CD40 interaction had been previously recognized in the generation of some primary (Bennett et al., 1998; Ridge et al., 1998; Schoenberger et al., 1998) and memory $CD8^+$ T cell responses (Janssen et al., 2003; Shedlock and Shen, 2003; Sun and Bevan, 2003), which may be related to inducing the expression on dendritic cells of CD70, the ligand for CD27. The possible central roles of CD70- and CD27-mediated responses in the IL-2-independent clonal expansion of the $CD8^+$ T cell are discussed in more detail in Section 4. The potential clinical relevance of a role for $CD4^+$ T cells in persistent viral infections is, of course, the loss of control by $CD8^+$ T cells of viral replication in CD4$^+$ T cell-deficient patients with acquired immune deficiency syndrome secondary to HIV infection.

There is an interesting possible contrasting requirement for IL-15 in the responses of memory and persistently stimulated CD8$^+$ T cells. While this cytokine is needed to maintain normal numbers of antigen-experienced CD8$^+$ T cells during the memory phase between infections (Becker et al., 2002; Goldrath et al., 2002; Tan et al., 2002), it may not be necessary to maintain the generation of sufficient T$_{EFF}$ for control of persistent γ-herpesvirus or HSV infections (Obar et al., 2004; Sheridan et al., 2006). This finding emphasizes the need to identify the antigen-experienced CD8$^+$ T cell pool that is IL-15 dependent. Although it might be argued that persistent antigenic stimulation obviates the need for sustaining antigen-experienced CD8$^+$ T cells during antigen-free period of a classical memory response, an alternative explanation may be that cells contained within the T$_{CM}$ pool, which mediate the continual generation of new T$_{EFF}$ (Stock et al., 2006), do not require IL-15. This possibility would be consistent with the finding of a quantitatively normal secondary response of memory CD8$^+$ T cells to LCMV in IL-15-deficient mice (Becker et al., 2002). This issue has implications for determining the transcription factors that are necessary to establish the self-renewing subset of antigen-experienced CD8$^+$ T cells. T-bet and eomesodermin (eomes) have been considered to be required for memory CD8$^+$ T cell maintenance based on their role in increasing expression of CD122, the IL-2βR (Intlekofer et al., 2005) that mediates signaling by IL-15. If the cells in the T$_{CM}$ subset that maintain the continuous production of new T$_{EFF}$ do not need IL-15 signaling, then eomes and T-bet may not necessarily be involved in the development of this important population in for either classical memory or persistent viral responses.

3. CLARIFYING THE ROLE OF IL-2 IN THE CLONAL EXPANSION AND EFFECTOR DIFFERENTIATION OF THE CD8$^+$ T CELL

3.1. Is IL-2 "the" or "a" mediator of CD8$^+$ T cell clonal expansion?

IL-2 has been considered to be the principle mediator of the clonal expansion of T cells since its initial identification as the "T cell growth factor" 20 years ago (Cantrell and Smith, 1984; Gillis and Smith, 1977; Morgan et al., 1976). Ligation of the TCR induces expression of CD25, the IL-2Rα chain, that enables high-affinity binding of IL-2 to the IL-2R and transcription of the IL-2 gene, perhaps with the assistance of signals from CD28. The apparent simplicity of this system and its ability to maintain the long-term growth *in vitro* of murine $CD8^+$ T cell clones supported its candidacy as the mediator of the clonal expansion of the antigen-stimulated T cell, a obligatory cellular response in the clonal selection principle of adaptive immunity. More recent reports showed that even if the $CD8^+$ T cell did not itself provide IL-2 for autocrine stimulation, relatively transient ligation of the TCR "programmed" the cell for a paracrine IL-2 response that extended for 7–10 cell cycles (Kaech and Ahmed, 2001; Wong and Pamer, 2001). Observations such as these seemed to confirm suggestions made for many years that $CD4^+$ T cell "help" for $CD8^+$ T cell responses was mediated by paracrine IL-2, although the role of $CD4^+$ T cells in $CD8^+$ T cell responses had been shown to be via the activation of dendritic cells (Bennett *et al.*, 1998; Ridge *et al.*, 1998; Schoenberger *et al.*, 1998). It is interesting to note that an emphasis on the role of IL-2 in clonal expansion leads logically to the linear differentiation pathway in which T_{EFF} are the generated directly from naive cells because IL-2-induced $CD8^+$ T cell replication is coupled to effector differentiation, with the assistance of additional cytokine signals (Mescher *et al.*, 2006).

Other findings suggest that prior IL-2R signaling may have negative or positive effects on subsequent antigen-dependent $CD8^+$ T cell proliferation, with reasons for the differing outcomes not being evident. For example, IL-2 was shown to program T cells for cell death in a process that has been termed, antigen- or activation-induced cell death (AICD), such that repetitive ligation of the TCR on T cells that had been stimulated by IL-2 caused an apoptotic response (Lenardo, 1991). In contrast, memory $CD8^+$ T cells that had received IL-2R signals during a primary response expanded better on rechallenge than those that had not, when assessed in the same mouse (Williams *et al.*, 2006). This finding has not been reconciled with the prior demonstration of normal secondary expansion of memory $CD8^+$ T cells when all memory cells lacked prior IL-2R signals (Yu *et al.*, 2003). Of course, even being able to evaluate memory $CD8^+$ T cells that have not received IL-2R signals indicates that IL-2 is not required for primary clonal expansion or for the development of T_{CM}. This outcome would not have been predicted by the linear differentiation model in which T_{CM} develop from T_{EFF} via T_{EM}, since in the absence of IL-2, the development of T_{EFF} does not occur (Yu *et al.*, 2003), and, of even greater significance, it is not consistent with IL-2 being the essential driver of clonal expansion.

3.2. $CD8^+$ T cell clonal expansion without IL-2R signaling

In F5 TCR-transgenic, IL-2-deficient mice, administration of antigenic peptide induced the expansion of the transgenic $CD8^+$ T cells but did not cause them to develop CTL activity (Kramer *et al.*, 1994). Thus, more than 10 years ago, immunologists were confronted with the possibility that in the absence of IL-2, clonal expansion without effector differentiation occurs in the antigenically stimulated CD8$^+$ T cell. The nonredundant role of IL-2 in effector differentiation, in contrast to its apparently nonessential role in proliferation in this study, was also demonstrated by the development of CTLs when IL-2 was coadministered with antigenic peptide. Perhaps this study was not considered to be definitive because it did not involve a "physiological" stimulus for the activation of CD8$^+$ T cells, a microbial infection, but in retrospect, it is of great interest.

Other investigators did evaluate the CD8$^+$ T cell response in IL-2-deficient mice with more ambiguous outcomes, which may have been related to the autoimmunity that is caused by the absence of IL-2-dependent regulatory T cells and to difficulties associated with assessing clonal expansion before pMHC tetramers were available, which required measuring antigen-specific CD8$^+$ T cells based on their effector function, an obvious problem in IL-2-deficient conditions where effector differentiation may not occur. However, the question was addressed more recently by two groups who overcame these problems either by introducing the deficiency of IL-2Rα chain into a TCR-transgenic, Rag-deficient background (D'Souza and Lefrancois, 2003), or by restricting the deficiency of the IL-2Rβ chain to postthymic cells, thereby permitting development of the IL-2-dependent regulatory T cells (Yu *et al.*, 2003). Viral stimulation of cells unable to respond to IL-2, or even to IL-15 in the IL-2R$\beta^{-/-}$ mice, caused normal primary clonal expansion in secondary lymphoid organs. Interestingly, expansion of the IL-2R$\alpha^{-/-}$ CD8$^+$ T cells was impaired in peripheral, nonlymphoid tissues in which differentiated T$_{EFF}$ would be expected to accumulate, suggesting a two-step process of clonal expansion with IL-2 being required only for a later phase that is associated with the accumulation of effector cells in peripheral tissues. This possibility was directly demonstrated by the absence of effector functions in *ex vivo* assays in the expanded IL-2R$\beta^{-/-}$, virus-specific CD8$^+$ T cells, demonstrating again the nonredundant role of IL-2 in T$_{EFF}$ generation. Furthermore, the memory cells that developed in the IL-2R$\beta^{-/-}$ mice were capable of a quantitatively normal response to secondary viral infection, which suggests a developmental pathway for T$_{CM}$ that does not involve prior differentiation to T$_{EFF}$.

Therefore, IL-2R signaling is not required for clonal expansion of the CD8$^+$ T cell or for the generation of the subset of memory cells that mediates secondary expansion, but is required for the development of T$_{EFF}$. These findings would fit easily with the model of antigen-dependent CD8$^+$ T cell development that proposes the occurrence of asymmetrical division of undifferentiated, self-renewing cells, as it presents the possibility of two pathways for clonal expansion, an IL-2-independent pathway that does not cause effector differentiation, enabling expansion through a process of self-renewal, and an IL-2-dependent pathway that is coupled to differentiation. Although these findings of IL-2-independent clonal expansion do not exclude a linear differentiation pathway, they imply a means for maintaining persistently stimulated CD8$^+$ T cell clones because avoiding IL-2-dependent effector differentiation during clonal expansion provides a means for evading replicative senescence and AICD.

4. CORECEPTORS MEDIATING IL-2-INDEPENDENT CD8$^+$ T CELL CLONAL EXPANSION

Knowing that quantitatively normal primary clonal expansion of CD8$^+$ T cells occurs without IL-2R signaling allows one to infer that if abnormal expansion is observed when signaling through coreceptor on CD8$^+$ T cells is interrupted, that coreceptor may mediate IL-2-independent CD8$^+$ T cell proliferation. CD27 is the best example of the result of such reasoning.

4.1. CD27

CD27 and its ligand, CD70, have been known to promote CD8$^+$ T cell proliferation *in vitro* for many years (Lens *et al.*, 1998), but a nonredundant role has become evident only relatively recently (Borst *et al.*, 2005). CD27 is expressed on all naive CD8$^+$ T cells and appears to be lost only when they become highly differentiated. CD70 is expressed by dendritic cells that have been activated by both innate and adaptive immune signals, as is discussed below, and also on activated B cells. An important advance occurred when CD27-deficient mice were shown to have impaired primary and secondary expansion of CD8$^+$ T cells in response to infection with influenza (Hendriks *et al.*, 2000). CD27 has also been found more recently to be necessary for the long-term CD8$^+$ T cell response to persistent polyomaviral infection (Kemball *et al.*, 2006).

The possibility that CD27 drives IL-2-independent responses of the CD8$^+$ T cell *in vivo* is supported by finding that stimulating IL-2$^{-/-}$ CD8$^+$ T cells *in vitro* with repetitive antigen and a recombinant form of soluble CD70 caused marked clonal expansion, no change in the CD62Lhigh status, and no effector differentiation (Carr *et al.*, 2006). Thus, the expanding cells more closely resembled the T$_{CM}$ of Sallusto *et al.* (1999) than of Wherry *et al.* (2003) in that they had not acquired a capacity for rapid synthesis of IFN-γ. The effect of CD27 on cell expansion was the result of both enhanced cell cycling and survival, with the latter being dependent on the ability of ligated CD27 to maintain the expression of IL-7Rα on TCR-stimulated cells. Since IL-7Rα expression contributes to the viability of activated cells after resolution of the acute phase of clonal expansion (Schluns *et al.*, 2000), this effect of CD27 costimulation may be especially important for long-term clonal expansion in persistent viral infections. Moreover, in contrast to IL-2R-stimulated cells (Lenardo, 1991), repetitive TCR ligation of CD27-stimulated CD8$^+$ T cells did not induce AICD or cause the loss of *in vivo* replicative function (Gattinoni *et al.*, 2005), but instead maintained the cellular response to CD70 *in vitro* and a capability for clonal expansion and effector differentiation after adoptive transfer and viral challenge *in vivo*. The additional observation that stimulation through CD27 selectively suppressed IL-2R-induced effector differentiation, while not impairing the proliferative response to IL-2 suggests that CD27 could mediate self-renewal of the CD8$^+$ T cell even in the presence of IL-2. Taken together, these two studies of Hendriks *et al.* (2000) and Carr *et al.* (2006) make CD27 a reasonable candidate for a coreceptor that drives TCR-dependent, IL-2-independent generation of the nondifferentiating, self-renewing subset of antigen-experienced CD8$^+$ T cells.

Recent findings of the role of CD70 on dendritic cells support a critical function for CD27 stimulation of the CD8$^+$ T cell. The ability of agonistic anti-CD40 antibody to promote CD8$^+$ T cell clonal expansion was inhibited by blocking antibody to CD70 (Rowley and Al-Shamkani, 2004), and the effect of agonistic anti-CD40 antibody was shown to be on the dendritic cell (Bullock and Yagita, 2005; Sanchez *et al.*, 2007; Schildknecht *et al.*, 2007; Taraban *et al.*, 2004, 2006). Thus, earlier studies of the ability of agonistic anti-CD40 antibody to replace the function of CD4$^+$ T cells in persistent γ-herpesvirus infection (Sarawar *et al.*, 2001) and of the role of CD4$^+$ T cells in CD8$^+$ T cell responses in general may be related to inducing dendritic cell-associated CD70 to maintain the IL-2-independent pool of undifferentiated, antigen-experienced CD8$^+$ T cells. This TCR/CD27 pathway of CD8$^+$ T cell clonal expansion may also mediate the effect of CD70-expressing antigen-presenting cells in the laminia propria, which contributes to mucosal immune responses to Listeria (Laouar *et al.*, 2005), and be the basis of the efficacy of blocking anti-CD70 antibody in preventing cardiac allograft rejection (Yamada *et al.*, 2005).

The study by Carr *et al.* (2006) indicates that a cell's response to CD70 requires repetitive TCR ligation, which is consistent with this being a pathway for clonal expansion in secondary lymphoid tissue where both antigen and CD70 would be available as long as a microbial infection persists and dendritic cells continue to receive TLR and CD40 signals. However, the T cell may need to receive other signals in addition to TCR and CD27 for effective clonal expansion because in mice with a transgene directing constitutive expression of CD70 on B cells, excessive T cell activation leads eventually both to B and, paradoxically, T cell depletion (Arens *et al.*, 2001; Tesselaar *et al.*, 2003). These studies did not examine the effect of the CD70 transgene in the context of a transgenic TCR that responds poorly to environmental antigens, so that the role of inappropriately "weak" TCR signaling, such as that which drives homeostatic expansion, was not evaluated. The unusual CD70-dependent immunodeficiency syndrome was at least partially explained by the subsequent finding that instead of T cell depletion that occurs in wild-type mice with the CD70 transgene, a T cell proliferative abnormality was observed in transgenic mice lacking CD95 (Arens *et al.*, 2005). This finding identifies Fas–FasL interactions as an essential control for CD27-dependent lymphocyte proliferation and prompts the question of how the presumed self-renewing, CD27-stimulated $CD8^+$ T cell responding to a microbial infection circumvents Fas-mediated apoptosis. Other costimulatory signals delivered by an appropriately activated dendritic cell, which would be absent from B cells constitutively expressing the transgenic CD70, may have a role.

4.2. Other coreceptors

If impaired clonal expansion of $CD8^+$ T cells does identify coreceptors for mediating an IL-2-independent response, then CD28 must also be considered as a candidate for this function. CD28-deficient mice have diminished clonal expansion of $CD8^+$ T cells in primary and secondary responses to influenza (Bertram *et al.*, 2002, 2004; Hendriks *et al.*, 2003, 2005), and mice lacking both CD27 and CD28 have essentially no primary or secondary $CD8^+$ T cell expansion (Hendriks *et al.*, 2003). Although CD28 is known to promote the production of IL-2 through transcriptional and posttranscriptional means, the normal proliferation of $CD8^+$ T cells in the absence of IL-2 excludes this as the basis for the impaired expansion associated with CD28 deficiency.

In some of these studies (Bertram *et al.*, 2002, 2004; Hendriks *et al.*, 2005), deficiency of 4–1BB, which like CD27 is a member of the tumor necrosis factor receptor superfamily, was found to diminish clonal expansion, but the defect was more prominent in the secondary than in the primary response. 4–1BB is not expressed on naive $CD8^+$ T cells, and the precise stage of antigen-dependent development of the $CD8^+$ T cell at which 4–1BB expression occurs is not clear. It may share with CD27 a capacity for IL-2-independent proliferation, but possibly at a later stage of development following IL-2-induced differentiation.

5. MODIFYING THE ANTIPROLIFERATIVE EFFECTS OF TYPES I AND II IFN

The $CD8^+$ T cell must proliferate rapidly in the presence of types I and II IFN produced by plasmacytoid dendritic cells, NK cells and NKT cells. Since IFNs are generally antiproliferative for all other cell types

(Balkwill and Oliver, 1977; Balkwill and Taylor-Papadimitriou, 1978; Lin et al., 1986), this capability is perhaps unique. Remarkably, CD8$^+$ T cells not only overcome the antiproliferative effects of IFNs, but even respond to them with enhanced clonal expansion. Furthermore, CD8$^+$ T cells use IFN-γ for differentiation, in that signaling through the IFN-γ receptor (IFN-γR) induces the expression of T-bet (Glimcher et al., 2004) and, since IL-2-stimulated CD8$^+$ T cells may acquire a capacity for producing IFN-γ, the cytokine has the potential for mediating an autocrine loop that induces terminal differentiation. For these reasons, it is important to evaluate how CD8$^+$ T cells regulate their responses to types I and II IFN.

5.1. The effects of types I and II IFN on CD8$^+$ T cells

IFN-γR signaling promotes apoptosis of antigen-stimulated CD8$^+$ T cells during the acute (Lohman and Welsh, 1998) and contraction phases of the primary response (Badovinac et al., 2000, 2004). In an apparently opposite outcome, IFN-γ has also been observed to promote the expansion of CD8$^+$ T cells (Sercan et al., 2006; Whitmire et al., 2005). Similarly, type I IFN also can enhance clonal expansion of CD8$^+$ T cells by maintaining their viability (Marrack et al., 1999) and proliferation *in vitro* (Curtsinger et al., 2005) and *in vivo* (Ahonen et al., 2004; Honda et al., 2005). Most importantly, this effect of type I IFN is known to be on the CD8$^+$ T cell itself because the expansion of IFNAR$^{-/-}$ CD8$^+$ T cells in wild-type mice infected with LCMV is diminished 100-fold (Kolumam et al., 2005; Thompson et al., 2006).

The capacity of IFN-γR signaling, but possibly not IFNAR signaling (Lighvani et al., 2001), to induce T-bet in the CD8$^+$ T cell may indicate that IFN-γ also has a unique role in differentiation. T-bet$^{-/-}$ CD8$^+$ T cells secrete less IFN-γ, have lower CTL activity (Sullivan et al., 2003), and have impaired effector function in a model of type 1 diabetes (Juedes et al., 2004). However, the precise role for T-bet in the function of CD8$^+$ T cells is unclear as its expression is not required for protective CD8$^+$ T cell immunity in all microbial infections (Way and Wilson, 2004), and some functions of T-bet may be replaced by its paralog, eomes (Intlekofer et al., 2005; Pearce et al., 2003). Perhaps T-bet induces a stage in effector development of the CD8$^+$ T cell that is not mediated by eomes, as suggested by nonredundant functions of T-bet in the expression of the IL-12Rβ2 chain (Afkarian et al., 2002; Pearce et al., 2003) and in the development of NK and NKT cells (Townsend et al., 2004).

5.2. Regulating IFN-γR expression

The biological importance of controlling IFN-γR signaling is suggested by the finding that fully differentiated CD4$^+$ TH1 cells and CD8$^+$ T cells do not express IFN-γR2, the signal transducing subunit of the heterodimeric receptor complex (Bach et al., 1995; Pernis et al., 1995; Tau et al., 2001). If IFN-γR2 is ectopically expressed in $CD4^+$ T cells, the development of TH1 cells is impaired (Tau et al., 2000); ecotopic expression of IFN-γR2 in $CD8^+$ T cells also inhibits the development of CTLs (Tau et al., 2001). Therefore, the transcriptional downregulation of IFN-γR2 with its attendant suppression of IFN-γR signaling is required for normal development of effector T cells of both the $CD4^+$ and $CD8^+$ T cell lineages. This transcriptional downregulation occurs not only in $CD8^+$ T cell clones generated through *in vitro* culture, but also during their primary clonal expansion during acute Listeria infection (Haring et al., 2005). Interestingly, IFN-γR2 expression in the antigen-stimulated cells returns after resolution of the infection in contrast to the apparently permanent repression of its expression in TH1 and CTL clones. Since genetic deletion of IFN-γR1 expression enables clonally expanding $CD8^+$ T cells to avoid *ex vivo*-induced AICD (Lohman and Welsh, 1998), decreased IFN-γR2 may be a developmentally regulated response to enhance $CD8^+$ T cell expansion. However, the means by which IFN-γR signaling promotes AICD is not known, and a previous suggestion that it was through the induction of caspase-8 (Refaeli et al., 2002) is not supported by the occurrence of AICD in caspase-8-deficient T cells (Salmena et al., 2003). Since AICD in $CD8^+$ T cells requires B lymphocyte-induced maturation protein-1 (Blimp-1) (Kallies et al., 2006), IFN-γR signaling may cause the expression of Blimp-1 (see Section 6).

The second means for controlling IFN-γR signaling is cell biological. While IFN-γR1 resides mainly at the plasma membrane, most of IFN-γR2 is in an intracellular compartment that has not been fully characterized, with only a few hundred copies of IFN-γR2 present at the cell surface (Rigamonti et al., 2000). A dipeptide motif in the cytoplasmic domain of IFN-γR2 possibly regulates trafficking to the plasma membrane (Rosenzweig et al., 2004), raising the possibility that cellular signals could acutely increase or decrease the cell's potential for responding to IFN-γ. A third means of regulating IFN-γR signaling is the redistribution of IFN-γR1 to the immunological synapse (Maldonado et al., 2004). If IFN-γ secretion induced by TCR ligation is also directed to this site, this redistribution of IFN-γR1 potentially could promote autocrine responses to the cytokine leading to T-bet expression and further differentiation of the $CD8^+$ T cell.

5.3. Stat1 as a "Switch" determining the effects of types I and II IFN on proliferation

The anti- and pro-proliferative effects of IFNs on $CD8^+$ T cells and other cell types suggest that a "switch" exists that determines which of these two opposing effects of the IFNs will occur. Such a switch was identified

10 years ago when the expression of Stat1 by fibroblasts was found to be required for type I and type II IFN to suppress serum-induced proliferation (Bromberg et al., 1996). Also remarkable was the finding that in Stat1-sufficient fibroblasts, IFN-γ suppressed the induction of c-Myc by platelet-derived growth factor (PDGF), while in Stat1-deficient cells IFN-γ no longer inhibited this growth factor response, and actually transiently induced c-Myc (Ramana et al., 2000). A gene profiling study showed that in Stat1$^{-/-}$ fibroblasts, IFN-γ and PDGF induced many of the same genes (Ramana et al., 2001), which may help explain how IFN-γ enhances the survival and proliferation of macrophage-colony stimulating factor-stimulated Stat1$^{-/-}$ bone marrow-derived macrophages while suppressing these responses in Stat1$^{+/+}$ cells (Gil et al., 2001). These findings have recently been extended to T cells with the demonstration that type I IFN suppressed the proliferation of wild-type murine T cells stimulated with phorbol ester and IL-2 but enhanced the proliferation of similarly stimulated Stat1$^{-/-}$ or Stat2$^{-/-}$ T cells (Gimeno et al., 2005).

These studies point to the possibility that if the CD8$^+$ T cell had a mechanism by which it could control the level of Stat1, it could control the nature of its growth response to type I and type II IFN. Regulation of Stat1 expression by the T cell has not been reported, but Stat1 in IFN-γ-stimulated fibroblasts is subject to ubiquitin- and proteosome-mediated degradation (Kim and Maniatis, 1996), which has also been shown to occur in osteopontin-treated macrophages (Gao et al., 2007). A nuclear E3 ubiquitin ligase, termed SLIM, has also been found to suppress Stat1-dependent signaling (Tanaka et al., 2005). In this respect, since serine phosphorylation often targets proteins for ubiquitin modification, it is interesting that ligation of either TCR or CD28 induces phosphorylation of serine 727 in Stat1 (Gamero and Larner, 2000; Lafont et al., 2000). Although phosphorylation of serine 727 in the transactivation domain is necessary for the transcriptional activity of Stat1, this or other serines that are phosphorylated by the kinase(s) involved in these responses (Tenoever et al., 2007) could also cause ubiquitination and trigger rapid degradation of Stat1. Thus, there may be a means by which TCR-stimulation of the CD8$^+$ T cell could induce a posttranslational Stat1 deficiency to enable IFNs to promote rather than suppress CD8$^+$ T cell expansion.

6. TRANSCRIPTIONAL CONTROL OF REPLICATIVE SENESCENCE: BMI-1, BLIMP-1, AND BCL6/BCL6b

The molecular determinants, other than telomerase, of whether the CD8$^+$ T cell maintains cell cycling capability or has a senescent phenotype have not been described. The ability of Bmi-1, a member of the

Polycomb-group complex, to prevent senescence of the hematopoietic stem cell (Park et al., 2003) and its expression in splenic T cell lymphocytes (Zhang et al., 2004) suggests that it may have a role in this process. Bmi-1 was discovered as a cooperating oncogene in Eμ-myc transgenic mice (Haupt et al., 1991; van Lohuizen et al., 1991). It maintains self-renewing hematopoietic, cerebellar, and leukemic stem cells (Lessard and Sauvagneau, 2003; Molofsky et al., 2003; Park et al., 2003) by suppressing transcription of the INK4b-Arf-INK4a tumor suppressor locus whose protein products regulate pRb and p53 (Jacobs et al., 1999a). Bmi-1$^{-/-}$ mice have reduced T and B cells secondary to impaired early development (van der Lugt et al., 1994), and the few mature T cells that are present have diminished *in vitro* proliferative function after TCR signaling.

Several findings suggest that expression of Bmi-1 is relevant to the proliferative response of the CD8$^+$ T cell. In parallel with the increase in Bmi-1 expression in B cells responding to ligated membrane immunoglobulin (Hasegawa et al., 1998), TCR stimulation has been shown to increase Bmi-1 mRNA and protein levels in murine CD8$^+$ T cells (Heffner and Fearon, 2007). The increase in Bmi-1 is likely to be related to the replication by the TCR-stimulated CD8$^+$ T cell because "knocking-down" Bmi-1 with a lentiviral vector expressing an appropriate shRNA suppresses CD8$^+$ T cell proliferation, and ectopic expression of Bmi-1 promotes expansion of CD8$^+$ T cells both *in vitro* and *in vivo*. Thus, Bmi-1 expression may be linked to the proliferative capability of the antigen-stimulated CD8$^+$ T cell, just as it is to the self-renewing hematopoietic stem cell.

The means by which Bmi-1 is shut off to cause replicative senescence may be related to c-Myc. c-Myc can bind to the *bmi-1* promoter and drive transcription, and haploinsufficient *c-myc*$^{+/-}$ fibroblasts have reduced Bmi-1 levels and display INK4a-dependent senescence (Guney et al., 2006). Also, there is defective homeostatic expansion of *c-myc*$^{+/-}$ memory CD8$^+$ T cells (Bianchi et al., 2006), which may reflect impaired cycling secondary to diminished Bmi-1. However, appropriate studies have not been done to determine whether the replicative abnormalities of lymphocytes with diminished c-Myc are caused by effects on the expression of Bmi-1. [Not relevant to this discussion, but noted for completeness, is the apparently paradoxical finding that nonphysiologically high levels of c-Myc, as occurs in Eμ-myc transgenic mice, drive *Ink4a* transcription, overcoming Bmi-1 transcriptional repression and inducing apoptosis or senescence (Jacobs et al., 1999b).]

The implication of these findings in the context of the development and differentiation of the antigen-stimulated CD8$^+$ T cell is that the transcriptional repressor, Blimp-1 (Turner et al., 1994), also termed PRDI based on its inhibition of the transcription of IFN-β (Keller and Maniatis, 1991), represses *c-myc* transcription in terminally differentiated plasma cells and mononuclear phagocytes (Chang et al., 2000; Lin et al., 1997). Therefore, Blimp-1 may indirectly repress transcription of Bmi-1 in these cells, in which Bmi-1 mRNA has been shown to be absent (Zhang et al., 2004), and in senescent, terminally differentiated $CD8^+$ T cells, in which Bmi-1 expression is diminished (Heffner and Fearon, 2007). Consistent with this possibility are the findings that Blimp-1 is expressed in "effector memory" $CD8^+$ T cells, mediates AICD, and suppresses the expansion of pMHC- and homeostatically stimulated $CD8^+$ T cells *in vivo* and *in vitro* (Kallies et al., 2006; Martins et al., 2006). Although a decrease in c-Myc was not seen when Blimp-1 was induced in T cells *in vitro*, this might be explained by cells with low c-Myc levels being selected against during culture. If Blimp-1 is found to repress *c-myc* transcription in the $CD8^+$ T cell as it does in other cells, then terminal differentiation of the $CD8^+$ T cell would be mediated by a mechanism that is remarkably similar to the B cell lineage. Furthermore, a negative regulatory role for Blimp-1 in the expression of Bmi-1 would provide a direct link between terminal differentiation, Blimp-1, and INK4a-mediated replicative senescence.

These possibilities emphasize the importance of determining the signals that induce the expression of Blimp-1. There may be several pathways for this, as Blimp-1 can be induced in a transformed B cell line solely by IL-2R signaling (Reljic et al., 2000), in myeloid cell lines by macrophage-colony stimulating factor (Chang et al., 2000), and in myeloid and B cell lines by the unfolded protein stress response (Doody et al., 2006). The coupling of effector differentiation of the $CD8^+$ T cell to stimulation by IL-2 and IFN-γ, the occurrence of AICD in T cells stimulated by these two cytokines (Lohman and Welsh, 1998; Refaeli et al., 2002), and the dependence of AICD in the $CD8^+$ T cell on Blimp-1 expression (Kallies et al., 2006) suggest that IL-2 and IFN-γ may mediate the induction of Blimp-1. However, as $CD4^+$ TH2 cells, which have differentiated in response to IL-4 rather than IFN-γ signaling, also can become Blimp-1^+ (Kallies et al., 2006; Martins et al., 2006), there is likely to be more than one pathway to Blimp-1 transcription in the $CD8^+$ T cell.

The ability of BCL6 to repress the expression of Blimp-1 in the B cell (Reljic et al., 2000; Shaffer et al., 2000) and prevent plasma cell differentiation in germinal center B cells (Dent et al., 1997; Fukuda et al., 1997; Ye et al., 1997), when coupled with its role and that of its paralog, BCL6b, in enhancing the generation of memory $CD8^+$ T cells and promoting the magnitude of the secondary $CD8^+$ T cell response (Ichii et al., 2002, 2004; Manders et al., 2005) suggest that these transcriptional repressors may suppress Blimp-1-induced terminal differentiation and loss of Bmi-1 expression in the $CD8^+$ T cell. Although neither has been reported to do this, the probable role of IL-2 in contributing to Blimp-1 expression in T cells and the ability of BCL6b to suppress the proliferative response of the $CD8^+$ T cell to IL-2 (Manders et al., 2005) make this at least plausible.

However, it is not possible to discuss the cellular interactions that would favor the expression of Blimp-1 versus BCL6/BCL6b because the signals that induce the expression of Blimp-1, BCL6, or BCL6b in the $CD8^+$ T cell have not been fully defined.

7. A REFINED MODEL FOR $CD8^+$ T CELL CLONAL EXPANSION: SEQUENTIAL PHASES OF CD27-DEPENDENT SELF-RENEWAL AND IL-2-DEPENDENT DIFFERENTIATION

The analysis of the $CD8^+$ T cell response to persistent viral infections, especially those caused by human and murine CMV, is informative because it reveals capabilities of the antigen-experienced $CD8^+$ T cell that are not evident in analyses of classical memory responses of this cell. Persistent antigenic stimulation of the $CD8^+$ T cell causes continual, "inflationary" clonal expansion (Karrer et al., 2003) with most antigen-specific cells having a senescent and highly differentiated T_{EFF} phenotype (Holtappels et al., 2000; Munks et al., 2006; Sierro et al., 2005). Adoptive transfer experiments showed that these differentiated T_{EFF} were not able to generate additional T_{EFF}, whereas antigen-experienced $CD8^+$ T cells with a less differentiated, T_{CM} phenotype and residing in secondary lymphoid organs did have this function (Pahl-Seibert et al., 2005). These findings, combined with the demonstration of long-term clonal persistence of $CD8^+$ T cells specific for continually presented viral epitopes (Cohen et al., 2002; Khan et al., 2002; Weekes et al., 1999), lead to the conclusion that persistently stimulated clones are maintained by a process of self-renewal, with asymmetrical division yielding both undifferentiating progeny and daughter cells that become committed to effector differentiation. The quantitatively normal clonal expansion without differentiation that occurs with antigen-stimulated, $IL-2R^{-/-}$ $CD8^+$ T cells (D'Souza and Lefrancois, 2003; Yu et al., 2003) demonstrates that IL-2-independent expansion is robust and that IL-2 has a nonredundant role in effector differentiation. Therefore, to avoid clonal senescence, antigen-stimulated $CD8^+$ T cells must establish a self-renewing, nondifferentiating pool that is capable of IL-2-independent expansion and that avoids IL-2-induced differentiation. This pool would serve as the source of cells that have the potential of entering a phase of IL-2-dependent expansion and effector differentiation when antigenic stimulation indicates the need for additional T_{EFF}.

Two sets of findings provide evidence that an IL-2-independent phase of clonal expansion can be mediated by CD27: first, unlike IL-2R-deficient mice, CD27-deficient mice show impaired expansion of antigen-specific $CD8^+$ T cells in acute (Hendriks et al., 2000) and persistent viral infections (Kemball et al., 2006), and second, ligation of CD27 on repetitively TCR-stimulated $CD8^+$ T cells in vitro causes IL-2-independent expansion without effector differentiation (Carr *et al.*, 2006). The expanded cells retain the potential for infection-induced expansion and differentiation *in vivo*. In addition, costimulation through CD27 suppresses differentiation caused by IL-2 *in vitro*. Although other coreceptors, such as CD28, may also have this capability, the importance of these findings with CD27 is that they establish the principle of IL-2-independent $CD8^+$ T cell clonal expansion without differentiation, that is, self-renewal. The recent demonstration of asymmetrical division of antigen-stimulated $CD8^+$ T cells (Chang *et al.*, 2007) is consistent with this view of two means for clonal expansion, one that does not initiate differentiation and the other that does. These findings do not accommodate the linear differentiation model, which envisions effector differentiation with loss, even if only temporary, of replicative function as being the first step in the development of the antigen-stimulated $CD8^+$ T cell (Wherry *et al.*, 2003).

Superimposed on this basic, underlying process are two additional themes: first, a remarkable switch in the nature of the response of the antigen-stimulated $CD8^+$ T cell to type I and possibly type II IFN from antiproliferative to proproliferative, the mechanism for which is suggested to be related to posttranslational regulation of Stat1, and second, the control of terminal differentiation and senescence. The latter may involve the expression of Blimp-1 in the IL-2-stimulated $CD8^+$ T cell (Kallies *et al.*, 2006; Martins *et al.*, 2006); Blimp-1 is not induced by repetitive TCR and CD27 signaling. Blimp-1 may indirectly suppress the expression of Bmi-1, which may be required for the $CD8^+$ T cell to prevent replicative senescence in the antigen-experienced $CD8^+$ T cell (Heffner and Fearon, 2007) as it is in the hematopoietic stem cell. Senescence may be delayed by BCL6 (Ichii *et al.*, 2002) or BCL6b (Manders *et al.*, 2005), which, by analogy to the function of BCL6 in the germinal center B cell, may suppress the induction of Blimp-1 by inhibiting transcriptional events downstream of IL-2R signaling. These views are summarized in Fig. 3.1.

8. CLINICAL EXTENSIONS OF THE TCR/CD27 PATHWAY: ADOPTIVE $CD8^+$ T CELL THERAPY

The definition of a means for expanding antigen-specific $CD8^+$ T cells *in vitro* without causing replicative senescence after adoptive transfer and *in vivo* challenge may increase the clinical utility of adoptive $CD8^+$ T cell therapy. Two general clinical situations have been examined for adoptive $CD8^+$ T cell therapy: the treatment of persistent viral diseases that occur in individuals rendered immunodeficient by HIV infection or during the course of bone marrow transplantation and in patients with cancer. Disseminated CMV infection has been successfully treated by adoptive transfer of CMV-specific $CD8^+$ T cells, with a recent example

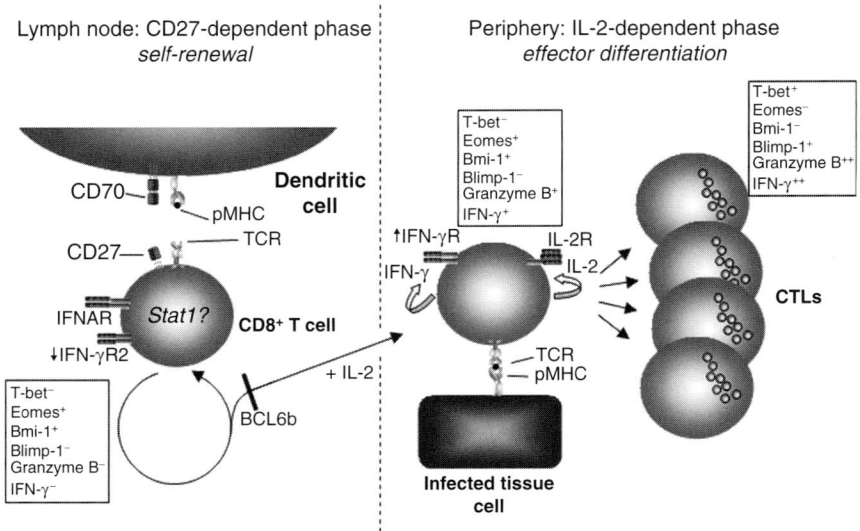

FIGURE 3.1 The two phases of central and peripheral CD8$^+$ T cell clonal expansion. (1) Dendritic cells "fully" activated by TLR ligands and CD40L present pMHC and CD70 to ligate TCR and CD27 on antigen-specific CD8$^+$ T cells. (2) TCR/CD27 signals clonal expansion. IL-2R signals may occur, but CD27 suppresses IL-2R-induced effector differentiation. (3) Repetitively ligated TCR on CD27-stimulated CD8$^+$ T cells may switch their response to type I IFN from growth inhibitory to growth enhancing, possibly through ubiquitin-mediated regulation of Stat1 transcriptional activity. (4) Unknown signals suppress IFN-γR2 expression to avoid AICD. (5) Expansion of the pool of self-renewing, antigen-specific CD8$^+$ T cells and competition for pMHC and CD70 on dendritic cells allow some T cells to initiate differentiation in response to IL-2 and eomes. These cells change their homing receptors and migrate to peripheral inflamed tissues. (6) Transcriptional repression of IL-2R signaling by BCL6 or BCL6b expands the central pool of replicating cells. (7) Encounter with pMHC in peripheral tissue causes secretion of IL-2, which maintains expansion and drives further differentiation. As autocrine and paracrine IFN-γ is produced and IFN-γR2 is reexpressed, T-bet is induced, which completes differentiation. (8) Blimp-1 levels rise as the cells differentiate, leading to suppression of Bmi-1, possibly indirectly, and cell cycle arrest. (See Plate 6 in Color Plate Section.)

using CMV-specific cells purified by cell sorting based on the binding of pMHC complexes bearing the relevant peptide (Cobbold *et al.*, 2005). No attempt was made to expand the cells by *in vitro* stimulation so that excessive differentiation with loss of *in vivo* replicative function did not occur. However, if expansion *in vitro* of IE1-specific CD8$^+$ T cells with maintenance of replicative function could be made possible through the TCR/CD27 pathway, a bank of CMV-specific CD8$^+$ T cells from normal individuals of differing HLA haplotypes could be established for adoptive transfer therapy. This would avoid the need to sort antigen-specific cells acutely for immediate adoptive transfer and facilitate the use of this therapy. Such an approach could be extended to the clonal expansion of $CD8^+$ T cells specific for multiple viruses for use in immunodeficient patients who not infrequently have uncontrolled infections involving more than a single virus (Leen et al., 2006). The ultimate success of this strategy would be enhanced if a similar approach could be developed for the *in vitro* expansion of $CD4^+$ T cells that will be required for the maintenance of the $CD8^+$ T cell response.

Immunological therapy of tumors has followed two general approaches (Blattman and Greenberg, 2004; Gattinoni et al., 2006): first, active immunization with tumor-associated antigens in combination with other immunopotentiating agents for therapeutic treatment of patients with clinically evident tumor (Hodi et al., 2003), or immunization for prophylactic treatment of patients following apparent total resection of the tumor, but in whom undetectable micrometastases may be present (Jager et al., 2006); and second, adoptive T cell therapy after *in vitro* expansion of T cells specific for tumor-associated antigens (e.g., Dudley et al., 2002). The advantage of therapeutic immunization is its relative simplicity, but the expansion of $CD8^+$ T cells that are specific for tumor-associated antigens may not overcome the hurdle of a local, immunosuppressive environment within the tumor itself (Willimsky and Blankenstein, 2005). Therefore, a major advantage of adoptive T cell therapy is the opportunity to alter the T cells during *in vitro* culture so that they may become resistant to immunosuppressive mediators in the tumor, such as TGF-β (Chen et al., 2005; Gorelik and Flavell, 2001). However, repetitive stimulation of $CD8^+$ T cells with antigen and IL-2 ablates their antitumor effects because of loss of *in vivo* replicative function (Gattinoni et al., 2005; Klebanoff et al., 2005), so that genetic modification during *in vitro* clonal expansion is not feasible using the standard means for $CD8^+$ T cell proliferation. As proposed for developing antigen-specific $CD8^+$ T cells to employ in adoptive therapy for viral diseases, the use of the TCR/CD27 pathway could potentially overcome this technical problem and enhance the efficacy of this approach to cancer immunotherapy.

ACKNOWLEDGMENTS

The work described in this manuscript was supported by grants from the Wellcome Trust, the National Institutes of Health, and the Medical Research Council. Thanks to Dr. James Carr for making the figure and to members of the Fearon lab for helpful comments.

REFERENCES

Afkarian, M., Sedy, J. R., Yang, J., Jacobson, N. G., Cereb, N., Yang, S. Y., Murphy, T. L., and Murphy, K. M. (2002). T-bet is a STAT1-induced regulator of IL-12R expression in naive $CD4^+$ T cells. *Nat. Immunol.* **3**, 549–557.

Ahonen, C. L., Doxsee, C. L., McGurran, S. M., Riter, T. R., Wade, W. F., Barth, R. J., Vasilakos, J. P., Noelle, R. J., and Kedl, R. M. (2004). Combined TLR and CD40 triggering induces potent CD8$^+$ T cell expansion with variable dependence on type I IFN. *J. Exp. Med.* **199,** 775–784.

Arens, R., Tesselaar, K., Baars, P. A., van Schijndel, G. M., Hendriks, J., Pals, S. T., Krimpenfort, P., Borst, J., van Oers, M. H., and van Lier, R. A. (2001). Constitutive CD27/CD70 interaction induces expansion of effector-type T cells and results in IFN-gamma-mediated B cell depletion. *Immunity* **15,** 801–812.

Arens, R., Baars, P. A., Jak, M., Tesselaar, K., van der Valk, M., van Oers, M. H., and van Lier, R. A. (2005). Cutting edge: CD95 maintains effector T cell homeostasis in chronic immune activation. *J. Immunol.* **174,** 5915–5920.

Bach, E. A., Szabo, S. J., Dighe, A. S., Ashkenazi, A., Aguet, M., Murphy, K. M., and Schreiber, R. D. (1995). Ligand-induced autoregulation of IFN-gamma receptor beta chain expression in T helper cell subsets. *Science* **270,** 1215–1218.

Bachmann, M. F., Wolint, P., Schwarz, K., Jager, P., and Oxenius, A. (2005). Functional properties and lineage relationship of CD8$^+$ T cell subsets identified by expression of IL-7 receptor alpha and CD62L. *J. Immunol.* **175,** 4686–4696.

Badovinac, V. P., Tvinnereim, A. R., and Harty, J. T. (2000). Regulation of antigen-specific CD8$^+$ T cell homeostasis by perforin and interferon-gamma. *Science* **290,** 1354–1358.

Badovinac, V. P., Porter, B. B., and Harty, J. T. (2004). CD8$^+$ T cell contraction is controlled by early inflammation. *Nat. Immunol.* **5,** 809–817.

Balkwill, F., and Taylor-Papadimitriou, J. (1978). Interferon affects both G1 and S$^+$ G2 in cells stimulated from quiescence to growth. *Nature* **274,** 798–800.

Balkwill, F. R., and Oliver, R. T. (1977). Growth inhibitory effects of interferon on normal and malignant human haemopoietic cells. *Int. J. Cancer* **20,** 500–505.

Barber, D. L., Wherry, E. J., Masopust, D., Zhu, B., Allison, J. P., Sharpe, A. H., Freeman, G. J., and Ahmed, R. (2006). Restoring function in exhausted CD8 T cells during chronic viral infection. *Nature* **439,** 682–687.

Becker, T. C., Wherry, E. J., Boone, D., Murali-Krishna, K., Antia, R., Ma, A., and Ahmed, R. (2002). Interleukin 15 is required for proliferative renewal of virus-specific memory CD8 T cells. *J. Exp. Med.* **195,** 1541–1548.

Bennett, S. R., Carbone, F. R., Karamalis, F., Flavell, R. A., Miller, J. F., and Heath, W. R. (1998). Help for cytotoxic-T-cell responses is mediated by CD40 signalling. *Nature* **393,** 478–480.

Bertram, E. M., Lau, P., and Watts, T. H. (2002). Temporal segregation of 4-1BB versus CD28-mediated costimulation: 4-1BB ligand influences T cell numbers late in the primary response and regulates the size of the T cell memory response following influenza infection. *J. Immunol.* **168,** 3777–3785.

Bertram, E. M., Dawicki, W., Sedgmen, B., Bramson, J. L., Lynch, D. H., and Watts, T. H. (2004). A switch in costimulation from CD28 to 4-1BB during primary versus secondary CD8 T cell response to influenza *in vivo*. *J. Immunol.* **172,** 981–988.

Bianchi, T., Gasser, S., Trumpp, A., and MacDonald, H. R. (2006). c-Myc acts downstream of IL-15 in the regulation of memory CD8 T-cell homeostasis. *Blood* **107,** 3992–3999.

Blattman, J. N., and Greenberg, P. D. (2004). Cancer immunotherapy: A treatment for the masses. *Science* **305,** 200–205.

Borst, J., Hendriks, J., and Xiao, Y. (2005). CD27 and CD70 in T cell and B cell activation. *Curr. Opin. Immunol.* **17,** 275–281.

Bouneaud, C., Garcia, Z., Kourilsky, P., and Pannetier, C. (2005). Lineage relationships, homeostasis, and recall capacities of central- and effector-memory CD8 T cells *in vivo*. *J. Exp. Med.* **201,** 579–590.

Brenchley, J. M., Karandikar, N. J., Betts, M. R., Ambrozak, D. R., Hill, B. J., Crotty, L. E., Casazza, J. P., Kuruppu, J., Migueles, S. A., Connors, M., Roederer, M., Douek, M., *et al.*

(2003). Expression of CD57 defines replicative senescence and antigen-induced apoptotic death of CD8$^+$ T cells. *Blood* **101,** 2711–2720.

Bromberg, J. F., Horvath, C. M., Wen, Z., Schreiber, R. D., and Darnell, J. E. (1996). Transcriptionally active Stat1 is required for the antiproliferative effects of both interferon alpha and interferon gamma. *Proc. Natl. Acad. Sci. USA* **93,** 7673–7678.

Brooks, D. G., Trifilo, M. J., Edelmann, K. H., Teyton, L., McGavern, D. B., and Oldstone, M. B. (2006). Interleukin-10 determines viral clearance or persistence *in vivo*. *Nat. Med.* **12,** 1301–1309.

Bullock, T. N., and Yagita, H. (2005). Induction of CD70 on dendritic cells through CD40 or TLR stimulation contributes to the development of CD8$^+$ T cell responses in the absence of CD4$^+$ T cells. *J. Immunol.* **174,** 710–717.

Cantrell, D. A., and Smith, K. A. (1984). The interleukin-2 T-cell system: A new cell growth model. *Science* **224,** 1312–1316.

Cardin, R. D., Brooks, J. W., Sarawar, S. R., and Doherty, P. C. (1996). Progressive loss of CD8$^+$ T cell-mediated control of a gamma-herpesvirus in the absence of CD4$^+$ T cells. *J. Exp. Med.* **184,** 863–871.

Carr, J. M., Carrasco, M. J., Thaventhiran, J. E., Bambrough, P. J., Kraman, M., Edwards, A. D., Al-Shamkhani, A., and Fearon, D. T. (2006). CD27 mediates interleukin-2-independent clonal expansion of the CD8$^+$ T cell without effector differentiation. *Proc. Natl. Acad. Sci. USA* **103,** 19454–19459.

Chang, D. H., Angelin-Duclos, C., and Calame, K. (2000). BLIMP-1: Trigger for differentiation of myeloid lineage. *Nat. Immunol.* **1,** 169–176.

Chang, J. T., Palanivel, V. R., Kinjyo, I., Schambach, F., Intlekofer, A. M., Banerjee, A., Longworth, S. A., Vinup, K. E., Mrass, P., Oliaro, J., Killeen, N., Orange, J. S., *et al*. (2007). Asymmetric T Lymphocyte division in the initiation of adaptive immune responses. *Science* **315,** 1687–1691.

Chao, C. C., Jensen, R., and Dailey, M. O. (1997). Mechanisms of L-selectin regulation by activated T cells. *J. Immunol.* **159,** 1686–1694.

Chen, M. L., Pittet, M. J., Gorelik, L., Flavell, R. A., Weissleder, R., von Boehmer, H., and Khazaie, K. (2005). Regulatory T cells suppress tumor-specific CD8 T cell cytotoxicity through TGF-beta signals *in vivo*. *Proc. Natl. Acad. Sci. USA* **102,** 419–424.

Cobbold, M., Khan, N., Pourgheysari, B., Tauro, S., McDonald, D., Osman, H., Assenmacher, M., Billingham, L., Steward, C., Crawley, C., Olavarria, E., Goldman, J., *et al*. (2005). Adoptive transfer of cytomegalovirus-specific CTL to stem cell transplant patients after selection by HLA-peptide tetramers. *J. Exp. Med.* **202,** 379–386.

Cohen, G. B., Islam, S. A., Noble, M. S., Lau, C., Brander, C., Altfeld, M. A., Rosenberg, E. S., Schmitz, J. E., Cameron, T. O., and Kalams, S. A. (2002). Clonotype tracking of TCR repertoires during chronic virus infections. *Virology* **304,** 474–484.

Curtsinger, J. M., Valenzuela, J. O., Agarwal, P., Lins, D., and Mescher, M. F. (2005). Cutting edge: Type I IFNs provide a third signal to CD8 T cells to stimulate clonal expansion and differentiation. *J. Immunol.* **174,** 4465–4469.

Dent, A. L., Shaffer, A. L., Yu, X., Allman, D., and Staudt, L. M. (1997). Control of inflammation, cytokine expression, and germinal center formation by BCL-6. *Science* **276,** 589–592.

Doody, G. M., Stephenson, S., and Tooze, R. M. (2006). BLIMP-1 is a target of cellular stress and downstream of the unfolded protein response. *Eur. J. Immunol.* **36,** 1572–1582.

D'Souza, W. N., and Lefrancois, L. (2003). IL-2 is not required for the initiation of CD8 T cell cycling but sustains expansion. *J. Immunol.* **171,** 5727–5735.

Dudley, M. E., Wunderlich, J. R., Robbins, P. F., Yang, J. C., Hwu, P., Schwartzentruber, D. J., Topalian, S. L., Sherry, R., Restifo, N. P., Hubicki, A. M., Robinson, M. R., Raffeld, M., *et al*. (2002). Cancer regression and autoimmunity in patients after clonal repopulation with antitumor lymphocytes. *Science* **298,** 850–854.

Fearon, D. T., Manders, P., and Wagner, S. D. (2001). Arrested differentiation, the self-renewing memory lymphocyte, and vaccination. *Science* **293**, 248–250.

Fearon, D. T., Carr, J. M., Telaranta, A., Carrasco, M. J., and Thaventhiran, J. E. (2006). The rationale for the IL-2-independent generation of the self-renewing central memory $CD8^+$ T cells. *Immunol. Rev.* **211**, 104–118.

Fukuda, T., Yoshida, T., Okada, S., Hatano, M., Miki, T., Ishibashi, K., Okabe, S., Koseki, H., Hirosawa, S., Taniguchi, M., Miyasaka, N., and Tokuhisa, T. (1997). Disruption of the Bcl6 gene results in an impaired germinal center formation. *J. Exp. Med.* **186**, 439–448.

Gamero, A. M., and Larner, A. C. (2000). Signaling via the T cell antigen receptor induces phosphorylation of Stat1 on serine 727. *J. Biol. Chem.* **275**, 16574–16578.

Gao, C., Guo, H., Mi, Z., Grusby, M. J., and Kuo, P. C. (2007). Osteopontin induces ubiquitin-dependent degradation of STAT1 in RAW264.7 murine macrophages. *J. Immunol.* **178**, 1870–1881.

Gattinoni, L., Klebanoff, C. A., Palmer, D. C., Wrzesinski, C., Kerstann, K., Yu, Z., Finkelstein, S. E., Theoret, M. R., Rosenberg, S. A., and Restifo, N. P. (2005). Acquisition of full effector function *in vitro* paradoxically impairs the *in vivo* antitumor efficacy of adoptively transferred $CD8^+$ T cells. *J. Clin. Invest.* **115**, 1616–1626.

Gattinoni, L., Powell, D. J., Jr., Rosenberg, S. A., and Restifo, N. P. (2006). Adoptive immunotherapy for cancer: Building on success. *Nat. Rev. Immunol.* **6**, 383–393.

Gil, M. P., Bohn, E., O'Guin, A. K., Ramana, C. V., Levine, B., Stark, G. R., Virgin, H. W., and Schreiber, R. D. (2001). Biologic consequences of Stat1-independent IFN signaling. *Proc. Natl. Acad. Sci. USA* **98**, 6680–6685.

Gillis, S., and Smith, K. A. (1977). Long term culture of tumour-specific cytotoxic T cells. *Nature* **268**, 154–156.

Gimeno, R., Lee, C. K., Schindler, C., and Levy, D. E. (2005). Stat1 and Stat2 but not Stat3 arbitrate contradictory growth signals elicited by alpha/beta interferon in T lymphocytes. *Mol. Cell. Biol.* **25**, 5456–5465.

Glimcher, L. H., Townsend, M. J., Sullivan, B. M., and Lord, G. M. (2004). Recent developments in the transcriptional regulation of cytolytic effector cells. *Nat. Rev. Immunol.* **4**, 900–911.

Goldrath, A. W., Sivakumar, P. V., Glaccum, M., Kennedy, M. K., Bevan, M. J., Benoist, C., Mathis, D., and Butz, E. A. (2002). Cytokine requirements for acute and basal homeostatic proliferation of naive and memory $CD8^+$ T cells. *J. Exp. Med.* **195**, 1515–1522.

Gorelik, L., and Flavell, R. A. (2001). Immune-mediated eradication of tumors through the blockade of transforming growth factor-beta signaling in T cells. *Nat. Med.* **7**, 1118–1122.

Guney, I., Wu, S., and Sedivy, J. M. (2006). Reduced c-Myc signaling triggers telomere-independent senescence by regulating Bmi-1 and p16(INK4a). *Proc. Natl. Acad. Sci. USA* **103**, 3645–3650.

Haring, J. S., Corbin, G. A., and Harty, J. T. (2005). Dynamic regulation of IFN-gamma signaling in antigen-specific $CD8^+$ T cells responding to infection. *J. Immunol.* **174**, 6791–6802.

Hasegawa, M., Tetsu, O., Kanno, R., Inoue, H., Ishihara, H., Kamiyasu, M., Taniguchi, M., and Kanno, M. (1998). Mammalian Polycomb group genes are categorized as a new type of early response gene induced by B-cell receptor cross-linking. *Mol. Immunol.* **35**, 559–563.

Haupt, Y., Alexander, W. S., Barri, G., Klinken, S. P., and Adams, J. M. (1991). Novel zinc finger gene implicated as myc collaborator by retrovirally accelerated lymphomagenesis in E mu-myc transgenic mice. *Cell* **65**, 753–763.

Heffner, M., and Fearon, D. T. (2007). Loss of T cell receptor-induced Bmi-1 in the $KLRG1^+$ senescent $CD8^+$ T lymphocyte. *Proc. Natl. Acad. Sci. USA* **104**, 13414–13419.

Hendriks, J., Gravestein, L. A., Tesselaar, K., van Lier, R. A., Schumacher, T. N., and Borst, J. (2000). CD27 is required for generation and long-term maintenance of T cell immunity. *Nat. Immunol.* **1**, 433–440.

Hendriks, J., Xiao, Y., and Borst, J. (2003). CD27 promotes survival of activated T cells and complements CD28 in generation and establishment of the effector T cell pool. *J. Exp. Med.* **198**, 1369–1380.

Hendriks, J., Xiao, Y., Rossen, J. W., van der Sluijs, K. F., Sugamura, K., Ishii, N., and Borst, J. (2005). During viral infection of the respiratory tract, CD27, 4–1BB, and OX40 collectively determine formation of CD8$^+$ memory T cells and their capacity for secondary expansion. *J. Immunol.* **175**, 1665–1676.

Hodi, F. S., Mihm, M. C., Soiffer, R. J., Haluska, F. G., Butler, M., Seiden, M. V., Davis, T., Henry-Spires, R., MacRae, S., Willman, A., Padera, R., Jaklitsch, M. T., *et al.* (2003). Biologic activity of cytotoxic T lymphocyte-associated antigen 4 antibody blockade in previously vaccinated metastatic melanoma and ovarian carcinoma patients. *Proc. Natl. Acad. Sci. USA* **100**, 4712–4717.

Holtappels, R., Pahl-Seibert, M. F., Thomas, D., and Reddehase, M. J. (2000). Enrichment of immediate-early 1 (m123/pp89) peptide-specific CD8 T cells in a pulmonary CD62L(lo) memory-effector cell pool during latent murine cytomegalovirus infection of the lungs. *J. Virol.* **74**, 11495–11503.

Honda, K., Yanai, H., Negishi, H., Asagiri, M., Sato, M., Mizutani, T., Shimada, N., Ohba, Y., Takaoka, A., Yoshida, N., and Taniguchi, T. (2005). IRF-7 is the master regulator of type-I interferon-dependent immune responses. *Nature* **434**, 772–777.

Ichii, H., Sakamoto, A., Hatano, M., Okada, S., Toyama, H., Taki, S., Arima, M., Kuroda, Y., and Tokuhisa, T. (2002). Role for Bcl-6 in the generation and maintenance of memory CD8$^+$ T cells. *Nat. Immunol.* **3**, 558–563.

Ichii, H., Sakamoto, A., Kuroda, Y., and Tokuhisa, T. (2004). Bcl6 acts as an amplifier for the generation and proliferative capacity of central memory CD8$^+$ T cells. *J. Immunol.* **173**, 883–891.

Intlekofer, A. M., Takemoto, N., Wherry, E. J., Longworth, S. A., Northrup, J. T., Palanivel, V. R., Mullen, A. C., Gasink, C. R., Kaech, S. M., Miller, J. D., Gapin, L., Ryan, K., *et al.* (2005). Effector and memory CD8$^+$ T cell fate coupled by T-bet and eomesodermin. *Nat. Immunol.* **6**, 1236–1244.

Jacobs, J. J., Kieboom, K., Marino, S., DePinho, R. A., and van Lohuizen, M. (1999a). The oncogene and Polycomb-group gene bmi-1 regulates cell proliferation and senescence through the ink4a locus. *Nature* **397**, 164–168.

Jacobs, J. J., Scheijen, B., Voncken, J. W., Kieboom, K., Berns, A., and van Lohuizen, M. (1999b). Bmi-1 collaborates with c-Myc in tumorigenesis by inhibiting c-Myc-induced apoptosis via INK4a/ARF. *Genes Dev.* **13**, 2678–2690.

Jager, E., Karbach, J., Gnjatic, S., Neumann, A., Bender, A., Valmori, D., Ayyoub, M., Ritter, E., Ritter, G., Jager, D., Panicali, D., Hoffman, E., *et al.* (2006). Recombinant vaccinia/fowlpox NY-ESO-1 vaccines induce both humoral and cellular NY-ESO-1-specific immune responses in cancer patients. *Proc. Natl. Acad. Sci. USA* **103**, 14453–14458.

Janssen, E. M., Lemmens, E. E., Wolfe, T., Christen, U., von Herrath, M. G., and Schoenberger, S. P. (2003). CD4$^+$ T cells are required for secondary expansion and memory in CD8$^+$ T lymphocytes. *Nature* **421**, 852–856.

Juedes, A. E., Rodrigo, E., Togher, L., Glimcher, L. H., and von Herrath, M. G. (2004). T-bet controls autoaggressive CD8 lymphocyte responses in type 1 diabetes. *J. Exp. Med.* **199**, 1153–1162.

Jung, T. M., Gallatin, W. M., Weissman, I. L., and Dailey, M. O. (1988). Down-regulation of homing receptors after T cell activation. *J. Immunol.* **141**, 4110–4117.

Kaech, S. M., and Ahmed, R. (2001). Memory CD8$^+$ T cell differentiation: Initial antigen encounter triggers a developmental program in naive cells. *Nat. Immunol.* **2**, 415–422.

Kaech, S. M., Tan, J. T., Wherry, E. J., Konieczny, B. T., Surh, C. D., and Ahmed, R. (2003). Selective expression of the interleukin 7 receptor identifies effector CD8 T cells that give rise to long-lived memory cells. *Nat. Immunol.* **4**, 1191–1198.

Kallies, A., Hawkins, E. D., Belz, G. T., Metcalf, D., Hommel, M., Corcoran, L. M., Hodgkin, P. D., and Nutt, S. L. (2006). Transcriptional repressor Blimp-1 is essential for T cell homeostasis and self-tolerance. *Nat. Immunol.* **7**, 466–474.

Karrer, U., Sierro, S., Wagner, M., Oxenius, A., Hengel, H., Koszinowski, U. H., Phillips, R. E., and Klenerman, P. (2003). Memory inflation: Continuous accumulation of antiviral $CD8^+$ T cells over time. *J. Immunol.* **170**, 2022–2029.

Karrer, U., Wagner, M., Sierro, S., Oxenius, A., Hengel, H., Dumrese, T., Freigang, S., Koszinowski, U. H., Phillips, R. E., and Klenerman, P. (2004). Expansion of protective $CD8^+$ T-cell responses driven by recombinant cytomegaloviruses. *J. Virol.* **78**, 2255–2264.

Keller, A. D., and Maniatis, T. (1991). Identification and characterization of a novel repressor of beta-interferon gene expression. *Genes Dev.* **5**, 868–879.

Kemball, C. C., Lee, E. D., Szomolanyi-Tsuda, E., Pearson, T. C., Larsen, C. P., and Lukacher, A. E. (2006). Costimulation requirements for antiviral $CD8^+$ T cells differ for acute and persistent phases of polyoma virus infection. *J. Immunol.* **176**, 1814–1824.

Khan, N., Shariff, N., Cobbold, M., Bruton, R., Ainsworth, J. A., Sinclair, A. J., Nayak, L., and Moss, P. A. (2002). Cytomegalovirus seropositivity drives the CD8 T cell repertoire toward greater clonality in healthy elderly individuals. *J. Immunol.* **169**, 1984–1992.

Khanna, K. M., Bonneau, R. H., Kinchington, P. R., and Hendricks, R. L. (2003). Herpes simplex virus-specific memory $CD8^+$ T cells are selectively activated and retained in latently infected sensory ganglia. *Immunity* **18**, 593–603.

Kim, T. K., and Maniatis, T. (1996). Regulation of interferon-gamma-activated STAT1 by the ubiquitin-proteasome pathway. *Science* **273**, 1717–1719.

Klebanoff, C. A., Gattinoni, L., Torabi-Parizi, P., Kerstann, K., Cardones, A. R., Finkelstein, S. E., Palmer, D. C., Antony, P. A., Hwang, S. T., Rosenberg, S. A., Waldmann, T. A., and Restifo, N. P. (2005). Central memory self/tumor-reactive $CD8^+$ T cells confer superior antitumor immunity compared with effector memory T cells. *Proc. Natl. Acad. Sci. USA* **102**, 9571–9576.

Kolumam, G. A., Thomas, S., Thompson, L. J., Sprent, J., and Murali-Krishna, K. (2005). Type I interferons act directly on CD8 T cells to allow clonal expansion and memory formation in response to viral infection. *J. Exp. Med.* **202**, 637–650.

Kramer, S., Mamalaki, C., Horak, I., Schimpl, A., Kioussis, D., and Hung, T. (1994). Thymic selection and peptide-induced activation of T cell receptor-transgenic CD8 T cells in interleukin-2-deficient mice. *Eur. J. Immunol.* **24**, 2317–2322.

Lafont, V., Decker, T., and Cantrell, D. (2000). Antigen receptor signal transduction: Activating and inhibitory antigen receptors regulate STAT1 serine phosphorylation. *Eur. J. Immunol.* **30**, 1851–1860.

Laouar, A., Haridas, V., Vargas, D., Zhinan, X., Chaplin, D., van Lier, R. A., and Manjunath, N. (2005). $CD70^+$ antigen-presenting cells control the proliferation and differentiation of T cells in the intestinal mucosa. *Nat. Immunol.* **6**, 698–706.

Leen, A. M., Myers, G. D., Sili, U., Huls, M. H., Weiss, H., Leung, K. S., Carrum, G., Krance, R. A., Chang, C. C., Molldrem, J. J., Gee, A. P., Brenner, M. K., *et al.* (2006). Monoculture-derived T lymphocytes specific for multiple viruses expand and produce clinically relevant effects in immunocompromised individuals. *Nat. Med.* **12**, 1160–1165.

Lenardo, M. J. (1991). Interleukin-2 programs mouse alpha beta T lymphocytes for apoptosis. *Nature* **353**, 858–861.

Lens, S. M., Tesselaar, K., van Oers, M. H., and van Lier, R. A. (1998). Control of lymphocyte function through CD27-CD70 interactions. *Semin. Immunol.* **10**, 491–499.

Lessard, J., and Sauvageau, G. (2003). Bmi-1 determines the proliferative capacity of normal and leukaemic stem cells. *Nature* **423**, 255–260.

Lighvani, A. A., Frucht, D. M., Jankovic, D., Yamane, H., Aliberti, J., Hissong, B. D., Nguyen, B. V., Gadina, M., Sher, A., Paul, W. E., and O'Shea, J. J. (2001). T-bet is rapidly induced by interferon-gamma in lymphoid and myeloid cells. *Proc. Natl. Acad. Sci. USA* **98**, 15137–15142.

Lin, S. L., Kikuchi, T., Pledger, W. J., and Tamm, I. (1986). Interferon inhibits the establishment of competence in Go/S-phase transition. *Science* **233**, 356–359.

Lin, Y., Wong, K., and Calame, K. (1997). Repression of c-myc transcription by Blimp-1, an inducer of terminal B cell differentiation. *Science* **276**, 596–599.

Liu, T., Khanna, K. M., Chen, X., Fink, D. J., and Hendricks, R. L. (2000). $CD8^+$ T cells can block herpes simplex virus type 1 (HSV-1) reactivation from latency in sensory neurons. *J. Exp. Med.* **191**, 1459–1466.

Lohman, B. L., and Welsh, R. M. (1998). Apoptotic regulation of T cells and absence of immune deficiency in virus-infected gamma interferon receptor knockout mice. *J. Virol.* **72**, 7815–7821.

Maldonado, R. A., Irvine, D. J., Schreiber, R., and Glimcher, L. H. (2004). A role for the immunological synapse in lineage commitment of CD4 lymphocytes. *Nature* **431**, 527–532.

Manders, P. M., Hunter, P. J., Telaranta, A. I., Carr, J. M., Marshall, J. L., Carrasco, M., Murakami, Y., Palmowski, M. J., Cerundolo, V., Kaech, S. M., Ahmed, R., and Fearon, D. T. (2005). BCL6b mediates the enhanced magnitude of the secondary response of memory $CD8^+$ T lymphocytes. *Proc. Natl. Acad. Sci. USA* **102**, 7418–7425.

Marrack, P., Kappler, J., and Mitchell, T. (1999). Type I interferons keep activated T cells alive. *J. Exp. Med.* **189**, 521–530.

Martins, G. A., Cimmino, L., Shapiro-Shelef, M., Szabolcs, M., Herron, A., Magnusdottir, E., and Calame, K. (2006). Transcriptional repressor Blimp-1 regulates T cell homeostasis and function. *Nat. Immunol.* **7**, 457–465.

Marzo, A. L., Klonowski, K. D., Le Bon, A., Borrow, P., Tough, D. F., and Lefrancois, L. (2005). Initial T cell frequency dictates memory $CD8^+$ T cell lineage commitment. *Nat. Immunol.* **6**, 793–799.

Mescher, M. F., Curtsinger, J. M., Agarwal, P., Casey, K. A., Gerner, M., Hammerbeck, C. D., Popescu, F., and Xiao, Z. (2006). Signals required for programming effector and memory development by $CD8^+$ T cells. *Immunol. Rev.* **211**, 81–92.

Michie, C. A., McLean, A., Alcock, C., and Beverley, P. C. (1992). Lifespan of human lymphocyte subsets defined by CD45 isoforms. *Nature* **360**, 264–265.

Migueles, S. A., Laborico, A. C., Shupert, W. L., Sabbaghian, M. S., Rabin, R., Hallahan, C. W., Van Baarle, D., Kostense, S., Miedema, F., McLaughlin, M., Ehler, L., Metcalf, J., *et al.* (2002). HIV-specific $CD8^+$ T cell proliferation is coupled to perforin expression and is maintained in nonprogressors. *Nat. Immunol.* **3**, 1061–1068.

Molofsky, A. V., Pardal, R., Iwashita, T., Park, I. K., Clarke, M. F., and Morrison, S. J. (2003). Bmi-1 dependence distinguishes neural stem cell self-renewal from progenitor proliferation. *Nature* **425**, 962–967.

Morgan, D. A., Ruscetti, F. W., and Gallo, R. (1976). Selective in vitro growth of T lymphocytes from normal human bone marrows. *Science* **193**, 1007–1008.

Munks, M. W., Cho, K. S., Pinto, A. K., Sierro, S., Klenerman, P., and Hill, A. B. (2006). Four distinct patterns of memory $CD8^+$ T cell responses to chronic murine cytomegalovirus infection. *J. Immunol.* **177**, 450–458.

Obar, J. J., Crist, S. G., Leung, E. K., and Usherwood, E. J. (2004). IL-15-independent proliferative renewal of memory $CD8^+$ T cells in latent gammaherpesvirus infection. *J. Immunol.* **173**, 2705–2714.

Pahl-Seibert, M. F., Juelch, M., Podlech, J., Thomas, D., Deegen, P., Reddehase, M. J., and Holtappels, R. (2005). Highly protective in vivo function of cytomegalovirus IE1 epitope-specific memory $CD8^+$ T cells purified by T-cell receptor-based cell sorting. *J. Virol.* **79**, 5400–5413.

Papagno, L., Spina, C. A., Marchant, A., Salio, M., Rufer, N., Little, S., Dong, T., Chesney, G., Waters, A., Easterbrook, P., Dunbar, P. R., Shepherd, D., et al. (2004). Immune activation and CD8$^+$ T-cell differentiation towards senescence in HIV-1 infection. *PLoS Biol.* **2**, E20.

Park, I. K., Qian, D., Kiel, M., Becker, M. W., Pihalja, M., Weissman, I. L., Morrison, S. J., and Clarke, M. F. (2003). Bmi-1 is required for maintenance of adult self-renewing haematopoietic stem cells. *Nature* **423**, 302–305.

Pearce, E. L., Mullen, A. C., Martins, G. A., Krawczyk, C. M., Hutchins, A. S., Zediak, V. P., Banica, M., DiCioccio, C. B., Gross, D. A., Mao, C. A., Shen, H., Cereb, N., et al. (2003). Control of effector CD8$^+$ T cell function by the transcription factor *Eomesodermin*. *Science* **302**, 1041–1043.

Pernis, A., Gupta, S., Gollob, K. J., Garfein, E., Coffman, R. L., Schindler, C., and Rothman, P. (1995). Lack of interferon gamma receptor beta chain and the prevention of interferon gamma signaling in TH1 cells. *Science* **269**, 245–257.

Ramana, C. V., Grammatikakis, N., Chernov, M., Nguyen, H., Goh, K. C., Williams, B. R., and Stark, G. R. (2000). Regulation of c-myc expression by IFN-gamma through Stat1-dependent and -independent pathways. *EMBO J.* **19**, 263–272.

Ramana, C. V., Gil, M. P., Han, Y., Ransohoff, R. M., Schreiber, R. D., and Stark, G. R. (2001). Stat1-independent regulation of gene expression in response to IFN-gamma. *Proc. Natl. Acad. Sci. USA* **98**, 6674–6679.

Reddehase, M. J., Rothbard, J. B., and Koszinowski, U. H. (1989). A pentapeptide as minimal antigenic determinant for MHC class I-restricted T lymphocytes. *Nature* **337**, 651–653.

Refaeli, Y., Van Parijs, L., Alexander, S. I., and Abbas, A. K. (2002). Interferon gamma is required for activation-induced death of T lymphocytes. *J. Exp. Med.* **196**, 999–1005.

Reljic, R., Wagner, S. D., Peakman, L. J., and Fearon, D. T. (2000). Suppression of signal transducer and activator of transcription 3-dependent B lymphocyte terminal differentiation by BCL-6. *J. Exp. Med.* **192**, 1841–1848.

Ridge, J. P., Di Rosa, F., and Matzinger, P. (1998). A conditioned dendritic cell can be a temporal bridge between a CD4$^+$ T-helper and a T-killer cell. *Nature* **393**, 474–478.

Rigamonti, L., Ariotti, S., Losana, G., Gradini, R., Russo, M. A., Jouanguy, E., Casanova, J. L., Forni, G., and Novelli, F. (2000). Surface expression of the IFN-gamma R2 chain is regulated by intracellular trafficking in human T lymphocytes. *J. Immunol.* **164**, 201–207.

Rosenzweig, S. D., Schwartz, O. M., Brown, M. R., Leto, T. L., and Holland, S. M. (2004). Characterization of a dipeptide motif regulating IFN-gamma receptor 2 plasma membrane accumulation and IFN-gamma responsiveness. *J. Immunol.* **173**, 3991–3999.

Rowley, T. F., and Al-Shamkhani, A. (2004). Stimulation by soluble CD70 promotes strong primary and secondary CD8$^+$ cytotoxic T cell responses *in vivo*. *J. Immunol.* **172**, 6039–6046.

Sallusto, F., Lenig, D., Forster, R., Lipp, M., and Lanzavecchia, A. (1999). Two subsets of memory T lymphocytes with distinct homing potentials and effector functions. *Nature* **401**, 708–712.

Salmena, L., Lemmers, B., Hakem, A., Matysiak-Zablocki, E., Murakami, K., Au, P. Y., Berry, D. M., Tamblyn, L., Shehabeldin, A., Migon, E., Wakeham, A., Bouchard, D., et al. (2003). Essential role for caspase 8 in T-cell homeostasis and T-cell-mediated immunity. *Genes Dev.* **17**, 883–895.

Sanchez, P. J., McWilliams, J. A., Haluszczak, C., Yagita, H., and Kedl, R. M. (2007). Combined TLR/CD40 stimulation mediates potent cellular immunity by regulating dendritic cell expression of CD70 *in vivo*. *J. Immunol.* **178**, 1564–1572.

Sarawar, S. R., Lee, B. J., Reiter, S. K., and Schoenberger, S. P. (2001). Stimulation via CD40 can substitute for CD4 T cell function in preventing reactivation of a latent herpesvirus. *Proc. Natl. Acad. Sci. USA* **98**, 6325–6329.

Schildknecht, A., Miescher, I., Yagita, H., and van den Broek, M. (2007). Priming of CD8$^+$ T cell responses by pathogens typically depends on CD70-mediated interactions with dendritic cells. *Eur. J. Immunol.* **37**, 716–728.

Schluns, K. S., Kieper, W. C., Jameson, S. C., and Lefrancois, L. (2000). Interleukin-7 mediates the homeostasis of naive and memory CD8 T cells *in vivo*. *Nat. Immunol.* **1**, 426–432.

Schoenberger, S. P., Toes, R. E., van der Voort, E. I., Offringa, R., and Melief, C. J. (1998). T-cell help for cytotoxic T lymphocytes is mediated by CD40-CD40L interactions. *Nature* **393**, 480–483.

Sercan, O., Hammerling, G. J., Arnold, B., and Schuler, T. (2006). Innate immune cells contribute to the IFN-gamma-dependent regulation of antigen-specific CD8$^+$ T cell homeostasis. *J. Immunol.* **176**, 735–739.

Shaffer, A. L., Yu, X., He, Y., Boldrick, J., Chan, E. P., and Staudt, L. M. (2000). BCL-6 represses genes that function in lymphocyte differentiation, inflammation, and cell cycle control. *Immunity* **13**, 199–212.

Shedlock, D. J., and Shen, H. (2003). Requirement for CD4 T cell help in generating functional CD8 T cell memory. *Science* **300**, 337–339.

Sheridan, B. S., Khanna, K. M., Frank, G. M., and Hendricks, R. L. (2006). Latent virus influences the generation and maintenance of CD8$^+$ T cell memory. *J. Immunol.* **177**, 8356–8364.

Sierro, S., Rothkopf, R., and Klenerman, P. (2005). Evolution of diverse antiviral CD8$^+$ T cell populations after murine cytomegalovirus infection. *Eur. J. Immunol.* **35**, 1113–1123.

Simon, C. O., Holtappels, R., Tervo, H. M., Bohm, V., Daubner, T., Oehrlein-Karpi, S. A., Kuhnapfel, B., Renzaho, A., Strand, D., Podlech, J., Reddehase, M. J., and Grzimek, N. K. (2006). CD8 T cells control cytomegalovirus latency by epitope-specific sensing of transcriptional reactivation. *J. Virol.* **80**, 10436–10456.

Smith, C. M., Wilson, N. S., Waithman, J., Villadangos, J. A., Carbone, F. R., Heath, W. R., and Belz, G. T. (2004). Cognate CD4$^+$ T cell licensing of dendritic cells in CD8$^+$ T cell immunity. *Nat. Immunol.* **5**, 1143–1148.

Stock, A. T., Jones, C. M., Heath, W. R., and Carbone, F. R. (2006). Cutting edge: Central memory T cells do not show accelerated proliferation or tissue infiltration in response to localized herpes simplex virus-1 infection. *J. Immunol.* **177**, 1411–1415.

Sullivan, B. M., Juedes, A., Szabo, S. J., von Herrath, M., and Glimcher, L. H. (2003). Antigen-driven effector CD8 T cell function regulated by T-bet. *Proc. Natl. Acad. Sci. USA* **100**, 15818–15823.

Sun, J. C., and Bevan, M. J. (2003). Defective CD8 T cell memory following acute infection without CD4 T cell help. *Science* **300**, 339–342.

Tan, J. T., Ernst, B., Kieper, W. C., LeRoy, E., Sprent, J., and Surh, C. D. (2002). Interleukin (IL)-15 and IL-7 jointly regulate homeostatic proliferation of memory phenotype CD8$^+$ cells but are not required for memory phenotype CD4$^+$ cells. *J. Exp. Med.* **195**, 1523–1532.

Tanaka, T., Soriano, M. A., and Grusby, M. J. (2005). SLIM is a nuclear ubiquitin E3 ligase that negatively regulates STAT signaling. *Immunity* **22**, 729–736.

Taraban, V. Y., Rowley, T. F., and Al-Shamkhani, A. (2004). Cutting edge: A critical role for CD70 in CD8 T cell priming by CD40-licensed APCs. *J. Immunol.* **173**, 6542–6546.

Taraban, V. Y., Rowley, T. F., Tough, D. F., and Al-Shamkhani, A. (2006). Requirement for CD70 in CD4$^+$ Th cell dependent and innate receptor-mediated CD8$^+$ T cell priming. *J. Immunol.* **177**, 2969–2975.

Tau, G. Z., von der Weid, T., Lu, B., Cowan, S., Kvatyuk, M., Pernis, A., Cattoretti, G., Braunstein, N. S., Coffman, R. L., and Rothman, P. B. (2000). Interferon gamma signaling alters the function of T helper type 1 cells. *J. Exp. Med.* **192**, 977–986.

Tau, G. Z., Cowan, S. N., Weisburg, J., Braunstein, N. S., and Rothman, P. B. (2001). Regulation of IFN-gamma signaling is essential for the cytotoxic activity of CD8$^+$ T cells. *J. Immunol.* **167**, 5574–5582.

Tenoever, B. R., Ng, S. L., Chua, M. A., McWhirter, S. M., Garcia-Sastre, A., and Maniatis, T. (2007). Multiple functions of the IKK-related kinase IKKepsilon in interferon-mediated antiviral immunity. *Science* **315,** 1274–1278.

Tesselaar, K., Arens, R., van Schijndel, G. M., Baars, P. A., van der Valk, M. A., Borst, J., van Oers, M. H., and van Lier, R. A. (2003). Lethal T cell immunodeficiency induced by chronic costimulation via CD27-CD70 interactions. *Nat. Immunol.* **4,** 49–54.

Thimme, R., Appay, V., Koschella, M., Panther, E., Roth, E., Hislop, A. D., Rickinson, A. B., Rowland-Jones, S. L., Blum, H. E., and Pircher, H. (2005). Increased expression of the NK cell receptor KLRG1 by virus-specific CD8 T cells during persistent antigen stimulation. *J. Virol.* **79,** 12112–12116.

Thompson, L. J., Kolumam, G. A., Thomas, S., and Murali-Krishna, K. (2006). Innate inflammatory signals induced by various pathogens differentially dictate the IFN-I dependence of CD8 T cells for clonal expansion and memory formation. *J. Immunol.* **177,** 1746–1754.

Townsend, M. J., Weinmann, A. S., Matsuda, J. L., Salomon, R., Farnham, P. J., Biron, C. A., Gapin, L., and Glimcher, L. H. (2004). T-bet regulates the terminal maturation and homeostasis of NK and Valpha14i NKT cells. *Immunity* **20,** 477–494.

Turner, C. A., Jr., Mack, D. H., and Davis, M. M. (1994). Blimp-1, a novel zinc finger-containing protein that can drive the maturation of B lymphocytes into immunoglobulin-secreting cells. *Cell* **77,** 297–306.

van der Lugt, N. M., Domen, J., Linders, K., van Roon, M., Robanus-Maandag, E., te Riele, H., van der Valk, M., Deschamps, J., Sofroniew, M., van Lohuizen, M., and Berns, A. (1994). Posterior transformation, neurological abnormalities, and severe hematopoietic defects in mice with a targeted deletion of the *bmi-1* proto-oncogene. *Genes Dev.* **8,** 757–769.

van Lint, A. L., Kleinert, L., Clarke, S. R., Stock, A., Heath, W. R., and Carbone, F. R. (2005). Latent infection with herpes simplex virus is associated with ongoing $CD8^+$ T-cell stimulation by parenchymal cells within sensory ganglia. *J. Virol.* **79,** 14843–14851.

van Lohuizen, M., Verbeek, S., Scheijen, B., Wientjens, E., van der Gulden, H., and Berns, A. (1991). Identification of cooperating oncogenes in E mu-myc transgenic mice by provirus tagging. *Cell* **65,** 737–752.

Vezys, V., Masopust, D., Kemball, C. C., Barber, D. L., O'Mara, L. A., Larsen, C. P., Pearson, T. C., Ahmed, R., and Lukacher, A. E. (2006). Continuous recruitment of naive T cells contributes to heterogeneity of antiviral CD8 T cells during persistent infection. *J. Exp. Med.* **203,** 2263–2269.

Voehringer, D., Blaser, C., Brawand, P., Raulet, D. H., Hanke, T., and Pircher, H. (2001). Viral infections induce abundant numbers of senescent CD8 T cells. *J. Immunol.* **167,** 4838–4843.

Voehringer, D., Koschella, M., and Pircher, H. (2002). Lack of proliferative capacity of human effector and memory T cells expressing killer cell lectinlike receptor G1 (KLRG1). *Blood* **100,** 3698–3702.

Way, S. S., and Wilson, C. B. (2004). Cutting edge: Immunity and IFN-gamma production during Listeria monocytogenes infection in the absence of T-bet. *J. Immunol.* **173,** 5918–5922.

Weekes, M. P., Carmichael, A. J., Wills, M. R., Mynard, K., and Sissons, J. G. (1999). Human $CD28-CD8^+$ T cells contain greatly expanded functional virus-specific memory CTL clones. *J. Immunol.* **162,** 7569–7577.

Wherry, E. J., Teichgraber, V., Becker, T. C., Masopust, D., Kaech, S. M., Antia, R., von Andrian, U. H., and Ahmed, R. (2003). Lineage relationship and protective immunity of memory CD8 T cell subsets. *Nat. Immunol.* **4,** 225–234.

Whitmire, J. K., Tan, J. T., and Whitton, J. L. (2005). Interferon-gamma acts directly on $CD8^+$ T cells to increase their abundance during virus infection. *J. Exp. Med.* **201,** 1053–1059.

Williams, M. A., Tyznik, A. J., and Bevan, M. J. (2006). Interleukin-2 signals during priming are required for secondary expansion of $CD8^+$ memory T cells. *Nature* **441,** 890–893.

Willimsky, G., and Blankenstein, T. (2005). Sporadic immunogenic tumours avoid destruction by inducing T-cell tolerance. *Nature* **437,** 141–146.

Wong, P., and Pamer, E. G. (2001). Cutting edge: Antigen-independent CD8 T cell proliferation. *J. Immunol.* **166,** 5864–5868.

Yamada, A., Salama, A. D., Sho, M., Najafian, N., Ito, T., Forman, J. P., Kewalramani, R., Sandner, S., Harada, H., Clarkson, M. R., Mandelbrot, D. A., Sharpe, A. H., *et al.* (2005). CD70 signaling is critical for CD28-independent CD8$^+$ T cell-mediated alloimmune responses *in vivo*. *J. Immunol.* **174,** 1357–1364.

Ye, B. H., Cattoretti, G., Shen, Q., Zhang, J., Hawe, N., de Waard, R., Leung, C., Nouri-Shirazi, M., Orazi, A., Chaganti, R. S., Rothman, P., Stall, A. M., *et al.* (1997). The BCL-6 proto-oncogene controls germinal-centre formation and Th2-type inflammation. *Nat. Genet.* **16,** 161–170.

Yu, A., Zhou, J., Marten, N., Bergmann, C. C., Mammolenti, M., Levy, R. B., and Malek, T. R. (2003). Efficient induction of primary and secondary T cell-dependent immune responses in vivo in the absence of functional IL-2 and IL-15 receptors. *J. Immunol.* **170,** 236–242.

Zhang, P., Iwasaki-Arai, J., Iwasaki, H., Fenyus, M. L., Dayaram, T., Owens, B. M., Shigematsu, H., Levantini, E., Huettner, C. S., Lekstrom-Himes, J. A., Akashi, K., and Tenen, D. G. (2004). Enhancement of hematopoietic stem cell repopulating capacity and self-renewal in the absence of the transcription factor C/EBP alpha. *Immunity* **21,** 853–863.

CHAPTER 4

Inherited Complement Regulatory Protein Deficiency Predisposes to Human Disease in Acute Injury and Chronic Inflammatory States

The Examples of Vascular Damage in Atypical Hemolytic Uremic Syndrome and Debris Accumulation in Age-Related Macular Degeneration

Anna Richards, David Kavanagh, and **John P. Atkinson**

Contents		
	1. Altered Self Triggers Innate Immunity	143
	1.1. Acute injury	143
	1.2. Debris accumulation	144
	2. Regulation of the Alternative Complement Pathway	145
	2.1. Overview of activation	145
	2.2. Regulation of the alternative complement pathway	148
	2.3. Regulatory proteins	148

Washington University School of Medicine, St. Louis, Missouri

3. Lessons from Homozygous Complement Regulatory
   Protein Deficiencies   153
   3.1. Plasma proteins FH and FI   153
   3.2. Membrane proteins MCP and Crry   153
4. Complement and Atypical Hemolytic
   Uremic Syndrome   154
   4.1. Hemolytic uremic syndrome   154
   4.2. Factor H   155
   4.3. FH-related genes   157
   4.4. Membrane cofactor protein: CD46   157
   4.5. Factor I   159
   4.6. Factor B   159
   4.7. Disease penetrance   160
5. Complement and Age-Related Macular Degeneration   160
   5.1. Age-related macular degeneration   160
   5.2. Factor H   161
   5.3. Factor B/C2   163
6. Immunopathogenesis of aHUS and AMD   163
   6.1. Atypical HUS   163
   6.2. Age-related macular degeneration   166
7. Treatment of aHUS and AMD   167
   7.1. Treatment options for aHUS   167
   7.2. Treatment options for AMD   168
8. Conclusions: Lessons and Implications   168
References   169

Abstract

In this chapter, we examine the role of complement regulatory activity in atypical hemolytic uremic syndrome (aHUS) and age-related macular degeneration (AMD). These diseases are representative of two distinct types of complement-mediated injury, one being acute and self-limited, the other reflecting accumulation of chronic damage. Neither condition was previously thought to have a pathologic relationship to the immune system. However, alterations in complement regulatory protein genes have now been identified as major predisposing factors for the development of both diseases. In aHUS, heterozygous mutations leading to haploinsufficiency and function-altering polymorphisms in complement regulators have been identified, while in AMD, polymorphic haplotypes in complement genes are associated with development of disease. The basic premise is that a loss of function in a plasma or membrane inhibitor of the alternative complement pathway allows for excessive activation of complement on the endothelium of the kidney in aHUS and on retinal debris in AMD.

These associations have much to teach us about the host's innate immune response to acute injury and to chronic debris deposition. We all experience cellular injury and, if we live long enough, will deposit debris in blood vessel walls (atherosclerosis leading to heart attacks and strokes), the brain (amyloid proteins leading to Alzheimer's disease), and retina (lipofuscin pigments leading to AMD). These are three common causes of morbidity and mortality in the developed world. The clinical, genetic, and immunopathologic understandings derived from the two examples of aHUS and AMD may illustrate what to anticipate in related conditions. They highlight how a powerful recognition and effector system, the alternative complement pathway, reacts to altered self. A response to acute injury or chronic debris accumulation must be appropriately balanced. In either case, too much activation or too little regulation promotes undesirable tissue damage and human disease.

1. ALTERED SELF TRIGGERS INNATE IMMUNITY

1.1. Acute injury

The raison d'être for the development of the immune system is to protect the host from microbes (Barilla-LaBarca and Atkinson, 2003; Walport, 2001a,b). Its goal is the destruction of a foreign antigen and facilitation of an adaptive immune response. Individuals with deficient immune systems have as the primary consequence, increased infections. Complete deficiency of C3 leads to severe, recurrent, life-threatening, pyogenic infections with encapsulated bacteria (Reis *et al.*, 2006). Complete deficiency of membrane attack components (Figueroa *et al.*, 1993) or of the alternative pathway (AP) components (Sprong *et al.*, 2006) predisposes to recurrent Neisserial infections.

However, early on in the characterization of total complement component deficiency states in man came the surprising observation that individuals deficient in early components of the classical pathway (CP), for example, C1q, C4, or C2, develop autoimmunity, especially systemic lupus erythematosus (SLE) (Manderson *et al.*, 2004). Approximately 90% of C1q and 80% of C4-deficient individuals present with SLE. In attempts to explain this association, investigators initially focused on a failure of immune-complex clearance (Atkinson, 1986; Walport and Lachmann, 1988) and then on inappropriate handling of self-antigens (Pickering *et al.*, 2000).

Much has been discovered about the role of innate immunity and especially the complement system's response to apoptosis and ischemia-reperfusion injury (IRI) (Carroll and Holers, 2005; Gershov *et al.*, 2000; Kim *et al.*, 2003; Mevorach *et al.*, 1998; Navratil *et al.*, 2006; Pickering *et al.*, 2000; Stahl *et al.*, 2003). These topics have been covered in this series (Carroll and Holers, 2005; Pickering *et al.*, 2000). IRI and apoptosis represent conditions in which the immune system sees "altered self." In animal models, the complement system contributes to the magnitude of the final injury. The interplay of lectins, natural antibodies, and the AP appears to vary from species to species and from tissue to tissue (Carroll and Holers, 2005; Gershov et al., 2000; Kim et al., 2003; Mevorach et al., 1998; Navratil et al., 2006).

IRI is a major concern following strokes and myocardial infarction, but the host's reaction system likely evolved primarily for the repair of cutaneous trauma. The critical goals were to prevent an infection, to dispose of apoptotic and necrotic cells and damaged tissue and, in these and other ways, to assist in wound repair. The size of the eventual scar was not the key issue. In contrast, a goal in the current treatment of strokes and myocardial infarctions is to minimize tissue damage caused by complement activation and other players in innate immunity.

Apoptosis is the term used to describe the programmed death of cells. Billions of cells, particularly in the bone marrow, become apoptotic on a daily basis and the erythrocyte lineage extrudes nuclear material as part of its maturation process. The adaptive immune system usually responds minimally to this form of altered self. In SLE, however, individuals develop high titer antinuclear antibodies, and many of the antigens recognized by these antibodies are exposed during apoptosis. This source of antigen and the high frequency of individuals with a complete deficiency of C1q or C4 developing SLE make for an attractive hypothesis to explain this autoimmune disease; namely, that the complement system assists in the removal of altered self and this process is disturbed in autoimmunity.

These two situations therefore have distinct end points: in the first, destruction of a foreign antigen and facilitation of an adaptive immune system are the goals; in the second, it is the safe removal of antigenic material with avoidance of an adaptive immune response. These examples are pertinent to the discussion to follow, as the balance between activation and regulation of the complement system underlies the pathogenesis of atypical hemolytic uremic syndrome (aHUS).

1.2. Debris accumulation

The second type of altered self to be analyzed in this discussion is the deposition of "garbage" in vital tissue. The average life expectancy in most of the developed world is now approaching 80 years. An unfortunate consequence (but consider the alternative) is the development of chronic diseases featuring debris deposition and accumulation: examples being urate crystals in joints (gout), lipids in large vessel walls (atherosclerosis), amyloid protein in the brain (Alzheimer's), and lipofuscin pigment in the retina (age-related macular degeneration; AMD). In each, there is a collection in critical tissues of altered and variably proinflammatory debris derived from self. The respective consequences are immobility, heart attacks and strokes, dementia, and visual loss. Formation of urate crystals in a joint can lead to an acute inflammatory ("like a pyogenic bacterial infection") gouty attack, but in chronic tophaceous gout the innate immune response tends to be low grade. In the other three conditions, the debris tends to be less inflammatory. In these four situations, waste accumulates in vital organs and an inflammatory response plays out over decades. Urate crystals, oxidized lipids, amyloid proteins, and lipofuscin pigments interact with the innate immune system. For example, fragments of C3 as well as many other participants in innate immunity coat such substances. For several reasons, until recently, such data from immunohistochemistry analyses have been largely ignored. First, it was correctly envisioned that complement activation was not likely to be the primary cause of the process. Second, many considered the presence of C3 fragments as representing innate immune markers; in other words, a bystander process. The discovery in 2005 that $\sim$50% of the genetic risk in AMD was related to a polymorphic variation in a complement regulatory protein led to a substantial reevaluation of the role of innate immunity in AMD.

A simple hypothesis for both aHUS and AMD is that the innate immune system is responding to altered self tissue. If there is excessive activation of the AP, it predisposes to and accelerates disease development. In aHUS and AMD, we now have genetic analyses to firmly establish a critical role for innate immunity in pathogenesis. Consequently, they will be used as the key examples in this discussion to point out a specific role for a deficiency of regulation of the AP of complement in disease causation.

2. REGULATION OF THE ALTERNATIVE COMPLEMENT PATHWAY

2.1. Overview of activation

The involvement of the AP in animal models of immunologic disease and in clinical medicine has been authoritatively reviewed by Thurman and Holers (2006). Here, we will outline how this pathway is activated and particularly focus on its feedback loop whose goal is to deposit additional C3b on a target (Fig. 4.1).

An ancient complement system probably consisted of C3-like protein along with a receptor for an "activated" form of this protein (Lambris, 1989). Related to the ancient α-2 macroglobulin family, C3 contains an internal thioester bond that, upon rupture, forms an ester linkage to a hydroxyl group on a target. The secreted C3, because of this unstable thioester bond (1–2% turnover per hour), spontaneously "ticks" over.

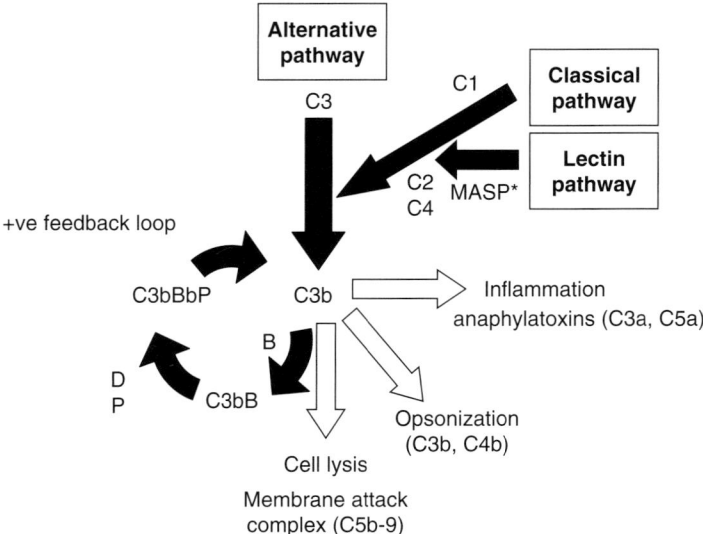

FIGURE 4.1 Positive feedback amplification loop of the alternative pathway (AP) of complement. The complement system has three activation pathways: the classical pathway (CP), lectin pathway (LP), and AP (dark arrows). The CP is activated by the binding of C1q subcomponent of C1 to IgM or IgG complexed with antigen. The LP is activated by the binding of lectins to repeating sugar motifs, present on the surface of many pathogens. The AP does not have a specific trigger but undergoes tick over or can be primed by the CP or LP. The C3b that is formed interacts with factor B (FB; B) which is then cleaved by factor D (FD; D) to form the AP C3 convertase (C3bBb). This enzyme complex is attached to the target covalently via C3b while Bb is the catalytic serine protease subunit. It is stabilized by binding properdin (P). Because C3 is the substrate for this convertase, a powerful feedback loop is created. *Mannan binding associated serine protease (C1 equivalent of the LP) (MASP).

One could envision its role being to attach to foodstuffs or microbes or activated proteases. To capture such targets, a recognition protein (receptor) on host cells was required. Such a system could operate for a single cell organism.

Subsequent evolutionary developments of the complement system can be viewed as providing a means to (1) amplify the process (put more C3b on target), and (2) increase the specificity and efficiency of the targeting process. Two such evolutionary developments were a cascade of proteases whose goals were to amplify the C3 activation process and more specific triggering mechanisms. The latter were antibodies in the case of the CP and lectins in the case of the lectin pathway (LP). In addition, the release of small proinflammatory peptides (C3a and C5a) and a lytic cascade (C5b-9) provided additional ingredients for an enhanced innate immune response.

For the alterative pathway, which does not have a specific trigger analogous to the CP or LP, the cascade employs a feedback loop. The target-bound C3b interacts with a 100-kDa zymogen serine protease of plasma known as factor B (FB) (Fig. 4.1). FB can then be activated by a small (25 kDa) plasma protease factor D (FD). FD cleaves FB into two fragments, Bb + Ba. The Bb piece remains bound to C3b to form the AP C3 convertase (C3bBb). This represents a powerful feedback loop because native C3 is the substrate for this heterodimeric enzyme. Newly generated C3bs also bind to the target and then form more C3 convertases. C3b anchors the heterodimeric protease to the membrane and Bb is the catalytic domain of the C3 convertase. This feedback loop can deposit several million C3bs on bacteria or a human cell in <2 min. Properdin (P) stabilizes the AP C3 convertase, extending its half-life from 20–30 s to 3–4 min. *In vivo*, it is probably required for efficient convertase activity.

A commonly asked question regarding the AP activation process relates to how the initial activated C3 is generated. There are several possibilities: (1) CP activation by antigen/antibody complexes; (2) LP activation via sugar/lectin complexes; or (3) spontaneous, continuous "tick over" of C3 to generate iC3 (C3 with a cleaved thioester bond which is analogous to C3b). In the most accepted rendition, originally worked out by groups led by Frank Austen and Hans Muller Eberhard, the newly activated C3 (C3 with a cleaved thioester bond) forms a transient fluid phase convertase to generate C3b (Fearon, 1979; Pangburn and Muller-Eberhard, 1984). Another likely mechanism is via the turnover process with direct binding to the target. Hourcade has recently pointed out a mechanism whereby P binds to a target and then interacts with spontaneously generated iC3 or C3b generated via the fluid phase convertase (Hourcade, 2006; Spitzer *et al.*, 2007). Once a small amount of activated C3 becomes bound to a target, it is then amplified via the feedback loop. The CP and LP accomplish C3 activation by proteolytic cleavage of C3 to C3b and C3a. The newly generated nascent C3b has a few microseconds to attach to a nearby target before it is hydrolyzed. This attachment mechanism largely restricts complement activation in time and space to the membrane on which the initial C3b is covalently bound.

In addition to instability of the thioester (Law *et al.*, 1979) and the fluid phase convertase of the AP convertase formation (Fearon, 1979; Pangburn and Muller-Eberhard, 1984), other evidence supports this C3 tick over concept. Upon purifying human albumin, a small fraction was noted to run on gels at ~100 kDa. It represented a C3d fragment covalently bound to albumin (Atkinson *et al.*, 1988). C3d is a small degradation fragment of C3b that remains attached to a target (immunologic scar) since it contains the site of thioester bond. Second, during their 120-day life span, human RBCs acquire fragments of C4d and C3d that are covalently attached to RBC membrane proteins. In the case of C4, which is an evolutionary cousin of C3 and also possesses a thioester bond, this C4d fragment is the basis for the Chido-Rodgers blood group antigen system because the two C4 genes are highly polymorphic (Atkinson *et al.*, 1988; Giles, 1988). Finally, in an observation that goes back over 100 years, serum left on bench top overnight no longer possesses complement functional (lytic) activity. C4 and C3 have turned over secondary to spontaneous activation of their thioester bond. The C3 and C4 proteins are present but are nonfunctional.

2.2. Regulation of the alternative complement pathway

Regulation of the AP C3 convertase is by two processes, decay accelerating activity (DAA) and cofactor activity (CA) (Fig. 4.2). The former refers to the dissociation of the convertase. Specifically, the decay accelerator protein displaces the catalytic Bb from the target-bound C3b. However and critical to this discussion, this C3b can bind another FB and then reform the convertase. To prevent this, C3b is inactivated by limited proteolytic cleavage (Fig. 4.2). This CA requires two participants—a cofactor protein that binds to the C3b and a protease that then cleaves the C3b. Such cleaved C3b, iC3b, is *no longer* capable of binding FB and thus cannot form the AP convertase. DAA and CA are synergistic in controlling the AP C3 convertase (Brodbeck *et al.*, 2000). Host regulatory proteins with DAA and CA for the AP C3 convertase are abundant in plasma and on cell surfaces. While the division of labor is different on cells versus the fluid phase, the composite functional repertoire is the same. The plasma protein factor H (FH) binds C3b and has both DAA and CA for the AP C3 convertase. On cells, membrane cofactor protein (MCP, CD46) has CA for C3b while decay accelerating factor (DAF, CD55) possesses disassociating activity for the AP C3 convertase. The same plasma serine protease, factor I (FI), mediates proteolytic cleavage of C3b by either FH or MCP. DAF and MCP are called intrinsic inhibitors because they only "work" on C3b and C3 convertases bound to the *same* cell on which they are expressed (Medof *et al.*, 1984; Oglesby *et al.*, 1992).

2.3. Regulatory proteins

2.3.1. Factor H (CFH)

FH is an abundant 150-kDa plasma protein synthesized by the liver (Vik *et al.*, 1990). It consists solely of 20 contiguous modules called complement control protein (CCP) repeats (Fig. 4.3). These modules contain ~60 amino acids with four invariant cysteines and 10–15 other highly conserved residues (Barlow *et al.*, 1991). The first four repeats of FH (CCPs 1–4) possess the major C3b binding site and the *only* cofactor site (Alexander and Quigg, 2007). As shown in Fig. 4.3, additional C3 fragment binding

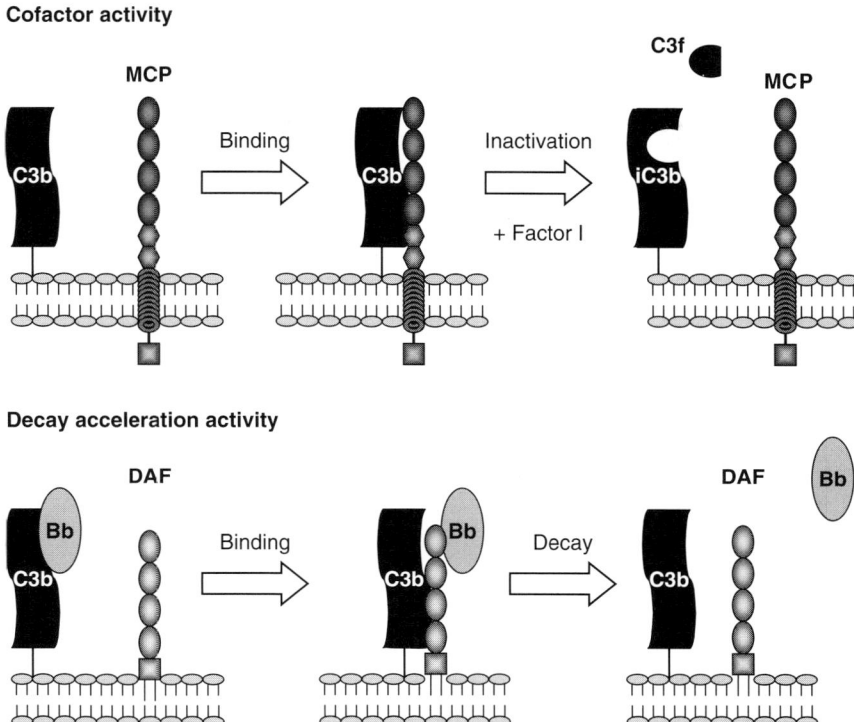

FIGURE 4.2 Regulation of the alternative pathway (AP) by cofactor activity (CA) and decay accelerating activating (DAA). CA results in the permanent inactivation of C3b to iC3b, such that it is no longer capable of binding FB and thus cannot form the AP convertase. CA requires both a cofactor protein and a protease, factor I (FI). MCP is the cofactor protein in this example. It is the deficiency of CA, either the cofactor protein (MCP or FH) or the protease FI, which predisposes an individual to aHUS. DAA is the dissociation of the C3/C5 convertases. Specifically, the decay accelerator protein, in this example DAF, displaces the catalytic Bb from target-bound C3b. However, this C3b can bind another FB and then reform the convertase.

sites are situated along this linear protein. Another major biologic activity scattered among the repeats are heparin (anionic) binding sites, with a major one in CCPs 19 and 20. These positively charged basic amino acid-rich binding sites allow this protein to attach to negatively charged, acidic extracellular tissues, such as matrixes and basement membranes, where they prevent these acellular materials from being attacked by the complement system. Healthy cells are protected by the constitutively expressed DAF, MCP, and an inhibitor of the membrane attack complex, CD59. However, in the setting of trauma, apoptosis, or necrosis, cell membranes may be damaged, turned inside out, or destroyed so that the outward

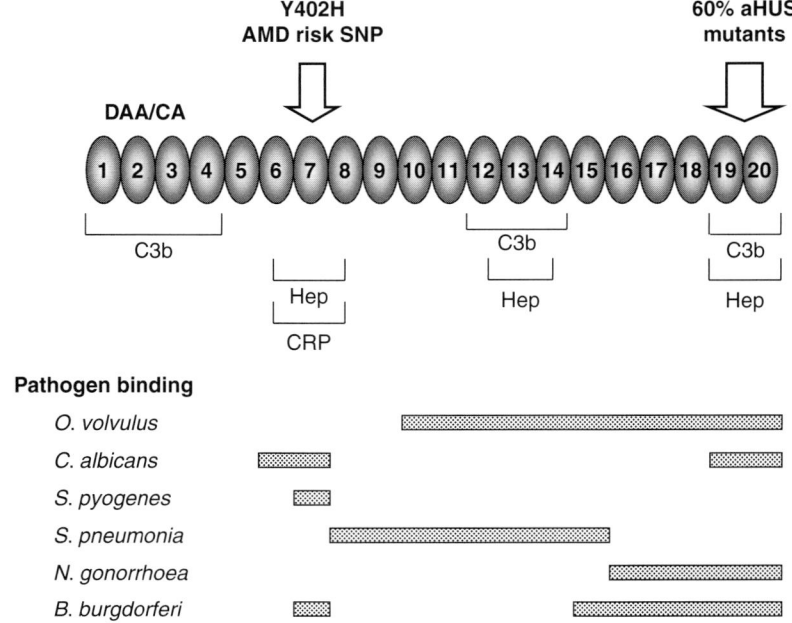

FIGURE 4.3 Functional domains of factor H (FH) and disease-associated genetic changes. FH is made up of 20 CCP modules. The binding sites for CRP, heparin (polyanions), and C3b are shown. The binding sites of microorganisms are shown by shaded boxes. The location of the AMD-associated SNP and the hotspot for aHUS-associated mutations are demonstrated by open arrows.

facing membrane proteins are no longer in place to guard against complement attack. Through binding of FH to such substances, activation by the complement system is controlled. As might be anticipated, multiple pathogens have evolved binding sites for FH in order to protect themselves from AP activation (Fig. 4.3).

2.3.2. Membrane cofactor protein

MCP (CD46) is a widely expressed membrane inhibitor of complement activation (Hourcade et al., 1989; Liszewski et al., 1996). It is a 55–65 kDa, type 1 transmembrane protein that binds C3b and C4b (Fig. 4.4). Upon such binding, FI can cleave C3b and C4b. The resulting fragments iC3b and C4d are not capable of forming convertases. This is particularly critical for regulation of the AP C3 convertase because C3b (but not iC3b) can bind FB and then reform the convertase. MCP is expressed on most cell types, as four isoforms, which vary in their juxtamembraneous O-linked sugar domain and cytoplasmic tail. The two distinct cytoplasmic tails possess CK and Src kinase sites which are phosphorylated upon MCP

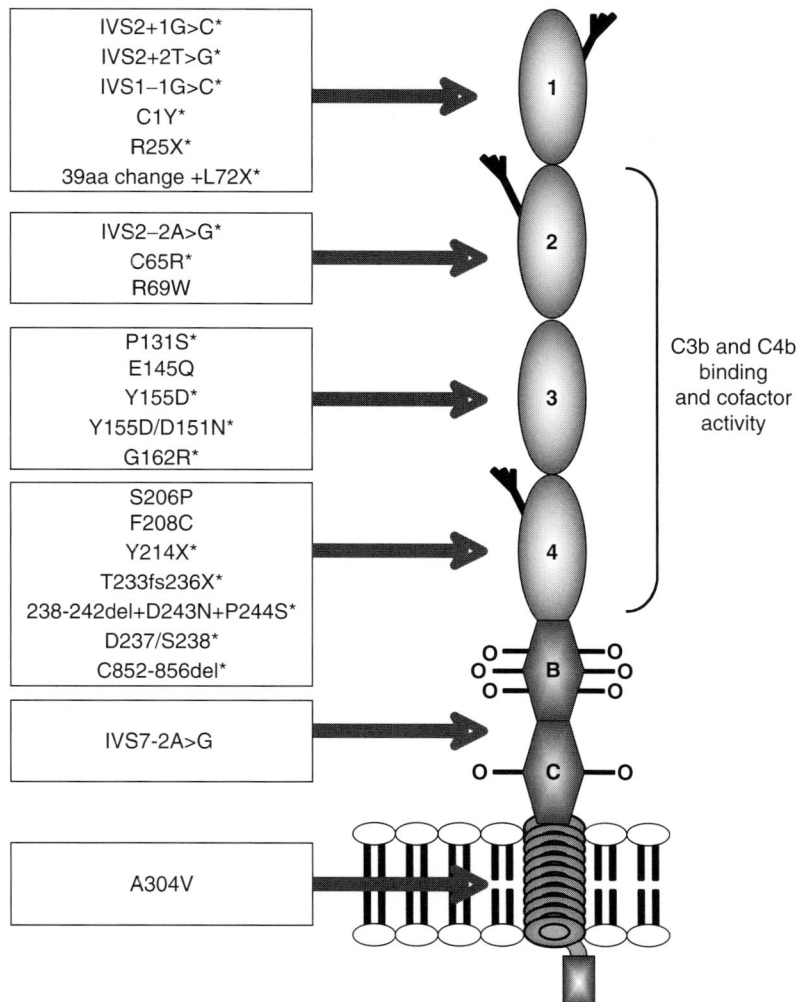

FIGURE 4.4 Diagram of membrane cofactor protein (MCP, CD46) structure and location of the initial mutations identified in aHUS. MCP is an ~65-kDa type 1 transmembrane glycoprotein. Beginning at the N-terminus, it consists of four ~60 amino acid complement control repeats (CCPs). CCPs 1, 2, and 4 each contain one N-glycosylation site. Following the CCPs is an alternatively spliced region, rich in serine, threonine, and proline (STP) that contain sites for O-glycosylation. The STP region is followed by a group of 12 amino acids of unknown function, a hydrophobic domain, a charged transmembrane anchor, and the alternatively spliced cytoplasmic tail (tail 1 or 2). The MCP-BC isoform is shown. Mutations associated with aHUS are clustered in the four extracellular CCPs of the molecule. Mutations associated with reduced expression levels are marked by an asterisk. (See Plate 7 in Color Plate Section.)

cross-linking (Kemper and Atkinson, 2007). In addition to its intrinsic complement regulatory activity, MCP is involved in sperm–egg interactions, being expressed on the inner acrosomal membrane (Riley-Vargas and Atkinson, 2003; Riley-Vargas et al., 2004, 2005). Cross-linking CD3 and CD46 on naïve human T cells leads to a T regulatory phenotype as evidenced by proliferation, IL-10 secretion, and granzyme B synthesis (Kemper and Atkinson, 2007). Finally, CD46 is a microbial magnet as eight human pathogens interact with this protein (Cattaneo, 2004; Gill and Atkinson, 2004; Liszewski et al., 2005; Riley-Vargas et al., 2004).

2.3.3. Factor I

FI is a two-chain serine protease of plasma (Fig. 4.5) that cleaves C3b and C4b but only in conjunction with a cofactor protein (Fearon, 1979; Goldberger et al., 1987). The gene encoding FI is at 4q25 and its genome covers 63 kb with 13 exons. FI is synthesized in the liver as a single chain precursor and is then proteolytically processed into a disulfide-linked heavy chain of 55 kDa and light chain of 38 kDa. The light chain consists of a typical catalytic serine protease domain while the heavy chain contains multiple modules whose functions are obscure. FI's serum concentration is 39–100 μg/ml (de Paula et al., 2003). It is present in most body secretions at ~10% of the serum level. This may derive from blood or be

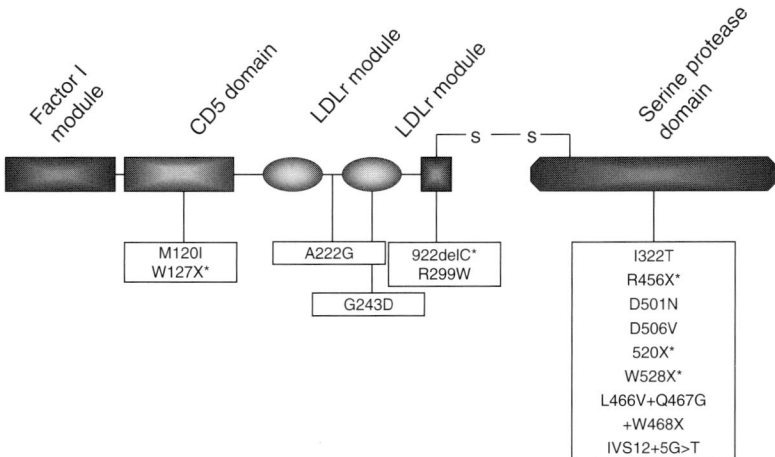

FIGURE 4.5 Diagram of factor I (FI) structure and location of the initial mutations identified in aHUS. Domains are illustrated by shaded figures and linked by nondomain regions (solid line). The disulfide bond linking heavy and light chains between C^{309} and C^{435} are shown. The position of CFI mutations associated with aHUS examined in this study are noted. The D-segment resides between LDLr domain and the serine protease domain. Asterisk denotes mutations causing reduced expression.

synthesized locally since multiple cell types including endothelial cells, monocytes/macrophages, and some epithelial cells synthesize FI (Dauchel *et al.*, 1990; Vyse *et al.*, 1996; Whaley, 1980).

3. LESSONS FROM HOMOZYGOUS COMPLEMENT REGULATORY PROTEIN DEFICIENCIES

3.1. Plasma proteins FH and FI

Homozygous FH deficiency has been described for humans (Ault *et al.*, 1997; Dragon-Durey *et al.*, 2004), the Norwegian Yorkshire pig (Hogasen *et al.*, 1995), and FH knockout mouse model (Pickering *et al.*, 2002). In man, there is an increased frequency of infection with encapsulated organisms and an association with the development of membranoproliferative glomerulonephritis type II (MPGN II). The pig and mouse models also develop MPGN II.

Over 30 families with complete FI deficiency have been described (Kavanagh *et al.*, 2005; Reis *et al.*, 2006). The clinical manifestations of increased susceptibility to recurrent infection with encapsulated microorganisms are present from early childhood. Two patients with complete FI deficiency have been reported to have renal disease. One patient had serological evidence of SLE and diffuses proliferative glomerulonephritis on renal biopsy (Amadei *et al.*, 2001). The other patient presented with a multisystem inflammatory disorder characterized by hepatitis, pneumonitis, myositis, and a microangiopathic vasculitis. This patient subsequently developed focal segmental glomerulosclerosis (Sadallah *et al.*, 1999).

A complete deficiency of either FH or FI therefore leads to uncontrolled activation of the amplification loop of the AP. Without fluid phase CA, there is spontaneous, unregulated C3 turnover to the point of exhaustion in plasma, such that initially, these patients were thought to have a primary C3 deficiency. In the FI-deficient patients, membrane CA mediated by MCP is also lacking. The secondary deficiency of C3 leads to a defect in opsonization, immune adherence, and phagocytosis, explaining the predisposition to infection with encapsulated organisms. These observations also support the concept of continuous AP-mediated low-grade turnover of C3 in the fluid phase and explain the finding by Chester Alper and colleagues three decades ago that infused plasma (containing donor FH and FI) was able to transiently correct the excessive C3 turnover (Ziegler *et al.*, 1975).

3.2. Membrane proteins MCP and Crry

In the mouse and rat, MCP is only expressed on the inner acrosomal membrane of spermatozoa (Riley-Vargas and Atkinson, 2003; Riley-Vargas *et al.*, 2005). This stands in contrast to its nearly ubiquitous expression pattern in other mammals including man. To replace the missing CA carried by MCP in these rodents, a related complement inhibitor known as Crry is expressed (Molina, 2002; Wong and Fearon, 1985). Molina knocked out this gene in the mouse and the result was embryonic lethality on ~day 7 (Xu et al., 2000). Immunohistochemistry demonstrated C3 fragments densely coating the developing placenta but not the maternally derived decidua. C3 deficiency in the mother rescued the phenotype, conclusively establishing that the maternal complement system was attacking the developing placenta to cause the mortality. More recently, deficiency of AP component FB has been shown to rescue the phenotype (X. Wu and J. A., unpublished data). Thus, in the mouse, Crry is absolutely essential to protect the embryo-derived placental trophoblast from attack by the mother's AP. This result, along with a substantial body of confirmatory evidence, establishes that all cells/tissues must have protection from the constantly turning over AP. It also points out that the functions of the plasma protein FH and Crry (or MCP) are not overlapping. FH is normal in these ($Crry^{-/-}$) mice and it did not protect the placenta (Xu et al., 2000). Likewise, Crry does not prevent fluid phase C3 turnover in $FH^{-/-}$ mice (Pickering et al., 2000).

Because of these data, a complete deficiency of MCP in man was conjectured to be embryonic lethal. However, two individuals were described with biallelic mutations and no detectable MCP on their peripheral blood cells (Fremeaux-Bacchi et al., 2006). Unfortunately, there is limited clinical data on both patients.

4. COMPLEMENT AND ATYPICAL HEMOLYTIC UREMIC SYNDROME

4.1. Hemolytic uremic syndrome

Hemolytic uremic syndrome (HUS) is a clinical triad of hemolytic anemia, thrombocytopenia, and acute renal failure. It is characterized by endothelial cell injury in the microvasculature of the kidney, and the host's subsequent innate immune response to damaged tissue (Kavanagh et al., 2008a). There are two main subtypes, diarrheal-associated/epidemic HUS (D+HUS) or nondiarrheal-associated or atypical HUS. In the epidemic form of HUS, a shiga toxin (Stx most typically derived from *Escherichia coli* O157:H7) mediates the damage to glomerular endothelial cells of the kidney (Tarr et al., 2005). No underlying genetic factors have been described to account for the 10–15% of individuals (mostly children) in these epidemics who develop HUS. In aHUS, other types of infections, for example, *Streptococcus pneumoniae*, drugs (e.g., mitomycin, quinine), and radiation act as the trigger resulting in endothelial cell injury

(Kavanagh et al., 2006), but there is usually no preceding diarrhea. aHUS is a disease with a 25% acute mortality and 50% develop end-stage renal disease (Noris and Remuzzi, 2005). In aHUS, haploinsufficiency of a complement regulatory protein of the AP predisposes to the disease, as do activating mutations in FB.

4.2. Factor H

A connection between mutations in FH and aHUS was first made by Warwicker et al. (1998). Using a candidate gene approach, they demonstrated linkage to 1q32, the site of the regulators of complement activation (RCA) gene cluster (Fig. 4.6), in three large pedigrees, and subsequently identified mutations in FH in one family and one sporadic case. These findings were subsequently confirmed in several large cohorts of aHUS patients (Buddles et al., 2000; Caprioli et al., 2001; Dragon-Durey et al., 2004; Neumann et al., 2003; Perez-Caballero et al., 2001; Richards et al., 2001). In these cohorts, mutations in FH accounted for between 15 and 30% of aHUS.

Approximately 60% of the independent mutational events cluster in CCPs 19–20 and another 20% are in CCPs 15–18 (Fig. 4.7). Other patients have mutations in additional parts of the gene leading to FH haploinsufficiency, as evidenced by ~50% of normal plasma levels. Only a few mutations in FH have been systematically studied but analyses of representative ones in CCP 20 have defined defects in binding to heparin, C3b, and endothelial cells (Heinen et al., 2006; Jokiranta et al., 2005; Jozsi et al., 2006; Manuelin et al., 2003; Sanchez-Corral et al., 2002, 2004; Vaziri-Sani et al., 2006).

Nuclear magnetic resonance (NMR) and crystal structures of CCPs 19 and 20 have been published (Herbert et al., 2006; Jokiranta et al., 2006).

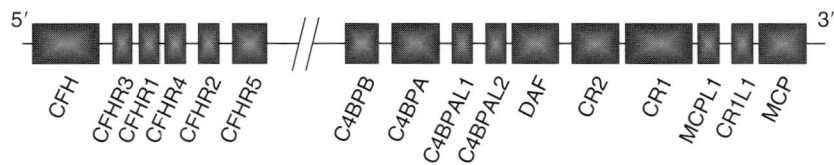

FIGURE 4.6 Representation of complement genes in the RCA cluster. Many complement genes reside in the RCA gene cluster on human chromosome 1q32. There are two main groups separated by ~14.5cM. Complement factor H (CFH) and CFH R1-R5, factor H (FH) and FH-related genes (FHRs): C4BPB and C4BPA, C4 binding protein β and α genes; C4BPAL1 and C4BPAL2, C4BP-like (partial duplicates) of α; DAF, decay accelerating factor (CD55); CR2, complement receptor 2 (CD21); CR1, complement receptor 1 (CD35); MCPL1, partial duplicate of membrane cofactor protein (MCP, CD46); CR1L1, partial duplicate of CR1; MCP, membrane cofactor protein (CD46).

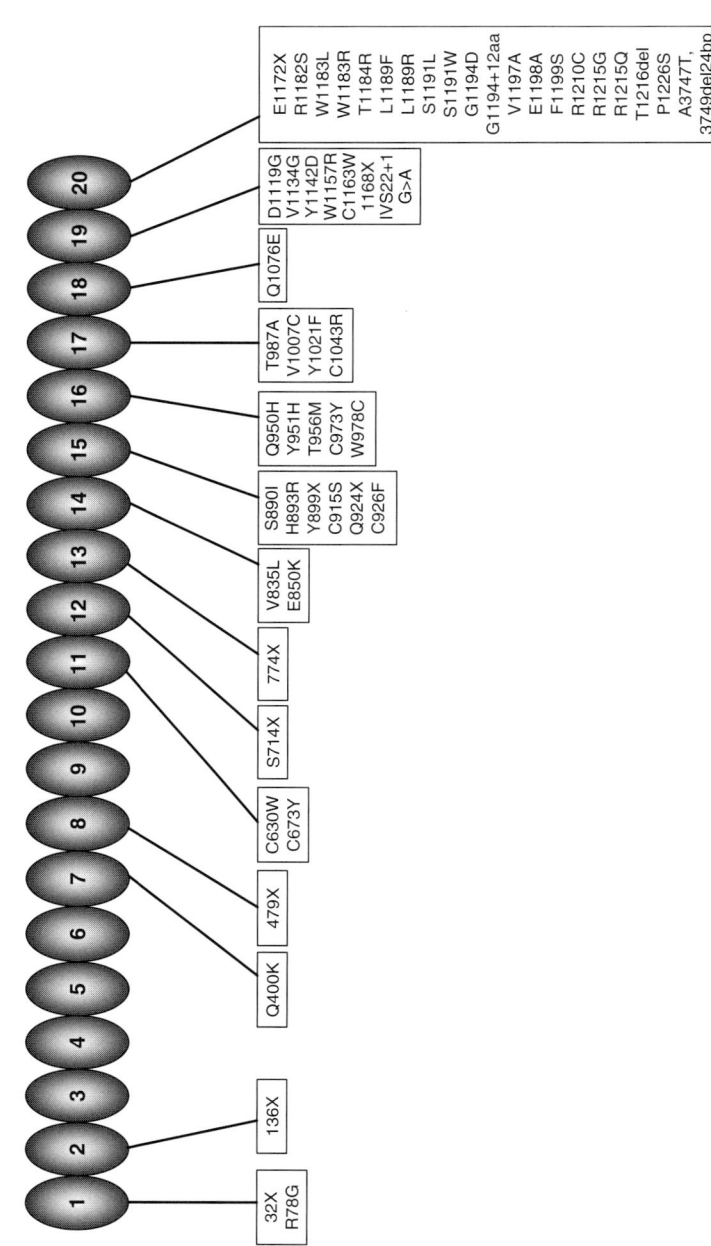

FIGURE 4.7 Location of the initial mutations identified in factor H (FH) in aHUS. Also, see Fig. 4.3.

Subsequent structure-based interpretation of the aHUS-associated mutations demonstrated the likely disruption of polyanion binding in the NMR structure (Herbert et al., 2006). However, analysis of the crystal structure suggested that it is binding affinity for C3d and C3b which is the critical activity perturbed by aHUS-linked mutations (Jokiranta et al., 2006).

A transgenic mouse model lacking the terminal five CCP domains of FH has been developed by Pickering et al. (2007). The mutant FH produced in these mice failed to bind to endothelial cells in a manner analogous to the mutations seen in aHUS individuals. This mouse regulated C3 activation in plasma but spontaneously developed aHUS.

Thus, the results of the functional studies, NMR and crystal structures, and the mouse model suggest that mutations in FH seen in aHUS interfere with ligand binding and thus prevent FH binding to host cell surfaces/basement membranes. This will prevent the control of AP amplification at these sites while fluid phase regulation remains unimpaired.

In addition to mutations in FH, single nucleotide polymorphisms (SNPs) in FH have also been associated with aHUS (Caprioli et al., 2003; Esparza-Gordillo et al., 2005; Fremeaux-Bacchi et al., 2005). One of these polymorphisms C-257T is located in a putative NF-κB binding site of the FH promoter. Although suggested to have a role in expression, there is no *in vivo* data to confirm this and the SNPs may simply be in linkage disequilibrium with other susceptibility alleles.

4.3. FH-related genes

In addition to FH, the RCA cluster contains five FH-related genes (FHR1–5) (Fig. 4.4). A haplotype containing a deletion of two of these genes, FHR1 and FHR3, has been demonstrated to increase the risk of aHUS (Zipfel et al., 2007). Although, a regulatory function is suggested by their ability to bind C3b and heparin, neither FHR1 nor FHR3 has intrinsic CA and DAA. FHR3 does have a cofactor-enhancing activity (Hellwage et al., 1999). Serum from aHUS patients lacking FHR1 and FHR3 showed an impaired ability to protect erythrocytes from complement activation (Zipfel et al., 2007). However, it is as yet unclear whether it is the absence of FHR1 and FHR3 that is responsible for the increased risk of aHUS per se, or whether this deletion is in linkage disequilibrium with other susceptibility alleles in FH.

4.4. Membrane cofactor protein: CD46

The initial linkage analysis to the RCA cluster in aHUS patients led to the discovery of mutations in FH in two of the three families used in the analysis (Venables et al., 2006; Warwicker et al., 1998). The cause of aHUS in the third family remained elusive until further linkage analysis by Richards et al. (2003) led to the discovery of mutations in MCP. Mutations in MCP account for 10–13% of aHUS (Caprioli et al., 2006; Esparza-Gordillo et al., 2005; Fremeaux-Bacchi et al., 2006; Noris et al., 2003). The initial 25 mutations in MCP in patients with aHUS have been reviewed (Richards et al., 2007). All but two occur in the four extracellular CCPs which contain the region critical for complement regulation (Fig. 4.4).

Mutations in MCP are of two general types, illustrated by the first two mutations identified in 2003 (Richards et al., 2003). In one family, a 6-bp deletion led to a 2 amino acid deletion. This results in a misfolded protein that is retained in the endoplasmic reticulum. The patients are haploinsufficient, expressing 50% of the normal levels as documented by fluorescence activated cell sorting (FACS) analysis of peripheral blood mononuclear cells (PBMCs) (Type 1 mutation). About 75% of aHUS-associated MCP mutations are of this type and are caused by a mixture of deletions as well as splice site, nonsense, and missense mutations (Richards et al., 2007).

In the remaining 25% of cases, the mutant is expressed normally but has absent or decreased complement regulatory activity (Type 2 mutation). For example, the S206P mutation in the initial report by Richards et al. (2003) has ~7% of the expected AP regulatory activity, if evaluated in the fluid phase or in situ. In two of these mutants (R69W, A304V), no detectable defect in C3b binding by the enzyme-linked immunosorbent assay (ELISA) or fluid-phase cofactor assays was observed. Subsequently, however, permanent Chinese Hamster Ovary (CHO) cell lines expressing equivalent copy numbers of these proteins allowed for their inhibitory activity to be quantitatively assessed in situ (Liszewski et al., 2007; Richards et al., 2007). They were both defective in CA on CHO cell lines when the AP was activated (Fang et al., 2007).

Approximately 75% of the aHUS patients with an MCP mutation are heterozygous (Richards et al., 2007). The other 25% represent an interesting mix in which there are biallelic mutations, two of which are compound heterozygotes and seven are homozygous. Of the latter, two individuals were shown to be null for MCP expression on the patients' peripheral blood cells. Unfortunately, the clinical information is limited on these two cases and their families. The others had reduced levels or expressed a mutant protein with detectable but reduced function. Because of MCP's potential role in autoimmunity, T regulatory cell development (Kemper and Atkinson, 2007; Kemper et al., 2003), and reproduction (Riley-Vargas and Atkinson, 2003; Riley-Vargas et al., 2005), their follow up will be of interest.

In addition to haploinsufficient predisposing to aHUS, Esparza-Gordillo et al. (2005) identified a haplotype they named MCPggaac that is increased two- to threefold in patients with aHUS (60%) versus controls (23%). The MCPggaac haplotype extends over a large part of the RCA gene cluster. The Madrid group analyzed the MCPggaac in a receptor gene assay system and demonstrated a reduction of 25% in transcriptional activity. Assuming these data can be transmitted to MCP expression *in vivo* and because most of these patients are also heterozygous for a mutation in FH, MCP, or FI, the functional level of normal MCP would be reduced 25% of that of normal individuals. The association between MCP SNPs, and aHUS has subsequently been confirmed in two independent cohorts of patients (Fremeaux-Bacchi *et al.*, 2005). These results suggest that it is the composite level of regulatory activity for the AP which accounts for the predisposition to aHUS.

4.5. Factor I

Mutations in FI account for 5–12% of the mutations in aHUS (Caprioli *et al.*, 2006; Esparza-Gordillo *et al.*, 2005; Fremeaux-Bacchi *et al.*, 2004; Kavanagh *et al.*, 2005). All FI mutations described so far in aHUS have been heterozygous (Fig. 4.5). Approximately 40% of the FI mutants associated with aHUS result in no or reduced secretion of FI (Type 1 mutations) (Kavanagh *et al.*, 2006). The remaining mutations produce a mutant protein that is secreted but is not functionally active.

We have analyzed the functional consequences of six mutated FI proteins where the protein is expressed normally (Kavanagh *et al.*, 2008b). Of those, three had no CA, one had ~30% of normal activity, and two had no detectable abnormality. The three with no activity (I322T, D501N, and D506V) had mutations in the serine protease domain. The other mutant-lacking activity (R299W) was in a domain of unknown function at the carboxyl terminus of the heavy chain, although it does contain the cysteine which forms the disulfide bridge linking the two chains. Two other examples, G243D and M120I, have also been studied in which no detectable functional deficiency was identified (Kavanagh *et al.*, 2008b). However, in none of these studies have all four cofactor proteins (CR1, FH, MCP, C4bp) been evaluated in the fluid phase and on C3b and C4b bound to a variety of targets.

4.6. Factor B

As this chapter was being written, gain of function mutations in FB were reported in aHUS (Goicoechea de Jorge *et al.*, 2007). One mutation (F286L) stabilized the convertase by increasing the affinity of its interaction with C3b (a more active enzyme) and was within a few amino acids of a previously reported mutant that also stabilized the convertase (Hourcade, 2006; Kuttner-Kondo *et al.*, 2003). The other mutation (K323E) made the enzyme complex more resistant to decay by FH. These data further illustrate that "just the right amount" of AP activation at a site of tissue injury in the renal endothelium is necessary to avoid this disease process. At least in the microvasculature of the kidney glomerulus, an *increase in AP activating capacity or a decrease in AP regulatory activity* predisposes to aHUS. These data raise the spectra of a gain of function mutations in C3, FD, or P as additional candidates for mutations in aHUS. Along these lines, it is surprising that a deficiency of DAF (examined so far in two series) has not been associated with aHUS (Goicoechea de Jorge *et al.*, 2007; Kavanagh *et al.*, 2007a).

4.7. Disease penetrance

The penetrance of mutations in FH, MCP, and FI is ~50% in all reported series with many patients having an unaffected parent. Thus, a mutation in a single complement regulatory gene by itself may not be sufficient to cause aHUS. In addition to being haploinsufficient for FI, FH, or MCP, other genetic factors related to AP activation and regulation may be required. For example, 10–20% of patients bearing an MCP mutation also carry a FH or FI mutation. Others carry the risk polymorphisms in promoter regions of the FH and MCP genes detailed earlier. Along these lines, a particularly informative pedigree was reported by Esparza-Gordillo *et al.* (2006) in which there were two mutations, one in MCP and one in FI, in addition to the MCPggaac haplotype. Only when all three risk factors came together in an individual did aHUS occur.

In the only family with an FB mutation (F286L) and aHUS, incomplete penetrance is again seen with only 7 of 11 developing aHUS. In this family, all seven individuals with aHUS had both the F286L mutation and the MCPggaac haplotype. Also, the single individual with K323E mutation in FB mutation has the MCPggaac haplotype. These results further emphasize that it requires the combined action of FH, MCP, and FI to control the amplification loop of the AP.

Even when genetic risk factors segregate together in an individual, often disease does not manifest until middle age. This suggests that a precipitating cause is needed to unmask these latent complement regulatory defects. In a recent series of patients with mutations in MCP, aHUS was precipitated in all cases by infection (Caprioli *et al.*, 2006). In FH-HUS, 70% of cases were preceded by infection while pregnancy and drugs both accounted for 4% (Caprioli *et al.*, 2006). In FI-HUS, 40% were preceded by pregnancy while 60% were precipitated by infection.

5. COMPLEMENT AND AGE-RELATED MACULAR DEGENERATION

5.1. Age-related macular degeneration

AMD is the leading cause of irreversible blindness in individuals over 60 years in the developed world (Gehrs *et al.*, 2006; van Leeuwen *et al.*, 2003). It affects 30–50 million people worldwide with 14 million severely visually impaired. The diagnosis of AMD rests on the finding of drusen (German for geode) in the macula. They appear as whitish to yellowish dots in the retina, a hallmark of AMD. The size and number of drusen correlate with severity of the visual loss. Their origin remains obscure but they consist in part of lipofuscin pigments derived from degenerating retinal pigmented epithelial (RPE) cells. This is thought to be the primary defect in AMD. They also contain locally synthesized and exogenously derived plasma components that are responding to this debris. Drusen accumulation is exacerbated by environmental and genetic factors. The primary environmental factor is smoking. A genetic influence has long been proposed because of the often familial nature (up to 30%) of this condition. In AMD, drusen number and size tend to progressively increase but at a variable pace from patient to patient.

AMD is commonly grouped into two clinical categories, namely "dry" atrophic type and the exudative, neovascular, or "wet" form. Dry AMD accounts for 90% of the disease and the 10% of wet AMD cases commonly develop on a background of dry disease. Wet AMD refers to ingrowth of choroidal vessels toward the fovea which is accompanied by serous or hemorrhagic fluid accumulation. It is responsible for most of the severe visual loss in AMD, but extensive drusen formation in dry AMD (geographic atrophy) may also cause visual loss.

In addition to drusen, there are pigmentary changes in the macula that are caused by dysfunctional (degenerative) changes in RPE cells. An important function of RPE cells is to regenerate visual pigments (rhodopsin). RPE cells phagocytose the shed tips of rods and cones. Probably because of incomplete digestion of lipofuscin pigments, they progressively accumulate in RPE cells throughout life and may be responsible for the eventual degenerative changes including apoptosis.

5.2. Factor H

An important role for innate immunity and, particularly the complement system, was proposed for AMD by Hageman *et al.* (2001). Using immunohistochemistry, this group demonstrated heavy coating of drusen with C3 fragments and several other complement-derived proteins including C5b-9 and FH. Although the basic observation was not in doubt, the pathophysiological significance was of unclear import. Most investigators felt that this was a bystander phenomenon and not directly or possibly even indirectly involved in the pathogenesis of AMD.

However, in 2005, four simultaneous reports of whole genome screens of AMD patients found an association with a SNP in the FH gene (Edwards *et al.*, 2005; Hageman *et al.*, 2005; Haines *et al.*, 2005; Klein *et al.*, 2005) leading to new agreement that innate immunity through the complement system was playing an important role in the pathogenesis. The SNP was in the coding region (CCP 7) of FH, and produced a change in amino acid 402 from tyrosine to histidine (Fig. 4.3). It accounts for ~50% of the genetic risk in AMD. Heterozygous individuals have a two- to threefold and homozygous individuals have a five- to sevenfold increased risk of developing AMD.

This polymorphism is present in 30–40% of the Caucasian and African populations but in less than 5% of Asians (Chen et al., 2006; Lau et al., 2006; Uka et al., 2006). CCP 7 is one of the three anionic binding sites in FH. CCP 7 is also where C-reactive protein (CRP) binds as well as four pathogens interact with FH (Fig. 4.3). The simplest scenario, analogous to the situation in aHUS, is that FH carrying the H402 variant does not as efficiently bind to components of a drusen as the Y402 FH variant. Consequently, in the individuals with the H402, the AP is *more active* on a drusen. Thus, there is more local inflammation and presumably more collateral damage in the retina. The H402 polymorphism is not the primary cause of AMD but is postulated to accelerate disease development.

The Y402H polymorphism affects binding affinity of FH to CRP, an acute phase protein. FH purified from sera of AMD patients homozygous for the 402H variant showed a significantly reduced binding to CRP compared to the Y402 variant. A recombinant fragment of FH (CCPs 5–7) containing the same amino acid change also showed reduced binding to CRP for the H402 variant (Laine et al., 2007; Sjoberg et al., 2007). It is hypothesized that, because the interaction of FH and CRP promotes complement-mediated clearance of cellular debris in a noninflammatory fashion, the H402 variant predisposes to an impaired targeting of FH to cellular debris. Consequently, in individuals with AMD, there will be a reduction in debris clearance and an increase in inflammation along the macular RPE–choroidal vessel interface.

Clark et al. (2006) showed that the H402Y polymorphism in CCP 7 of FH is adjacent to a heparin binding site, and that the variants differentially recognize heparin. The H402 variant eluted at a lower salt concentration from a HiTrap heparin-affinity column compared with Y402, indicating that the latter has a higher affinity for heparin/GAGs. This work was confirmed by Herbert et al. (2007). They also demonstrated that the Y402 variant binds more tightly to a heparin-affinity column and to defined-length sulfated heparin oligosaccharides used in gel-mobility shift assays than the H402 variant. These data suggest that the protective Y402 variant has a higher affinity for heparin sulfate residues in exposed basement membranes. It supports a causal link between H402Y and AMD, whereby a reduction of FH binding to age-related changes in the glycosaminoglycans composition and apoptotic activity of the macula predisposes to AP complement-mediated injury. One mechanism by which this may be mediated was proposed by Fernando *et al.* (2007) who analyzed the complexes formed between heparin and recombinant FH CCP 6–8 domains using analytical ultracentrifugation and X-ray scattering. They suggested that the H402 variant may self-associate more readily than Y402 allotype to form dimers. This may reduce its availability to interact with heparin and possibly CRP.

Of the microbes reported to bind FH, *Streptococcal pyogenes* seems most likely to have driven this evolutionary change. The hypothesis under evaluation is that the H402 variant provides a survival advantage against this pathogen. Specifically, this polymorphic variant *decreases* the ability of S. *pyogenes* to bind FH and therefore protects itself against complement attack. This amino acid change may have the unanticipated *negative consequence,* if one lives long enough, in that FH H402 also does not bind as well to a drusen. The result of this set of circumstances is that more complement activation and inflammation occurs for a given degree of RPE degeneration and drusen accumulation.

Further, other protective and risk forms of FH have now been described. For example, a protective haplotype consists of deletion at the end of the FH gene and encompassing all of FHR1/3 gene that occurs in ~5% of normal individuals but only in ~1–2% of AMD patients (Hageman *et al.*, 2006; Hughes *et al.*, 2006). The other major noncomplement genetic risk factor in AMD, LOC387715 (Jakobsdottir *et al.*, 2005; Rivera *et al.*, 2005), has been mapped to a little-known tissue protease.

5.3. Factor B/C2

In 2006, two protective haplotypes and a single risk haplotype of FB were identified in AMD (Gold *et al.*, 2006). One of the protective SNP haplotypes leads to an amino acid change in FB previously shown to result in reduced AP activity (Lokki and Koskimies, 1991). The concept here, analogous to aHUS, is that a gain of function mutation would be disease accelerating while a loss of function would be protective against the development of AMD.

6. IMMUNOPATHOGENESIS OF aHUS and AMD

6.1. Atypical HUS

In aHUS, the following sequence of events may be proposed: a toxin, commonly in association with an infectious illness, causes an injury to endothelial cells in the vasculature of the renal glomerulus. This leads to complement activation at this site because the injury induces apoptotic and/or necrotic cells, with or without exposure of the underlying basement membrane. Natural antibodies, for example, antiendothelial cell antibodies, lectins, or the AP itself could be involved in generating the initial C3b deposition. The AP feedback loop is then engaged by this C3b, leading to further C3b deposition. In the presence of a lack of normal regulatory activity on injured cells with MCP deficiency, or on cells and at exposed tissue sites with FH deficiency, this allows for excessive complement activation (Fig. 4.8). This in turn generates too much opsonization, C3a/C5a anaphylotoxin liberation, and C5b-9 formation—such that any one of or a combination of these factors then leads to a procoagulant state. The release of tissue factor in response to C5a release and of von Willebrand factor from damaged endothelial cells are examples of how this procoagulant state may be perpetuated (Kavanagh et al., 2008a).

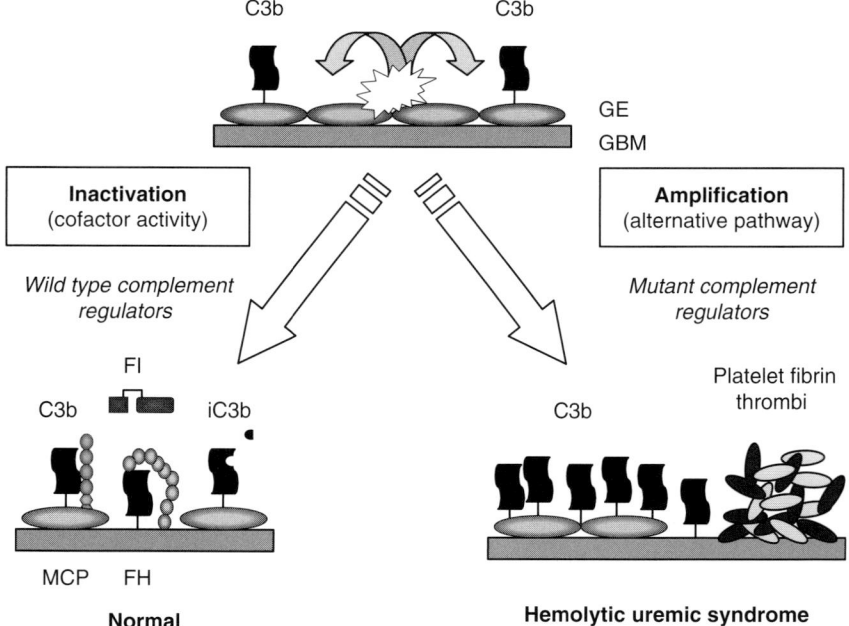

FIGURE 4.8 Outcome of C3b deposition on glomerular vasculature. On microbes, massive amplification is the desired result. On healthy self-tissue, inactivation is the desired result. On damaged or modified self-tissue, a delicate balance is presumably in place to allow for both repair and recovery and to limit the immune response.
If regulators such as factor H (FH) or MCP are deficient, excessive activation occurs, leading to further cell damage. GE, glomerular endothelium; GBM, glomerular basement membrane.

6.1.1. Related issues

6.1.1.1. Overlapping function of FH and MCP Mutations in aHUS unequivocally point out that these two regulatory proteins' functions are not overlapping *in vivo*. At least in the kidney glomerulus, a 50% reduction in one or the other predisposes to disease. These data are consistent with studies of Crry or FH-deficient mice in which neither one can protect against accelerated C3 turnover in a complete absence of the other (Pickering *et al.*, 2002; Xu *et al.*, 2000).

6.1.1.2. Complement activation and regulation on endothelial cells Endothelial cells abundantly express complement regulatory proteins. Typical copy numbers/cell on several types of human endothelial cells (including HMECS) are ~200,000 of DAF, ~400,000 of MCP, and ~1,400,000 of CD59 (A. R. and J. A., unpublished data). These are the highest copy number/cell that we are aware of for a normal human cell population. aHUS teaches us that this number of MCP/cell is required to guard against excessive activation on renal glomerular endothelial cells (Liszewski *et al.*, 2007). These complement regulatory MCP haploinsufficient individuals are healthy except for the development of aHUS. Other organ systems do not develop disease despite expressing 50% of the normal level. Also, aHUS points out that at least, in the renal endothelium, FH and MCP are both required. They are not redundant as MCP haploinsufficient aHUS patients have normal levels of FH and vice versa. Along this line, 50% of normal serum levels of FH or FI are also inadequate to protect against the development of aHUS. Thus, there is a reason why humans express a certain level of MCP on endothelial cells and why plasma concentrations of FH and FI are 500 μg/ml and not 250 μg/ml, and 50 μg/ml and not 25 μg/ml, respectively.

6.1.1.3. Binding of FH to human cells FH does not bind or at least it has been very hard to demonstrate that FH binds to normal human cells. A possible exception to this has been human endothelial cells in culture. Endothelial cells, particularly human umbilical vein endothelial cells (HUVECs) have been shown to bind FH in several studies, particularly by the Zipfel group (Jozsi *et al.*, 2007). Moreover, these investigators have convincingly shown that several FH mutations in CCP 20 reduce endothelial cell binding. Consequently, there may be a role for FH binding to normal human endothelial cells *in vivo*, but we are not aware that this has ever been demonstrated. However, what is clearer is that FH, through its anionic binding sites, can attach to injured cells, exposed basement membranes, and tissue matrices where it protects these materials from complement attack. If it were not for this capability, excessive complement activation would likely occur at sites of trauma, apoptosis, and necrosis.

FH does though bind to "normal" human cells once C3b is bound. However, on normal human cells, it requires a large amount of C3b (as it does on an ELISA plate) to be bound. In the systems we have investigated, FH minimally, if at all, inhibits the quantity of C3b deposited by the AP activated by antibody (Barilla-LaBarca et al., 2002; Liszewski et al., 2007). What FH does accomplish with remarkable efficiency is to convert the deposited C3b to iC3b, and thereby to provide another ligand for complement receptors (Barilla-LaBarca et al., 2002; Liszewski et al., 2007). Additional studies of injured cells are needed. Even in the absence of C3b deposition or in the setting of small amount of deposited C3b plus enhanced exposure to GAGs, FH binding may be efficient. In most situations of cellular injury, one would anticipate that FH and MCP combine forces to prevent an excessive/undesirable amount of AP activation.

6.2. Age-related macular degeneration

Two scenarios will be proposed to account for a role of the complement system in AMD. They are not mutually exclusive. The first is, in essence, identical to that in aHUS except that the injured (possibly degenerate) cell is the RPE cell rather than kidney microvascular endothelial cell. The process of RPE destruction, whether triggered by environmental factors (smoking, infection, hereditary), is accelerated by AP activation.

A second proposal concerns AP activation by drusen. C3 fragments, FH, and C5b-9 have all been shown to coat drusen (Hageman et al., 2001). Most complement-activating components are synthesized by RPE cells. Complement regulatory proteins including FH and FH-like and related proteins, MCP, DAF, and CD59 are also synthesized by RPE cells. It is unclear how much of the deposited complement activation fragments are derived from plasma versus locally synthesized. In any case, the proposal comes down to the hypothesis of excessive and undesirable AP activation on drusen. The local generation of C3 fragments, C3a/C5a, and C5b-C9 contribute to tissue damage by promoting inflammation, vascular ingress, and RPE degeneration.

The Y402H polymorphism may reduce the affinity of FH for the contents of a drusen. This reduced binding activity of H402 thus allows for greater AP activity in this extracellular site of drusen formation. Since CRP also binds to this same region of FH, involving CCPs 6–8, an alteration in its interaction with FH could also be a contributing factor. CRP binding can activate the CP but it also downmodulates subsequent loop amplification and C5b-9 formation. While other scenarios are certainly possible (such as FH binding influencing drusen formation or propagations independent of its role in AP regulation), the preceding hypothesis seems to best fit with what we have learned relative to complement FH deficiency in aHUS.

7. TREATMENT OF aHUS and AMD

7.1. Treatment options for aHUS

7.1.1. Plasma infusion/exchange
Plasma infusion replaces deficient serum complement regulatory proteins such as FH, FI, and FB and has been shown to be effective in managing chronic recurrent childhood aHUS. A volume of 20–30 ml/kg has been recommended for FH-HUS (Kavanagh et al., 2006). For those with aHUS due to anti-FH antibodies, plasma exchange is a logical therapy for removing the IgG antibodies (Kavanagh et al., 2006). Unfortunately, neither treatment addresses the underlying pathology, and relapse upon discontinuation of therapy is the expected outcome.

7.1.2. Renal transplantation
Since FH, FI, and FB are primarily synthesized in the liver, transplantation of a kidney will not correct this plasma protein deficiency. Consistent with this, renal transplantation in FH is associated with an ∼80% recurrence rate while in FI patients there is an ∼100% recurrence of aHUS in the donated allograft (Bresin et al., 2006; Kavanagh et al., 2006). In one patient transplanted with an FB mutation to date, there was again the expected recurrence of aHUS (Goicoechea de Jorge et al., 2007). In contrast, the transplanted kidney in aHUS patient carrying an MCP mutation expresses normal levels of MCP. In this case, the success rate is ∼90%, similar to that of most other kidney transplants (Richards et al., 2007). For this reason, it is recommended that all aHUS patients be screened for FH, FI, FB, and MCP mutations prior to transplantation (Kavanagh et al., 2007b). Renal transplant is contraindicated in an FH or an FI mutation but acceptable with an MCP mutation. If a family member is contemplated to be a donor, mutational screening of the donor will be necessary due to a report describing the development of *de novo* aHUS in living-related parental donors who carried a previously unsuspected FH mutation (Donne et al., 2002).

7.1.3. Liver/renal transplantation
As FH, FI, and FB are all synthesized in the liver; combined liver/renal transplantation represents a logical step in the efforts to treat the underlying cause of aHUS caused by mutations in these serum factors. It has now been used on four occasions in an attempt to treat FH-HUS, but unfortunately the first three cases had a very poor outcome with death occurring in two (Cheong et al., 2004; Remuzzi et al., 2002, 2005). However, the most recent case which used preemptive plasma infusion/exchange to elevate levels of wild-type complement regulators prior to the procedure was successful, suggesting that it is overcoming complement activation at

the time of the transplants which is the key factor in determining outcome and prognosis (Saland et al., 2006).

7.2. Treatment options for AMD

7.2.1. Antiangiogenic treatments

In addition to the standard measures of visual rehabilitation, blood pressure control, smoking cessation, a diet rich in vegetables and antioxidants, and the additional use of photodynamic laser for the neovascular (wet) type of AMD, new optimism now exists for treatment of AMD in the form of antiangiogenic therapies, namely vascular endothelial-derived growth factor (VEGF) inhibitors. VEGF is a potent mitogen and vascular permeability factor that plays a pivotal role in neovascularization. The role of VEGF in AMD is less clear but increased levels are present in neovascular membranes (Matsuoka et al., 2004). Putative mechanisms in the pathogenesis of AMD include induction of choroidal new vessels, increasing vascular permeability with the formation of subretinal fluid or acting as a proinflammatory agent causing leucocyte margination and damage to retinal endothelial cells. Currently, Anecortave acetate (Retaane, Alcon), a synthetic cortisone which blocks angiogenesis and inhibits VEGF; Pegaptanib (Macugen, EyeTech/Pfizer), a highly selective inhibitor of VEGF (VEGF-165); Ranibizumab, (Lucentis, Genentech), a mouse/human monoclonal antibody fragment that binds and blocks all forms of VEGF (at the VEGF receptor binding site); and Bevacizumab (Avastin, Roche), a whole humanized mouse antibody that binds and blocks all forms of VEGF (at the VEGF receptor binding site), are under investigation in clinical studies.

7.2.2. Complement inhibitors

The identification of polymorphic variants in FH and FB as risk factors for the development of AMD due to overactivation of the AP of complement has identified a new area for possible therapeutic intervention in AMD. The use of complement inhibitors such as the anti-C5 antibody is under consideration as a potential therapeutic strategy (Hillmen et al., 2004, 2006).

8. CONCLUSIONS: LESSONS AND IMPLICATIONS

HUS and AMD are diseases in which there were only a few hints that innate immunity and specifically the complement system might be playing a major role in disease causation. Subsequently, candidate gene studies identified genes that turned out to be predisposing in $\sim$50% of aHUS patients, most of whom are haploinsufficient for one of three complement regulatory proteins required for degradation of C3b (CA). A polymorphic variant in FH, present in about one-third of the American population, accounts for 50% of the genetic risk in AMD. *The common link between these syndromes is excessive AP activation secondary to decreased regulation of the AP.* These two examples have implications for identifying how innate immunity plays out in human disease. *First*, they beautifully illustrate the power of human genetics to provide unsuspected disease associations. *Second*, the association accounts for why the FH level in the blood is ~500 μg/ml and not ~250 μg/ml and why there are ~400,000 copies and not ~200,000 copies of MCP per human endothelial cell. *Third*, these associations point out our profound lack of knowledge of innate immune function in specialized sites such as on endothelial cells and in the retina. This raises the question of who are the key innate immune players in the white matter of the brain or in wall of a coronary vessel? We are largely ignorant about these issues which are so important to clinical medicine. *Fourth*, we are increasingly facing diseases featuring debris accumulation and yet we are just beginning to understand the innate response to this type of altered self. *Fifth*, low-grade innate/inflammatory immune responses in the retina, brain, and blood vessel walls occur over decades, a pace that is not easy for an immunologist to study. *Sixth*, inhibition of the AP and other innate immune recognition and effector/signaling cascades will likely become a frontier in drug development. For example, in AMD we need trials of the long-term downregulation of AP activation. *Seventh*, polymorphic variations leading to what appear to be modest changes in regulatory activity, like Y402H in AMD, may protect against a streptococcal infection in childhood. However, it becomes deleterious in the setting of our aging population where chronic diseases featuring debris deposition now cause much morbidity and mortality.

In conclusion, a role for innate immunity, particularly the AP of complement is now established in the pathogenesis of aHUS and AMD. Further challenges include development of effective therapeutic strategies for both conditions and to investigate the role of AP in other chronic conditions where "debris accumulation" is a key feature of the pathogenesis.

REFERENCES

Alexander, J. J., and Quigg, R. J. (2007). The simple design of complement factor H: Looks can be deceiving. *Mol. Immunol.* **44**, 123–132.

Amadei, N., Baracho, G. V., Nudelman, V., Bastos, W., Florido, M. P., and Isaac, L. (2001). Inherited complete factor I deficiency associated with systemic lupus erythematosus, higher susceptibility to infection and low levels of factor H. *Scand. J. Immunol.* **53**, 615–621.

Atkinson, J. P. (1986). Complement activation and complement receptors in systemic lupus erythematosus. *Springer Semin. Immunopathol.* **9**, 179–194.

Atkinson, J. P., Chan, A. C., Karp, D. R., Killion, C. C., Brown, R., Spinella, D., Shreffler, D. C., and Levine, R. P. (1988). Origin of the fourth component of complement related Chido and Rodgers blood group antigens. *Complement* **5,** 65–76.

Ault, B. H., Schmidt, B. Z., Fowler, N. L., Kashtan, C. E., Ahmed, A. E., Vogt, B. A., and Colten, H. R. (1997). Human factor H deficiency. Mutations in framework cysteine residues and block in H protein secretion and intracellular catabolism. *J. Biol. Chem.* **272,** 25168–25175.

Barilla-LaBarca, M. L., and Atkinson, J. P. (2003). Rheumatic syndromes associated with complement deficiency. *Curr. Opin. Rheumatol.* **15,** 55–60.

Barilla-LaBarca, M. L., Liszewski, M. K., Lambris, J. D., Hourcade, D., and Atkinson, J. P. (2002). Role of membrane cofactor protein (CD46) in regulation of C4b and C3b deposited on cells. *J. Immunol.* **168,** 6298–6304.

Barlow, P. N., Baron, M., Norman, D. G., Day, A. J., Willis, A. C., Sim, R. B., and Campbell, I. D. (1991). Secondary structure of a complement control protein module by two-dimensional 1H NMR. *Biochemistry (Moscow)* **30,** 997–1004.

Bresin, E., Daina, E., Noris, M., Castelletti, F., Stefanov, R., Hill, P., Goodship, T. H., and Remuzzi, G. (2006). Outcome of renal transplantation in patients with non-Shiga toxin-associated haemolytic uraemic syndrome: Prognostic significance of genetic background. *Clin. J. Am. Soc. Nephrol.* **1,** 88–99.

Brodbeck, W. G., Mold, C., Atkinson, J. P., and Medof, M. E. (2000). Cooperation between decay-accelerating factor and membrane cofactor protein in protecting cells from autologous complement attack. *J. Immunol.* **165,** 3999–4006.

Buddles, M., Donne, R., Richards, A., Goodship, J., and Goodship, T. H. (2000). Complement factor H gene mutation associated with autosomal recessive atypical hemolytic uraemic syndrome. *Am. J. Hum. Genet.* **66,** 1721–1722.

Caprioli, J., Bettinaglio, P., Zipfel, P. F., Amadei, B., Daina, E., Gamba, S., Skerka, C., Marziliano, N., Remuzzi, G., and Noris, M. (2001). The molecular basis of familial hemolytic uremic syndrome: Mutation analysis of factor H gene reveals a hot spot in short consensus repeat 20. *J. Am. Soc. Nephrol.* **12,** 297–307.

Caprioli, J., Castelletti, F., Bucchioni, S., Bettinaglio, P., Bresin, E., Pianetti, G., Gamba, S., Brioschi, S., Daina, E., Remuzzi, G., and Noris, M. (2003). Complement factor H mutations and gene polymorphisms in haemolytic uraemic syndrome: The C-257T, the A2089G and the G2881T polymorphisms are strongly associated with the disease. *Hum. Mol. Genet.* **12,** 3385–3395.

Caprioli, J., Noris, M., Brioschi, S., Pianetti, G., Castelletti, F., Bettinaglio, P., Mele, C., Bresin, E., Cassis, L., Gamba, S., Porrati, F., Bucchioni, S., *et al.* (2006). Genetics of HUS: The impact of MCP, CFH and IF mutations on clinical presentation, response to treatment, and outcome. *Blood* **108,** 1267–1279.

Carroll, M. C., and Holers, V. M. (2005). Innate autoimmunity. *Adv. Immunol.* **86,** 137–157.

Cattaneo, R. (2004). Four viruses, two bacteria, and one receptor: Membrane cofactor protein (CD46) as pathogens' magnet. *J. Virol.* **78,** 4385–4388.

Chen, L. J., Liu, D. T., Tam, P. O., Chan, W. M., Liu, K., Chong, K. K., Lam, D. S., and Pang, C. P. (2006). Association of complement factor H polymorphisms with exudative age-related macular degeneration. *Mol. Vis.* **12,** 1536–1542.

Cheong, H. I., Lee, B. S., Kang, H. G., Hahn, H., Suh, K. S., Ha, I. S., and Choi, Y. (2004). Attempted treatment of factor H deficiency by liver transplantation. *Pediatr. Nephrol.* **19,** 454–458.

Clark, S. J., Higman, V. A., Mulloy, B., Perkins, S. J., Lea, S. M., Sim, R. B., and Day, A. J. (2006). His-384 allotypic variant of factor H associated with age-related macular degeneration has different heparin binding properties from the non-disease-associated form. *J. Biol. Chem.* **281,** 24713–24720.

Dauchel, H., Julen, N., Lemercier, C., Daveau, M., Ozanne, D., Fontaine, M., and Ripoche, J. (1990). Expression of complement alternative pathway proteins by endothelial cells. Differential regulation by interleukin 1 and glucocorticoids. *Eur. J. Immunol.* **20,** 1669–1675.

de Paula, P., Barbosa, J., Junior, P., Ferriani, V., Latorre, M., Nudelman, V., and Isaac, L. (2003). Ontogeny of complement regulatory proteins—Concentrations of factor H, factor I, C4b-binding protein, properdin and vitronectin in healthy children of different ages and in adults. *Scand. J. Immunol.* **58,** 572–577.

Donne, R. L., Abbs, I., Barany, P., Elinder, C. G., Little, M., Conlon, P., and Goodship, T. H. (2002). Recurrence of hemolytic uremic syndrome after live related renal transplantation associated with subsequent *de novo* disease in the donor. *Am. J. Kidney Dis.* **40,** E22.

Dragon-Durey, M.-A., Fremeaux-Bacchi, V., Loirat, C., Blouin, J., Niaudet, P., Deschenes, G., Copp, P., Fridman, W. H., and Weiss, L. (2004). Heterozygous and homozygous factor H deficiencies associated with hemolytic uremic syndrome or membranoproliferative glomerulonephritis: Report and genetic analysis of 16 cases. *J. Am. Soc. Nephrol.* **15,** 787–795.

Edwards, A. O., Ritter, R., 3rd, Abel, K. J., Manning, A., Panhuysen, C., and Farrer, L. A. (2005). Complement factor H polymorphism and age-related macular degeneration. [See comment]. *Science* **308,** 421–424.

Esparza-Gordillo, J., Goicoechea de Jorge, E., Buil, A., Berges, L. C., Lopez-Trascasa, M., Sanchez-Corral, P., and Rodriguez de Cordoba, S. (2005). Predisposition to atypical hemolytic uremic syndrome involves the concurrence of different susceptibility alleles in the regulators of complement activation gene cluster in 1q32. *Hum. Mol. Genet.* **14,** 703–712.

Esparza-Gordillo, J., Jorge, E. G., Garrido, C. A., Carreras, L., Lopez-Trascasa, M., Sanchez-Corral, P., and de Cordoba, S. R. (2006). Insights into hemolytic uremic syndrome: Segregation of three independent predisposition factors in a large, multiple affected pedigree. *Mol. Immunol.* **43,** 1769–1775.

Fang, C. J., Fremeaux-Bacchi, V., Liszewski, M. K., Pianetti, G., Noris, M., Goodship, T. H. J., and Atkinson, J. P. (2007). Membrane cofactor protein mutations in atypical hemolytic uremic syndrome (aHUS), fatal Stx-HUS, C3 glomerulonephritis and the HEELP syndrome. *Blood,* Prepublished online October 3, 2007; DOI 10.1182/blood-2007-04-084533.

Fearon, D. T. (1979). Activation of the alternative complement pathway. *CRC Crit. Rev. Immunol.* **1,** 1–32.

Fernando, A. N., Furtado, P. B., Clark, S. J., Gilbert, H. E., Day, A. J., Sim, R. B., and Perkins, S. J. (2007). Associative and structural properties of the region of complement factor H encompassing the Tyr402His disease-related polymorphism and its interactions with heparin. *J. Mol. Biol.* **368,** 564–581.

Figueroa, J., Andreoni, J., and Densen, P. (1993). Complement deficiency states and meningococcal disease. *Immunol. Res.* **12,** 295–311.

Fremeaux-Bacchi, V., Dragon-Durey, M.-A., Blouin, J., Vigneau, C., Kuypers, D., Boudailiez, B., Loirat, C., Rondeau, E., and Fridman, W. H. (2004). Complement factor I: A susceptibility gene for atypical haemolytic uraemic syndrome. *J. Med. Genet.* **41,** e84.

Fremeaux-Bacchi, V., Kemp, E. J., Goodship, J. A., Dragon-Durey, M.-A., Strain, L., Loirat, C., Deng, H.-W., and Goodship, T. H. J. (2005). The development of atypical haemolytic-uraemic syndrome is influenced by susceptibility factors in factor H and membrane cofactor protein: Evidence from two independent cohorts. *J. Med. Genet.* **42,** 852–856.

Fremeaux-Bacchi, V., Moulton, E. A., Kavanagh, D., Dragon-Durey, M.-A., Blouin, J., Caudy, A., Arzouk, N., Cleper, R., Francois, M., Guest, G., Pourrat, J., Seligman, R., *et al.* (2006). Genetic and functional analyses of membrane cofactor protein (CD46) mutations in atypical hemolytic uremic syndrome. *J. Am. Soc. Nephrol.* **17,** 2017–2025.

Gehrs, K. M., Anderson, D. H., Johnson, L. V., and Hageman, G. S. (2006). Age-related macular degeneration-emerging pathogenetic and therapeutic concepts. *Ann. Med.* **38,** 450–471.

Gershov, D., Kim, S., Brot, N., and Elkon, K. B. (2000). C-reactive protein binds to apoptotic cells, protects the cells from assembly of the terminal complement components, and sustains an antiinflammatory innate immune response: Implications for systemic autoimmunity. [Erratum appears in *J. Exp. Med.* 2001; **193**(12), 1439]. *J. Exp. Med.* **192,** 1353–1364.

Giles, C. M. (1988). Antigenic determinants of human C4, Rodgers and Chido. *Exp. Clin. Immunogenet.* **5,** 99–114.

Gill, D., and Atkinson, J. (2004). CD46 in *Neisseria pathogenesis*. *Trends Mol. Med.* **10,** 459–465.

Goicoechea de Jorge, E., Harris, C. L., Esparza-Gordillo, J., Carreras, L., Arranz, E. A., Garrido, C. A., Lopez-Trascasa, M., Sanchez-Corral, P., Morgan, B. P., and Rodriguez de Cordoba, S. (2007). Gain-of-function mutations in complement factor B are associated with atypical hemolytic uremic syndrome. *Proc. Natl. Acad. Sci. USA* **104,** 240–245.

Gold, B., Merriam, J. E., Zernant, J., Hancox, L. S., Taiber, A. J., Gehrs, K., Cramer, K., Neel, J., Bergeron, J., Barile, G. R., Smith, R. T., Group, A. G. C. S., *et al.* (2006). Variation in factor B (BF) and complement component 2 (C2) genes is associated with age-related macular degeneration. *Nat. Genet.* **38,** 458–462.

Goldberger, G., Bruns, G. A., Rits, M., Edge, M. D., and Kwiatkowski, D. J. (1987). Human complement factor I: Analysis of cDNA-derived primary structure and assignment of its gene to chromosome 4. *J. Biol. Chem.* **262,** 10065–10071.

Hageman, G. S., Luthert, P. J., Victor Chong, N. H., Johnson, L. V., Anderson, D. H., and Mullins, R. F. (2001). An integrated hypothesis that considers drusen as biomarkers of immune-mediated processes at the RPE-Bruch's membrane interface in aging and age-related macular degeneration. *Prog. Retin. Eye Res.* **20,** 705–732.

Hageman, G. S., Anderson, D. H., Johnson, L. V., Hancox, L. S., Taiber, A. J., Hardisty, L. I., Hageman, J. L., Stockman, H. A., Borchardt, J. D., Gehrs, K. M., Smith, R. J., Silvestri, G., *et al.* (2005). A common haplotype in the complement regulatory gene factor H (HF1/CFH) predisposes individuals to age-related macular degeneration. *Proc. Natl. Acad. Sci. USA* **102,** 7227–7232.

Hageman, G. S., Hancox, L. S., Taiber, A. J., Gehrs, K. M., Anderson, D. H., Johnson, L. V., Radeke, M. J., Kavanagh, D., Richards, A., Atkinson, J., Meri, S., Bergeron, S., *et al.* (2006). Extended haplotypes in the complement factor H (CFH) and CFH-related (CFHR) family of genes protect against age-related macular degeneration: Characterization, ethnic distribution and evolutionary implications. *Ann. Med.* **38,** 592–604.

Haines, J. L., Hauser, M. A., Schmidt, S., Scott, W. K., Olson, L. M., Gallins, P., Spencer, K. L., Kwan, S. Y., Noureddine, M., Gilbert, J. R., Schnetz-Boutaud, N., Agarwal, N., *et al.* (2005). Complement factor H variant increases the risk of age-related macular degeneration. [See comment]. *Science* **308,** 419–421.

Heinen, S., Sanchez-Corral, P., Jackson, M. S., Strain, L., Goodship, J. A., Kemp, E. J., Skerka, C., Jokiranta, T. S., Meyers, K., Wagner, E., Robitaille, P., Esparza-Gordillo, P., *et al.* (2006). De novo gene conversion in the RCA gene cluster (1q32) causes mutations in complement factor H associated with atypical hemolytic uremic syndrome. *Hum. Mutat.* **27,** 292–293.

Hellwage, J., Jokiranta, T. S., Koistinen, V., Vaarala, O., Meri, S., and Zipfel, P. F. (1999). Functional properties of complement factor H-related proteins FHR-3 and FHR-4: Binding to the C3d region of C3b and differential regulation by heparin. *FEBS Lett.* **462,** 345–352.

Herbert, A. P., Uhrin, D., Lyon, M., Pangburn, M. K., and Barlow, P. N. (2006). Disease-associated sequence variations congregate in a polyanion recognition patch on human factor H revealed in three-dimensional structure. *J. Biol. Chem.* **281,** 16512–16520.

Herbert, A. P., Deakin, J. A., Schmidt, C. Q., Blaum, B. S., Egan, C., Ferreira, V. P., Pangburn, M. K., Lyon, M., Uhrin, D., and Barlow, P. N. (2007). Structure shows glycosaminoglycan- and protein-recognition site in factor H is perturbed by age-related macular degeneration-linked SNP. *J. Biol. Chem.* **282**, 18960–18968.

Hillmen, P., Hall, C., Marsh, J. C. W., Elebute, M., Bombara, M., Petro, B. E., Cullen, M. J., Richards, S. J., Rollins, S. A., Mojcik, C. F., and Rother, R. P. (2004). Effect of eculizumab on hemolysis and transfusion requirements in patients with paroxysmal nocturnal hemoglobinuria. *N. Engl. J. Med.* **350**, 552–559.

Hillmen, P., Young, N. S., Schubert, J., Brodsky, R. A., Socie, G., Muus, P., Roth, A., Szer, J., Elebute, M. O., Nakamura, R., Browne, P., Risitano, P., *et al.* (2006). The complement inhibitor eculizumab in paroxysmal nocturnal hemoglobinuria. *N. Engl. J. Med.* **355**, 1233–1243.

Hogasen, K., Jansen, J. H., Mollnes, T. E., Hovdenes, J., and Harboe, M. (1995). Hereditary porcine membranoproliferative glomerulonephritis type II is caused by factor H deficiency. *J. Clin. Invest.* **95**, 1054–1061.

Hourcade, D. (2006). The role of properdin in the assembly of the alternative pathway C3 convertases of complement. *J. Biol. Chem.* **281**, 2128–2132.

Hourcade, D., Holers, V. M., and Atkinson, J. P. (1989). The regulators of complement activation (RCA) gene cluster. *Adv. Immunol.* **45**, 381–416.

Hughes, A. E., Orr, N., Esfandiary, H., Diaz-Torres, M., Goodship, T., and Chakravarthy, U. (2006). A common CFH haplotype, with deletion of CFHR1 and CFHR3, is associated with lower risk of age-related macular degeneration. *Nat. Genet.* **38**, 1173–1177.

Jakobsdottir, J., Conley, Y. P., Weeks, D. E., Mah, T. S., Ferrell, R. E., and Gorin, M. B. (2005). Susceptibility genes for age-related maculopathy on chromosome 10q26. *Am. J. Hum. Genet.* **77**, 389–407.

Jokiranta, T. S., Cheung, Z., Seeberger, H., Jozsi, M., Heinen, S., Noris, M., Remuzzi, G., Ormsby, R., Gordon, D. L., Meri, S., Hellwage, J., and Zipfel, P. F. (2005). Binding of complement factor H to endothelial cells is mediated by the carboxy-terminal glycosaminoglycan binding site. *Am. J. Pathol.* **167**, 1171–1181.

Jokiranta, T. S., Jaakola, V. P., Lehtinen, M. J., Parepalo, M., Meri, S., and Goldman, A. (2006). Structure of complement factor H carboxyl-terminus reveals molecular basis of atypical haemolytic uremic syndrome. *EMBO J.* **25**, 1784–1794.

Jozsi, M., Heinen, S., Hartmann, A., Ostrowicz, C. W., Halbich, S., Richter, H., Kunert, A., Licht, C., Saunders, R. E., Perkins, S. J., Zipfel, P. F., and Skerka, C. (2006). Factor H and atypical hemolytic uremic syndrome: Mutations in the C-terminus cause structural changes and defective recognition functions. *J. Am. Soc. Nephrol.* **17**, 170–177.

Jozsi, M., Oppermann, M., Lambris, J. D., and Zipfel, P. F. (2007). The C-terminus of complement factor H is essential for host cell protection. *Mol. Immunol.* **44**, 2697–2706.

Kavanagh, D., Kemp, E. J., Mayland, E., Winney, R. J., Duffield, J., Warwick, G., Richards, A., Ward, R., Goodship, J. A., and Goodship, T. H. J. (2005). Mutations in complement factor I predispose to the development of atypical hemolytic uremic syndrome. *J. Am. Soc. Nephrol.* **16**, 2150–2155.

Kavanagh, D., Goodship, T. H., and Richards, A. (2006). Atypical haemolytic uraemic syndrome. *Br. Med. Bull.* **77/78**, 5–22.

Kavanagh, D., Burgess, R., Spitzer, D., Richards, A., Diaz-Torres, M. L., Goodship, J. A., Hourcade, D. E., Atkinson, J. P., and Goodship, T. H. (2007a). The decay accelerating factor mutation I197V found in hemolytic uraemic syndrome does not impair complement regulation. *Mol. Immunol.* **44**, 3162–3167.

Kavanagh, D., Richards, A., Fremeaux-Bacchi, V., Noris, M., Goodship, J. A., Remuzzi, G., and Atkinson, J. P. (2007b). Screening for complement system abnormalities in patients with atypical hemolytic uremic syndrome. *Clin. J. Am. Soc. Nephrol.* **2**, 591–596.

Kavanagh, D., Richards, A., and Atkinson, J. P. (2008a). Complement regulatory genes and hemolytic uremic syndromes. *Annu. Rev. Med.* **59,** 61–77.

Kavanagh, D., Richards, A., Noris, M., Hauhart, R., Liszewski, M. K., Karpman, D., Goodship, J. A., Fremeaux-Bacchi, V., Remuzzi, G., Goodship, T. H. J., and Atkinson, J. P. (2008b). Characterization of mutations in complement factor I (CFI) associated with hemolytic uremic syndrome. *Mol. Immunol.* **45,** 95–105.

Kemper, C., and Atkinson, J. P. (2007). T-cell regulation: With complements from innate immunity. *Nat. Rev. Immunol.* **7,** 9–18.

Kemper, C., Chan, A. C., Green, J. M., Brett, K. A., Murphy, K. M., and Atkinson, J. P. (2003). Activation of human $CD4^+$ cells with CD3 and CD46 induces a T-regulatory cell 1 phenotype. *Nature* **421,** 388–392.

Kim, S. J., Gershov, D., Ma, X., Brot, N., and Elkon, K. B. (2003). Opsonization of apoptotic cells and its effect on macrophage and T cell immune responses. *Ann. N.Y. Acad. Sci.* **987,** 68–78.

Klein, R. J., Zeiss, C., Chew, E. Y., Tsai, J. Y., Sackler, R. S., Haynes, C., Henning, A. K., Sangiovanni, J. P., Mane, S. M., Mayne, S. T., Bracken, M. B., Ferris, M. B., et al. (2005). Complement factor H polymorphism in age-related macular degeneration. [See comment]. *Science* **308,** 385–389.

Kuttner-Kondo, L. A., Dybvig, M. P., Mitchell, L. M., Muqim, N., Atkinson, J. P., Medof, M. E., and Hourcade, D. E. (2003). A corresponding tyrosine residue in the C2/factor B type A domain is a hot spot in the decay acceleration of the complement C3 convertases. *J. Biol. Chem.* **278,** 52386–52391.

Laine, M., Jarva, H., Seitsonen, S., Haapasalo, K., Lehtinen, M. J., Lindeman, N., Anderson, D. H., Johnson, P. T., Jarvela, I., Jokiranta, T. S., Hageman, G. S., Immonen, G. S., et al. (2007). Y402H polymorphism of complement factor H affects binding affinity to C-reactive protein. *J. Immunol.* **178,** 3831–3836.

Lambris, J. D. (1989). "Third Component of Complement." Springer-Verlag, New York.

Lau, L. I., Chen, S. J., Cheng, C. Y., Yen, M. Y., Lee, F. L., Lin, M. W., Hsu, W. M., and Wei, Y. H. (2006). Association of the Y402H polymorphism in complement factor H gene and neovascular age-related macular degeneration in Chinese patients. *Invest. Ophthalmol. Vis. Sci.* **47,** 3242–3246.

Law, S. K., Lichtenberg, N. A., and Levine, R. P. (1979). Evidence for an ester linkage between the labile binding site of C3b and receptive surfaces. *J. Immunol.* **123,** 1388–1394.

Liszewski, M. K., Farries, T. C., Lublin, D. M., Rooney, I. A., and Atkinson, J. P. (1996). Control of the complement system. *Adv. Immunol.* **61,** 201–283.

Liszewski, M. K., Kemper, C., Price, J. D., and Atkinson, J. P. (2005). Emerging roles and new functions of CD46. *Springer Semin. Immunopathol.* **27,** 345–358.

Liszewski, M. K., Leung, M. K., Schraml, B., Goodship, T. H., and Atkinson, J. P. (2007). Modeling how CD46 deficiency predisposes to atypical hemolytic uremic syndrome. *Mol. Immunol.* **44,** 1559–1568.

Lokki, M. L., and Koskimies, S. A. (1991). Allelic differences in hemolytic activity and protein concentration of BF molecules are found in association with particular HLA haplotypes. *Immunogenetics* **34,** 242–246.

Manderson, A. P., Botto, M., and Walport, M. J. (2004). The role of complement in the development of systemic lupus erythematosus. *Annu. Rev. Immunol.* **22,** 431–456.

Manuelin, T., Hellwage, J., Meri, S., Capriolo, J., Noris, M., Heinen, S., Jozsi, M., Neumann, H. P. H., Remuzzi, G., and Zipfel, P. F. (2003). Mutations in factor H reduce binding affinity to C3b and heparin and surface attachment to endothelial cells in haemolytic uraemic syndrome. *J. Clin. Invest.* **111,** 1181–1190.

Matsuoka, M., Ogata, N., Otsuji, T., Nishimura, T., Takahashi, K., and Matsumura, M. (2004). Expression of pigment epithelium derived factor and vascular endothelial growth factor

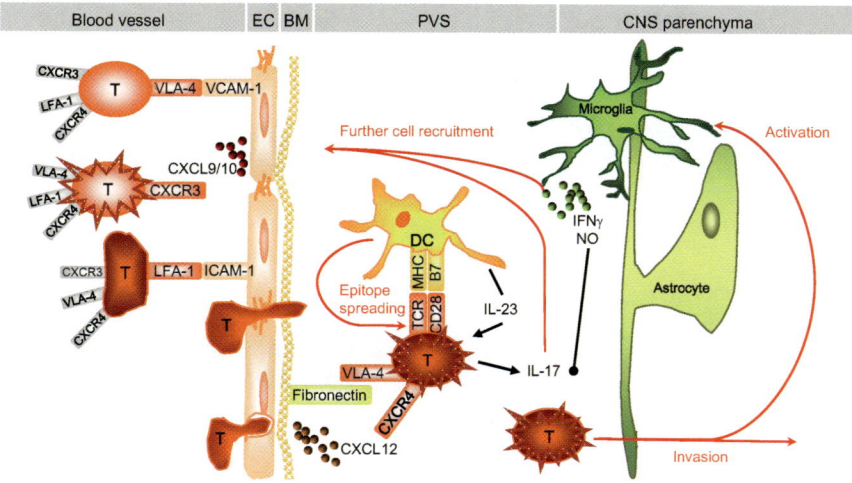

PLATE 1

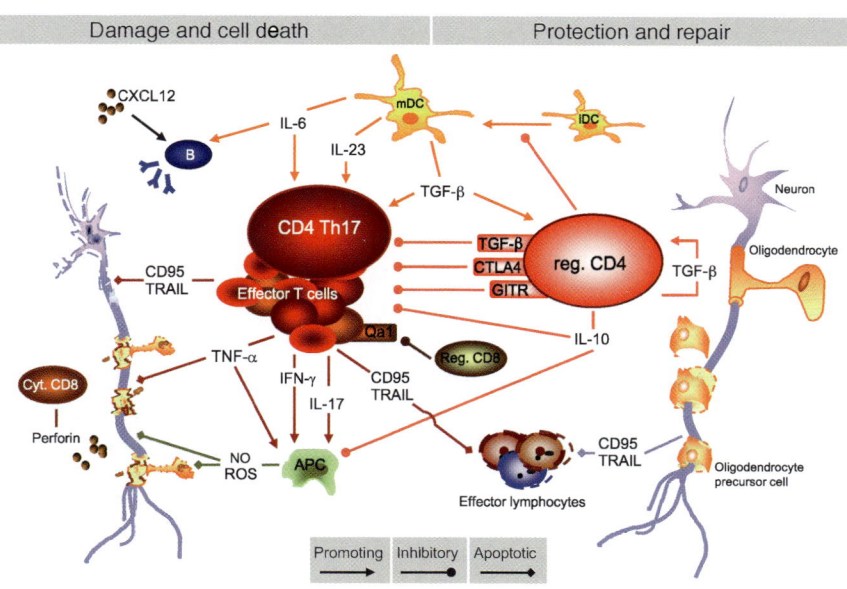

PLATE 2

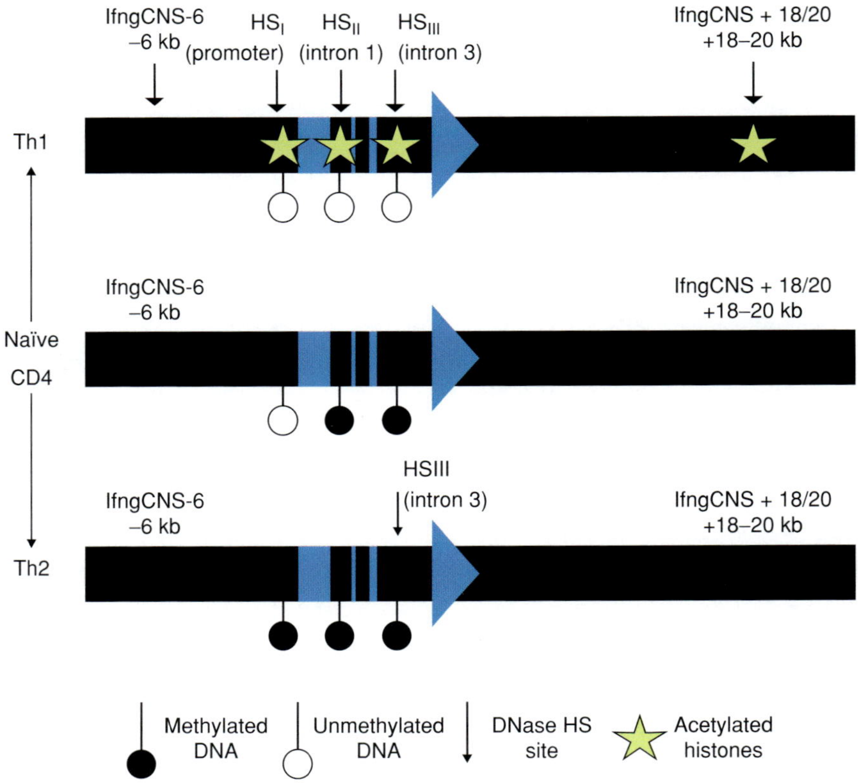

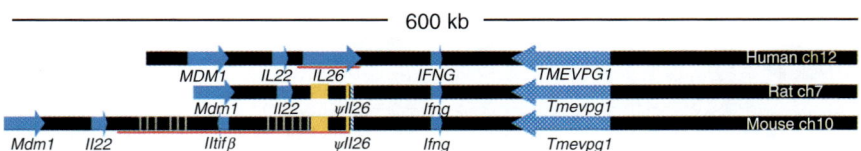

PLATE 3

PLATE 4

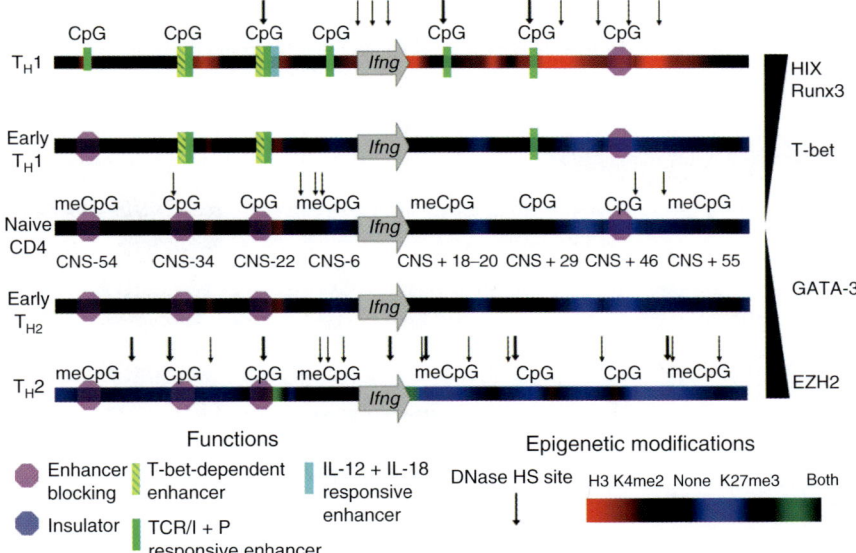

PLATE 5

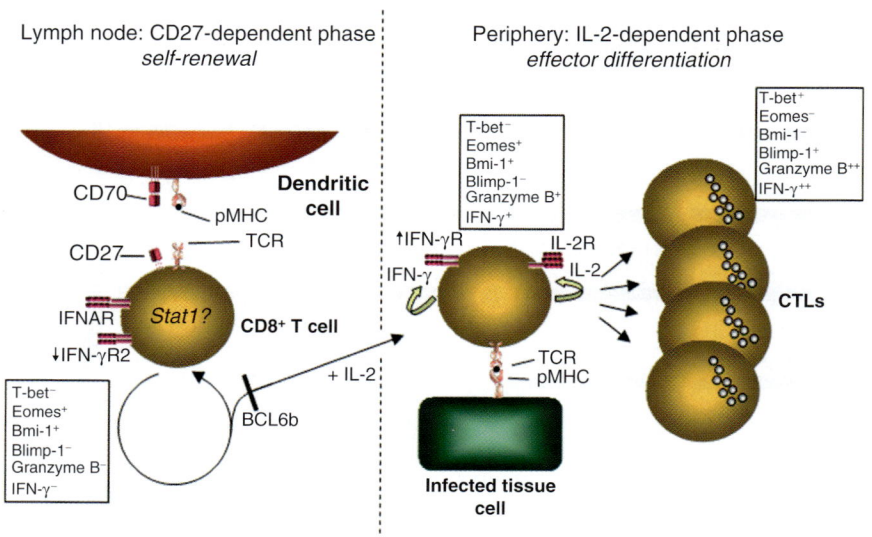

PLATE 6

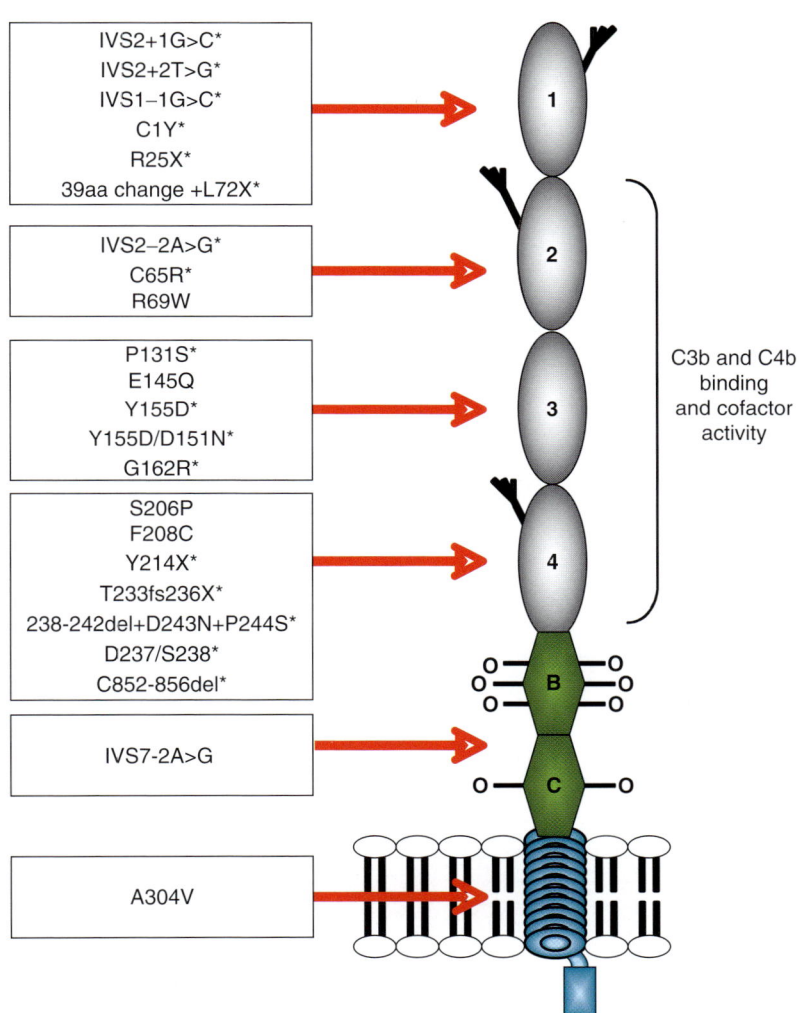

PLATE 7

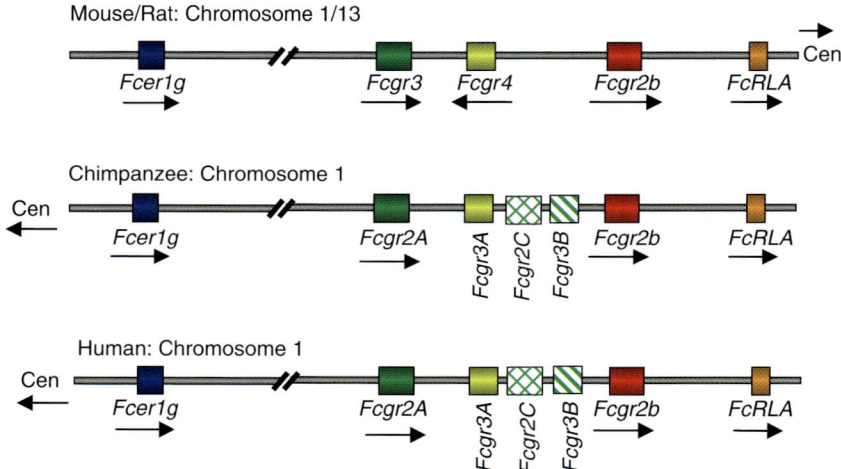

PLATE 8

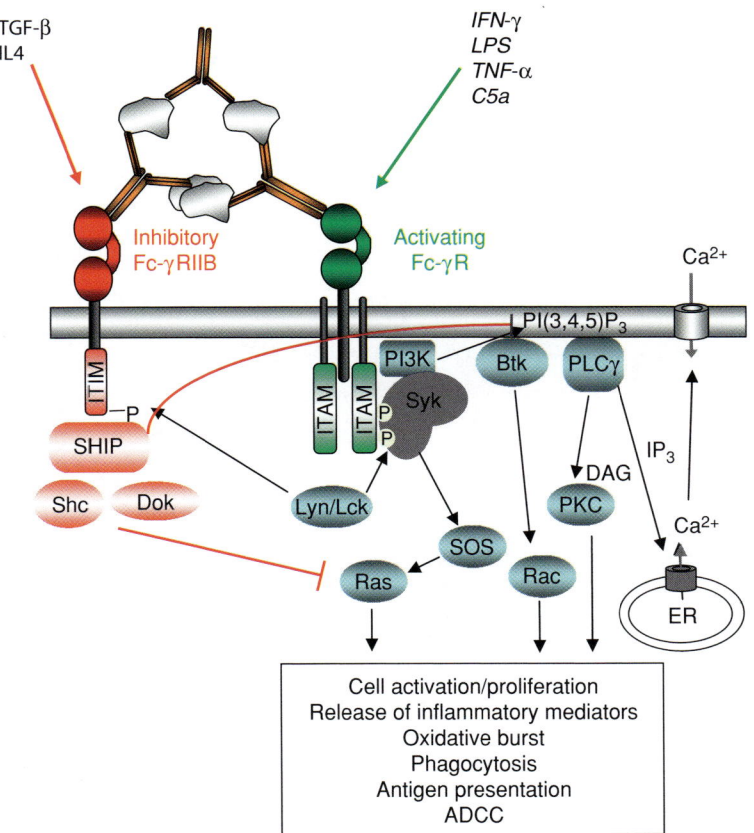

PLATE 9

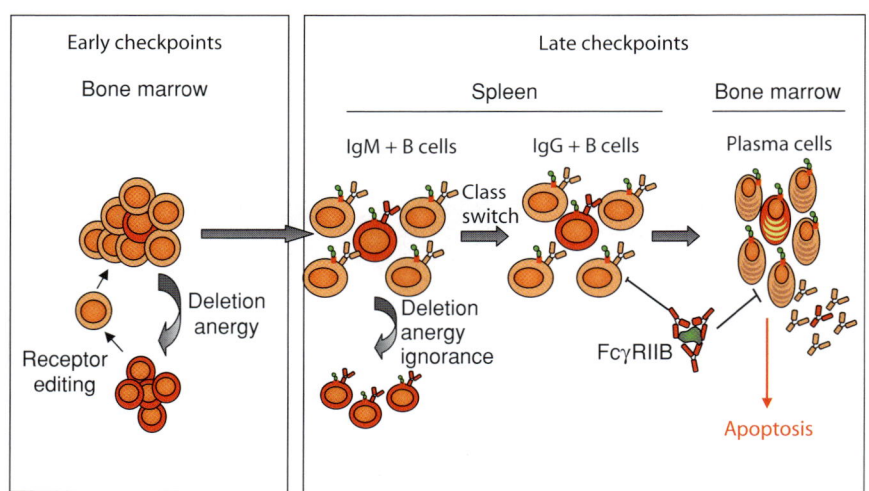

PLATE 10

in choroidal neovascular membranes and polypoidal choroidal vasculopathy. *Br. J. Ophthalmol.* **88,** 809–815.

Medof, M. E., Kinoshita, T., and Nussenzweig, V. (1984). Inhibition of complement activation on the surface of cells after incorporation of decay-accelerating factor (DAF) into their membranes. *J. Exp. Med.* **160,** 1558–1578.

Mevorach, D., Mascarenhas, J. O., Gershov, D., and Elkon, K. B. (1998). Complement-dependent clearance of apoptotic cells by human macrophages. *J. Exp. Med.* **188,** 2313–2320.

Molina, H. (2002). The murine complement regulator Crry: New insights into the immunobiology of complement regulation. *Cell. Mol. Life Sci.* **59,** 220–229.

Navratil, J. S., Liu, C. C., and Ahearn, J. M. (2006). Apoptosis and autoimmunity. *Immunol. Res.* **36,** 3–12.

Neumann, H. P. H., Salzmann, M., Bohnert-Iwan, B., Mannuelian, T., Skerka, C., Lenk, D., Bender, B. U., Cybulla, M., Priegler, P., Konigsrainer, A., Neyer, U., Bock, U., *et al.* (2003). Hemolytic uremic syndrome and mutations of the factor H gene: A registry-based study of German speaking countries. *J. Med. Genet.* **40,** 676–681.

Noris, M., and Remuzzi, G. (2005). Hemolytic uremic syndrome. *J. Am. Soc. Nephrol.* **16,** 1035–1050.

Noris, M., Brioschi, S., Caprioli, J., Todeschini, M., Bresin, E., Porrati, F., Gamba, S., and Remuzzi, G. (2003). Familial haemolytic uraemic syndrome and an MCP mutation. *Lancet* **362,** 1542–1547.

Oglesby, T. J., Allen, C. J., Liszewski, M. K., White, D. J. G., and Atkinson, J. P. (1992). Membrane cofactor protein (MCP; CD46) protects cells from complement-mediated attack by an intrinsic mechanism. *J. Exp. Med.* **175,** 1547–1551.

Pangburn, M. K., and Muller-Eberhard, H. J. (1984). The alternative pathway of complement. *Springer Semin. Immunopathol.* **7,** 163–192.

Perez-Caballero, D., Gonzalez-Rubio, C., Gallardo, M. E., Vera, M., Lopez-Trascasa, M., and de Cordoba, S. R. (2001). Clustering of missense mutations in the C-terminal region of factor H in atypical hemolytic uremic syndrome. *Am. J. Hum. Genet.* **68,** 478–484.

Pickering, M. C., Botto, M., Taylor, P. R., Lachmann, P. J., and Walport, M. J. (2000). Systemic lupus erythematosus, complement deficiency, and apoptosis. *Adv. Immunol.* **76,** 227–324.

Pickering, M. C., Cook, H. T., Warren, J., Bygrave, A. E., Moss, J., Walport, M. J., and Botto, M. (2002). Uncontrolled C3 activation causes membranoproliferative glomerulonephritis in mice deficient in complement factor H. *Nat. Genet.* **31,** 424–428.

Pickering, M. C., Goicoechea de Jorge, E., Martinez-Barricarte, R., Recalde, S., Garcia-Layana, K. L., Moss, J., Walport, M. J., Cook, H. T., Rodriguez de Cordoba, S., and Botto, M. (2007). Spontaneous haemolytic uraemic syndrome triggered by complement factor H lacking surface recognition domains. *J. Exp. Med.* **204,** 1249–1256.

Reis, E., Falcao, D. A., and Isaac, L. (2006). Clinical aspects and molecular basis of primary deficiencies of complement component C3 and its regulatory proteins factor I and factor H. *Scand. J. Immunol.* **63,** 155–168.

Remuzzi, G., Ruggenenti, P., Codazzi, D., Noris, M., Caprioli, J., Locatelli, G., and Gridelli, B. (2002). Combined kidney and liver transplantation for familial haemolytic uraemic syndrome. *Lancet* **359,** 1671–1672.

Remuzzi, G., Ruggenenti, P., Colledan, M., Gridelli, B., Bertani, A., Bettinaglio, P., Bucchioni, S., Sonzogni, A., Bonanomi, E., Sonzogni, V., Platt, J. L., Perico, J. L., *et al.* (2005). Hemolytic uremic syndrome: A fatal outcome after kidney and liver transplantation performed to correct factor H gene mutation. *Am. J. Transplant.* **5,** 1146–1150.

Richards, A., Buddles, M. R., Donne, R. L., Kaplan, B. S., Kirk, E., Venning, M. C., Tielemans, C. L., Goodship, J. A., and Goodship, T. H. (2001). Factor H mutations in

hemolytic uremic syndrome cluster in exons 18–20, a domain important for host cell recognition. *Am. J. Hum. Genet.* **68,** 485–490.
Richards, A., Kemp, E. J., Liszewski, M. K., Goodship, J. A., Lampe, A. K., Decorte, R., Muslumanolu, M. H., Kavukcu, S., Filler, G., Pirson, Y., Wen, L. S., Atkinson, L. S., *et al.* (2003). Mutations in human complement regulator, membrane cofactor protein (CD46), predispose to development of familial hemolytic uremic syndrome. *Proc. Natl. Acad. Sci. USA* **100,** 12966–12971.
Richards, A., Liszewski, M. K., Kavanagh, D., Fang, C. J., Moulton, E. A., Fremeaux-Bacchi, V., Remuzzi, G., Noris, M., Goodship, T. H. J., and Atkinson, J. P. (2007). Implications of the initial mutations in membrane cofactor protein (MCP; CD46) leading to atypical hemolytic uremic syndrome. *Mol. Immunol.* **44,** 111–122.
Riley-Vargas, R. C., and Atkinson, J. P. (2003). Expression of membrane cofactor protein (MCP; CD46) on spermatozoa: Just a complement inhibitor? *MAI* **3,** 75–78.
Riley-Vargas, R. C., Gill, D. B., Kemper, C., Liszewski, M. K., and Atkinson, J. P. (2004). CD46: Expanding beyond complement regulation. *Trends Immunol.* **25,** 496–503.
Riley-Vargas, R. C., Lanzendorf, S., and Atkinson, J. P. (2005). Targeted and restricted complement activation on acrosome-reacted spermatozoa. *J. Clin. Invest.* **115,** 1241–1249.
Rivera, A., Fisher, S. A., Fritsche, L. G., Keilhauer, C. N., Lichtner, P., Meitinger, T., and Weber, B. H. (2005). Hypothetical LOC387715 is a second major susceptibility gene for age-related macular degeneration, contributing independently of complement factor H to disease risk. *Hum. Mol. Genet.* **14,** 3227–3236.
Sadallah, S., Gudat, F., Laissue, J. A., Spath, P. J., and Schifferli, J. A. (1999). Glomerulonephritis in a patient with complement factor I deficiency. *Am. J. Kidney Dis.* **33,** 1153–1157.
Saland, J., Emre, S., Schneider, B., Benchimol, C., Ames, S., Bromberg, J., Remuzzi, G., Strain, L., and Goodship, T. H. J. (2006). Favorable long-term outcome after liver-kidney transplant for recurrent hemolytic uremic syndrome associated with a factor H mutation. *Am. J. Transplant.* **6,** 1948–1952.
Sanchez-Corral, P., Perez-Caballero, D., Huarte, O., Simckes, A., Goicoechea, E., Lopez-Trascasa, M., and Rodriguez de Cordoba, S. (2002). Structural and functional characterization of factor H mutations associated with atypical haemolytic uraemic syndrome. *Am. J. Hum. Genet.* **71,** 1285–1295.
Sanchez-Corral, P., Gonzalez-Rubio, C., Rodriguez de Cordoba, S., and Lopez-Trascasa, M. (2004). Functional analysis in serum from atypical Hemolytic Uremic Syndrome patients reveals impaired protection of host cells associated with mutations in factor H. *Mol. Immunol.* **41,** 81–84.
Sjoberg, A. P., Trouw, L. A., Clark, S. J., Sjolander, J., Heinegard, D., Sim, R. B., Day, A. J., and Blom, A. M. (2007). The factor H variant associated with age-related macular degeneration (His-384) and the non-disease-associated form bind differentially to C-reactive protein, fibromodulin, DNA, and necrotic cells. *J. Biol. Chem.* **282,** 10894–10900.
Spitzer, D., Mitchell, L. M., Atkinson, J. P., and Hourcade, D. E. (2007). Properdin can initiate complement activation by binding specific target surfaces and providing a plate form for *de novo* convertase assembly. *J. Immunol.* **179,** 2600–2608.
Sprong, T., Roos, D., Weemaes, C., Neeleman, C., Geesing, C. L., Mollnes, T. E., and van Deuren, M. (2006). Deficient alternative complement pathway activation due to factor D deficiency by 2 novel mutations in the complement factor D gene in a family with meningococcal infections. *Blood* **107,** 4865–4870.
Stahl, G. L., Xu, Y., Hao, L., Miller, M., Buras, J. A., Fung, M., and Zhao, H. (2003). Role for the alternative complement pathway in ischemia/reperfusion injury. *Am. J. Pathol.* **162,** 449–455.
Tarr, P. I., Gordon, C. A., and Chandler, W. L. (2005). Shiga-toxin-producing *Escherichia coli* and haemolytic uraemic syndrome. *Lancet* **365,** 1073–1086.
Thurman, J. M., and Holers, V. M. (2006). The central role of the alternative complement pathway in human disease. *J. Immunol.* **176,** 1305–1310.

Uka, J., Tamura, H., Kobayashi, T., Yamane, K., Kawakami, H., Minamoto, A., and Mishima, H. K. (2006). No association of complement factor H gene polymorphism and age-related macular degeneration in the Japanese population. *Retina* **26**, 985–987.

van Leeuwen, R., Klaver, C. C., Vingerling, J. R., Hofman, A., and de Jong, P. T. (2003). Epidemiology of age-related maculopathy: A review. *Eur. J. Epidemiol.* **18**, 845–854.

Vaziri-Sani, F., Holmberg, L., Sjoholm, A. G., Kristoffersson, A. C., Manea, M., Fremeaux-Bacchi, V., Fehrman-Ekholm, I., Raafat, R., and Karpman, D. (2006). Phenotypic expression of factor H mutations in patients with atypical hemolytic uremic syndrome. *Kidney Int.* **69**, 981–988.

Venables, J. P., Strain, L., Routledge, D., Bourn, D., Powell, H. M., Warwicker, P., Diaz-Torres, M., Sampson, A., Mead, P., Webb, M., Pirson, Y., Jackson, Y., *et al.* (2006). Atypical haemolytic uraemic syndrome associated with a hybrid complement gene. *PLoS Med.* **3**, 1957–1967.

Vik, D. P., Munoz-Canoves, P., Kozono, H., Martin, L. G., Tack, B. F., and Chaplin, D. D. (1990). Identification and sequence analysis of four complement factor H-related transcripts in mouse liver. *J. Biol. Chem.* **265**, 3193–3201.

Vyse, T. J., Morley, B. J., Bartok, I., Theodoridis, E. L., Davies, K. A., Webster, A. D., and Walport, M. J. (1996). The molecular basis of hereditary complement factor I deficiency. *J. Clin. Invest.* **97**, 925–933.

Walport, M. J. (2001a). Complement. Second of two parts. *N. Engl. J. Med.* **344**, 1140–1144.

Walport, M. J. (2001b). Complement. First of two parts. *N. Engl. J. Med.* **344**, 1058–1066.

Walport, M. J., and Lachmann, P. J. (1988). Erythrocyte complement receptor type 1, immune complexes, and the rheumatic diseases. *Arthritis Rheum.* **31**, 153–158.

Warwicker, P., Goodship, T. H. J., Donne, R. L., Pirson, Y., Nicholls, A., Ward, R. M., Turnpenny, P., and Goodship, J. A. (1998). Genetic studies into inherited and sporadic hemolytic uremic syndrome. *Kidney Int.* **53**, 836–844.

Whaley, K. (1980). Biosynthesis of the complement components and the regulatory proteins of the alternative complement pathway by human peripheral blood monocytes. *J. Exp. Med.* **151**, 501–516.

Wong, W. W., and Fearon, D. T. (1985). p65: A C3b-binding protein on murine cells that shares antigenic determinants with the human C3b receptor (CR1) and is distinct from murine C3b receptor. *J. Immunol.* **134**, 4048–4056.

Xu, C., Mao, D., Holers, V. M., Palanca, B., Cheng, A. M., and Molina, H. (2000). A critical role for the murine complement regulator Crry in fetomaternal tolerance. *Science* **287**, 498–501.

Ziegler, J. B., Alper, C. A., Rosen, R. S., Lachmann, P. J., and Sherington, L. (1975). Restoration by purified C3b inactivator of complement-mediated function *in vivo* in a patient with C3b inactivator deficiency. *J. Clin. Invest.* **55**, 668–672.

Zipfel, P. F., Edey, M., Heinen, S., Jozsi, M., Richter, H., Misselwitz, J., Hoppe, B., Routledge, D., Strain, L., Hughes, A. E., Goodship, J. A., Licht, J. A., *et al.* (2007). Deletion of complement factor H-related genes CFHR1 and CFHR3 is associated with atypical hemolytic uremic syndrome. *PLoS Genet.* **3**, e41.

CHAPTER 5

Fc-Receptors as Regulators of Immunity

Falk Nimmerjahn[*,†] and **Jeffrey V. Ravetch**[†]

Contents

	1. Introduction	180
	2. The Family of Activating and Inhibitory FcRs	181
	3. Activating and Inhibitory FcR Signaling Pathways	183
	4. The Role of Activating and Inhibitory FcRs on Innate Immune Effector Cells	185
	5. Modulation of Antibody Activity	188
	6. Activating and Inhibitory FcR Expression on DCs	190
	7. FcγRIIB as a Master Regulator of Humoral Tolerance and Plasma Cell Survival	192
	8. Summary and Outlook	195
	Acknowledgments	196
	References	196

Abstract

Receptors for immunoglobulins [Fc-receptors (FcRs)] are widely expressed throughout the immune system. By binding to the antibody Fc-portion, they provide a link between the specificity of the adaptive immune system and the powerful effector functions triggered by innate immune effector cells. By virtue of coexpression of activating and inhibitory FcRs on the same cell, they set a threshold for immune cell activation by immune complexes (ICs). Besides their involvement in the efferent phase of an immune response, they are also important for modulating adaptive immune responses by regulating B cell and dendritic cell (DC) activation. Deletion of the inhibitory FcR leads to the loss of tolerance in the humoral

---

[*] Laboratory for Experimental Immunology and Immunotherapy, Nikolaus-Fiebiger-Center for Molecular Medicine, University of Erlangen-Nuremberg, Erlangen 91054, Germany
[†] Laboratory for Molecular Genetics and Immunology, Rockefeller University, New York, New York

Advances in Immunology, Volume 96
ISSN 0065-2776, DOI: 10.1016/S0065-2776(07)96005-8

© 2007 Elsevier Inc.
All rights reserved.

immune system and the development of autoimmune disease. Uptake of ICs by FcRs on DCs and the concomitant triggering of activating and inhibitory signaling pathways will determine the strength of the initiated T-cell response. Loss of this balanced signaling results in uncontrolled responses that can lead to the damage of healthy tissues and ultimately to the initiation of autoimmune processes. In this chapter, we will discuss how coexpression of different activating and inhibitory receptors on different immune cells of the innate and adaptive immune system modulates cell activity. Moreover, we will focus on exogenous factors that can influence the balanced triggering of activating and inhibitory FcRs, such as the cytokine milieu and the role of differential antibody glycosylation.

1. INTRODUCTION

Autoimmune, infectious, and malignant diseases affect millions of people worldwide. Antibodies together with T cells and innate immune effector cells are crucial to defend the body from such threats, and a complex array of interactions between these different immune system players is necessary to ensure the success of an immune response. Cellular receptors for the different immunoglobulin isotypes (IgA, IgE, IgM, and IgG), so-called Fc-receptors (FcRs), are involved in regulating and executing antibody-mediated responses (Ravetch, 2003). By doing so, they link the specificity of the adaptive immune system to the powerful effector functions triggered by innate immune cells such as mast cells, neutrophils, monocytes, and macrophages. It is of utmost importance, however, that these proinflammatory reactions are tightly regulated to prevent destruction of healthy tissues. If this regulation fails, overwhelming responses and in the worst case chronic autoimmune diseases might be initiated (Dijstelbloem et al., 2001). There is convincing evidence that imbalanced immune responses are responsible for autoimmune diseases such as arthritis, multiple sclerosis, and systemic lupus erythematosus (SLE). It is widely accepted that many factors, including genetic and environmental components, are involved in the initiation and severity of autoimmune symptoms. Thus, identification of these components might be helpful to gain further insight into these diseases and to develop novel immunotherapeutic strategies to interfere with chronic inflammation. In the opposite scenario, as for example in immunotherapy of cancer or viral infections, this knowledge might be useful to enhance proinflammatory responses to clear pathogen-infected or malignant cells. This chapter will focus on the role of cellular FcRs for immunoglobulin G (IgG), the Fcγ-receptors (FcγR), in these different processes, including the maintenance of humoral tolerance and the regulation of adaptive and innate immune responses.

Moreover, we will include an overview of exogenous factors, such as pro- and antiinflammatory cytokines and differential antibody glycosylation that impact the resulting cellular response by changing the expression level or interaction with FcRs. There are several excellent reviews dealing with the important roles of FcRs for other antibody isotypes, such as FcαRs and FcεRs, which will not be covered in this chapter (Kraft and Novak, 2006; Wines and Hogarth, 2006; Woof and Kerr, 2006). Ultimately, we will briefly discuss how this information might be used for novel therapeutic approaches.

2. THE FAMILY OF ACTIVATING AND INHIBITORY FcRs

FcRs are widely expressed on cells of the immune system and select other cell types, such as endothelial cells, mesangial cells, and osteoclasts; one of the few hematopoietic cell types that do not show notable FcR expression are T cells (Daeron, 1997; Hulett and Hogarth, 1994; Ravetch, 2003). Four different classes of FcRs have been identified in rodents, which are called FcγRI, FcγRIIB, FcγRIII, and FcγRIV (Nimmerjahn and Ravetch, 2006; Nimmerjahn et al., 2005). FcγRs are well conserved between different mammals and orthologous proteins to these rodent receptors were found in most species. The corresponding human proteins are called FcγRIA, FcγRIIB (CD32B), FcγRIIA (CD32A), FcγRIIC, FcγRIIIA (CD16), and FcγRIIIB. Although the extracellular portion of FcγRIIA is highly homologous to mouse FcγRIII, the intracellular portion differs significantly. Other human FcR genes such as FcγRIB and FcγRIC do not code for functional proteins due to disrupted open reading frames. In addition, FcγRIIIB, a GPI-anchored FcR selectively expressed on neutrophils, is not found in mice. Structurally, FcRs as well as their ligands, the family of IgG molecules consisting of four members in mice (IgG1, IgGa, IgG2b, IgG3) and humans (IgG1–IgG4), belong to the large immunoglobulin superfamily. Whereas the majority of FcRs have two extracellular domains, FcγRI has an additional third domain which has been suggested to be important for the higher affinity of this receptor for IgG (Allen and Seed, 1989). Resolution of the crystal structure of human FcγRIIIA bound to IgG1 was crucial in defining the precise FcR–IgG interaction sites (Radaev et al., 2001; Sondermann et al., 2000). Thus, only one of the two extracellular domains makes contact with the CH2-domain of the antibody Fc-portion (Radaev and Sun, 2002). This interaction site is different from other IgG-binding proteins such as protein A/G, mannose-binding lectin (MBL), or the neonatal Fc-receptor (FcRn) (Jefferis and Lund, 2002). Moreover, this structural data in combination with results obtained from other methods suggests a 1:1 model of antibody–FcR interaction (Kato *et al.*, 2000; Zhang *et al.*, 2000). On a functional level, FcRs can be divided by two different ways: first, based on the affinity for their ligand and second, based on the type of signaling pathway that is initiated on FcR cross-linking. The majority of FcRs including FcγRIIB, FcγRIII, and FcγRIV as well as their corresponding human counterparts FcγRIIA/B/C and FcγRIIIA/B have a low affinity for the IgG Fc-portion in the micromolar range (Dijstelbloem *et al.*, 2001; Nimmerjahn and Ravetch, 2006; Nimmerjahn *et al.*, 2005). Only FcγRI displays a higher affinity (10^8–10^9 M^{-1}) enabling significant binding to monomeric antibodies. All other FcRs selectively interact with antibodies in the form of immune complexes (ICs), which usually consist of multiple antibodies bound to their target antigen.

FcRs differ in regard to the signaling pathways they initiate. Thus, there is one inhibitory receptor, FcγRIIB; all other FcRs with the exeption of human FcγRIIIB, which has no signaling function, trigger activating signaling pathways (Ravetch, 2003). On the genomic level the FcR genes are clustered in proximity with the novel family of FcR-like genes on chromosome 1 in mice, chimpanzees, and humans (Fig. 5.1). While the gene organization is highly conserved, FcR genes of other species such as rats, dogs, pigs, cows, and cats are localized on different chromosomes (Nimmerjahn and Ravetch, 2006). With the increasing availability of

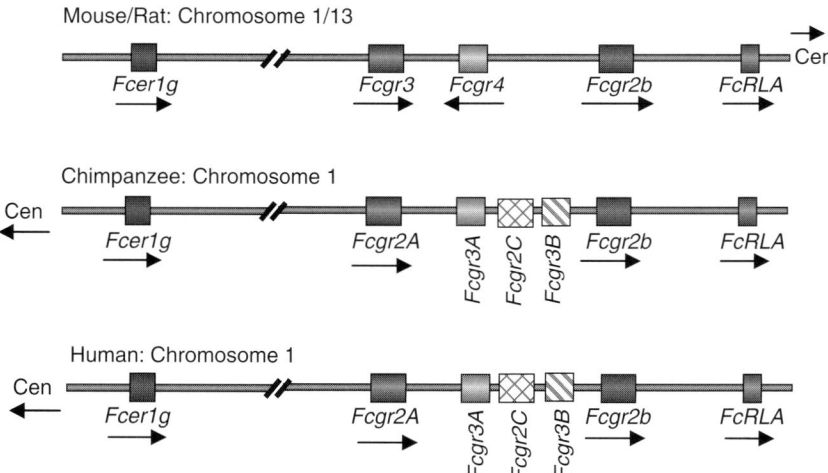

FIGURE 5.1 Organization of rodent and primate FcR genes. The genomic organization of the different FcR genes in humans, chimpanzees, and rodents (mouse and rat) is shown according to the ensembl database (www.ensembl.org). The different colors identify genes that are closely related and predicted to be orthologues. FcRLA (FcRX) is an FcR-like gene that is closely linked to the classical FcR genes. The position of the centromer (Cen) and the direction of transcription are indicated by the arrows. (See Plate 8 in Color Plate Section.)

genome sequence information, it is becoming clear that primates and humans have the greatest variety of FcR genes, most likely due to recent gene duplication and diversification processes, although some of the genes do not code for functional proteins (Qiu et al., 1990). Despite this higher complexity the underlying mechanisms and functions of these proteins defined in rodent animal models have been largely recapitulated by results obtained in humans supporting the value of these model systems. For the preclinical evaluation of novel antibody-based therapeutics, the development of novel humanized mouse models, such as mice transgenic for multiple human FcRs, will be of great value.

3. ACTIVATING AND INHIBITORY FcR SIGNALING PATHWAYS

One widely applicable rule with respect to the cellular expression pattern is that activating and inhibitory FcRs are coexpressed on the same cell (Ravetch, 2003). Thus, IC binding to cells will trigger both activating and inhibitory signaling pathways, thereby setting a threshold for cell activation, which will determine the magnitude of the ensuing response (Fig. 5.2). On innate immune effector cells such as mast cells, neutrophils, and macrophages, these dual signals regulate a variety of downstream responses such as cell degranulation, phagocytosis, antibody-dependent cellular cytotoxicity (ADCC), and antigen presentation (Ravetch, 2003). On B cells that do not express activating FcRs, FcγRIIB regulates activating signaling pathways initiated by the B-cell receptor (BCR) (Bolland and Ravetch, 1999; Ravetch and Lanier, 2000). In addition to the inhibitory FcγRIIB, B cells express several of the recently discovered FcR-like proteins (Davis et al., 2004, 2005; Ehrhardt et al., 2003, 2005; Facchetti et al., 2002). As efforts to demonstrate binding to immunoglobulins have not been successful to date, FcγRIIB remains the only molecule with Fc-fragment binding capacity that will be triggered if B cells bind to antigen presented in the form of ICs.

Another difference between activating and inhibitory FcRs is that in contrast to the single-chain inhibitory FcR that contains an immunoreceptor tyrosine-based inhibitory motif (ITIM) in its cytosolic domain, activating FcRs with the exception of human FcγRIIA/C cannot signal autonomously (Hulett and Hogarth, 1994). They have to associate with additional signaling adaptor molecules that might differ depending on the cell type. In NK cells, for example, the ζ-chain serves as an adaptor molecule, whereas the so-called FcR common γ-chain associates with activating receptors in the majority of other cell types (Fig. 5.2). In mast cells yet another adaptor molecule called the β-chain was found to be associated with FcγRIII and FcεRI (Kinet, 1999). All of these adaptors contain immunoreceptor tyrosine-based activation motifs (ITAM) in their

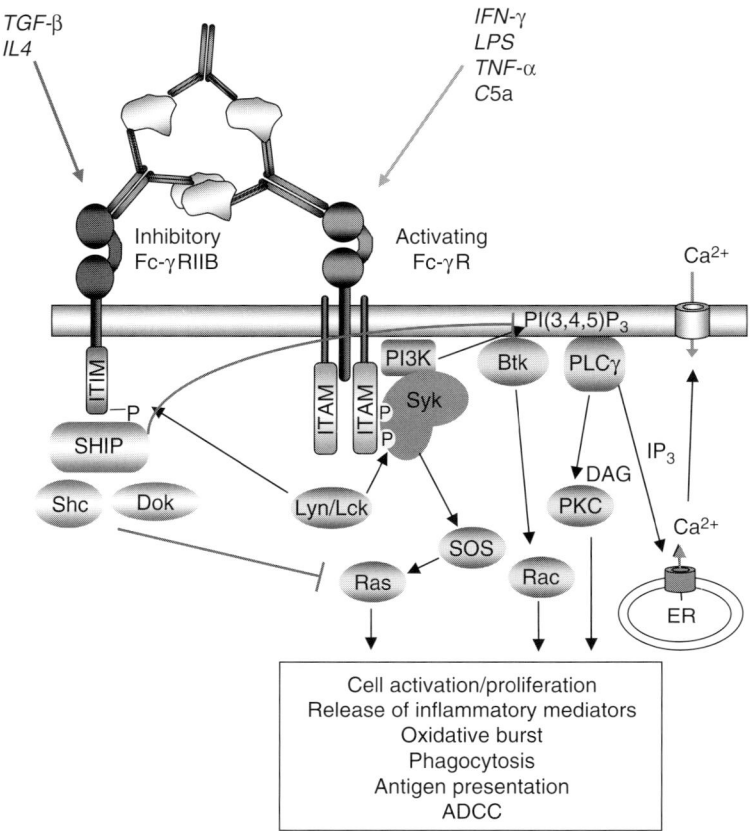

FIGURE 5.2 Coregulation of activating and inhibitory FcR signaling. IC binding to innate immune effector cells, such as monocytes, granulocytes, and macrophages, simultaneously triggers activating and inhibitory signaling pathways. Upon receptor cross-linking, Lyn phosphorylates both ITAM and ITIM in the cytoplasmic domain of the FcR common γ-chain and FcγRIIB, thereby initiating both signaling pathways. Red lines indicate points where inhibitory signaling pathways interfere with their activating counterparts. Moreover, factors that change the balanced expression of activating and inhibitory FcRs are shown at the top. See text for further details. (See Plate 9 in Color Plate Section.)

cytosolic portion that become tyrosine-phosphorylated by members of the Src family of kinases that may be associated with the receptor in an inactive form on FcR cross-linking. Phosphorylation of the ITAM sequences creates SH2 sites for docking and activation of Syk kinases. Importantly, low-affinity ligands result in nonproductive signaling complexes that fail to couple to downstream events and might even behave as antagonists (Torigoe *et al.*, 1998). Depending on the cell type and individual FcR, the involved Src kinase family members might vary. For example, Lyn is associated with the FcεRI pathway in mast cells, whereas Lck is associated with FcγRIIIA in NK cells. In macrophages both of these kinases as well as Hck are associated with FcγRI and FcγRIIA (Takai, 2002). Following the phosphorylation of the ITAM motif the recruitment and activation of Syk kinases ensues that leads to the recruitment of a variety of intracellular substrates, including PI3K, Btk and other Tec family kinases, phospholipase C-γ (PLCγ), and adaptor proteins such as SLP-76 and BLNK (Fig. 5.2) (Takai, 2002). Moreover, the Ras/Raf/MAP kinase pathway is activated through Sos bound to Grb2 that is recruited on phosphorylation of Shc. Another crucial step is the activation of PI3K by Syk, which results in the generation of phosphatidyl-inositol-3-phosphates. This leads to the recruitment of *Btk* and *PLCγ* that recognize PIP3 with their pleckstrin homology (PH) domains leading to the production of inositol triphosphate (IP3) and diacylglycerol (DAG), which are crucial for the mobilization of intracellular calcium and activation of protein kinase C (PKC).

The role of the inhibitory receptor signaling is to dampen these activating pathways by interfering with the generation of key intermediates such as PIP3 (Fig. 5.2). This is initiated by phosphorylation of the ITIM motif in the cytosolic portion of FcγRIIB by Lyn that leads to the recruitment and activation of the SH2-domain containing inositol 5′ phosphatase (SHIP) (Takai, 2002). The key function of activated SHIP is to hydrolyze phosphatidyl inositol intermediates, such as PIP3, and thereby to interfere with the membrane recruitment of *Btk* and *PLCγ*, thus dampening ITAM signaling-mediated calcium release and downstream effector functions such as ADCC, phagocytosis, cytokine secretion, and release of inflammatory mediators. The Ras pathway is also inhibited by recruitment of Shc and DOK to tyrosine-phosphorylated SHIP, which inhibits cell proliferation.

On B cells, another ITIM- and SHIP-independent signaling pathway that leads to apoptosis via an Abl-family kinase-dependent pathway has been described for selective cross-linking of FcγRIIB without concomitant triggering of the BCR (Pearse *et al.*, 1999; Tzeng *et al.*, 2005). This situation might arise during the germinal center reaction when somatic hypermutation generates BCRs that lose specificity for their cognate antigen presented on follicular dendritic cells (FDCs). The importance of this pathway for the maintenance of humoral tolerance and plasma cell homeostasis will be discussed in greater detail later.

4. THE ROLE OF ACTIVATING AND INHIBITORY FcRs ON INNATE IMMUNE EFFECTOR CELLS

As indicated before, virtually all innate immune effector cells with the exception of NK cells coexpress activating and inhibitory FcRs (Fig. 5.2). Thus, the magnitude of any response following IC binding is determined by the level of activating versus inhibitory signaling events. Besides FcγRs, several of these cell types such as basophils, mast cells, and monocytes express FcRs for other immunoglobulin isotypes including FcαRs, FcμRs, and FcεRs, which are also negatively regulated by the inhibitory FcγRIIB. Gene deletion studies were of great importance to gain insight into the role of FcγRs for antibody-mediated effector functions triggered by innate immune cells. Some examples for responses triggered by IC binding to these cell types include phagocytosis, release of inflammatory mediators, and ADCC (Ravetch, 2003).

In rodents, cell surface expression and signaling capacity of all activating FcγRs (FcγRI, FcγRIII, and FcγRIV) is dependent on the γ-chain. Therefore, genetic deletion of this subunit leads to the loss of cell surface expression and functional inactivation of all activating FcRs. Not surprisingly, γ-chain knockout animals had dramatically impaired antibody-mediated effector cell responses (Clynes and Ravetch, 1995; Clynes et al., 1998; Hamaguchi et al., 2006; Nimmerjahn and Ravetch, 2005; Park et al., 1998; Sylvestre and Ravetch, 1994; Takai et al., 1994; Uchida et al., 2004; Zhang et al., 2004). In contrast, studies using mice deficient in a variety of complement proteins such as CR2, C3, C4, or mannose-binding lectin (MBL) failed to demonstrate major defects in these efferent responses (Azeredo da Silveira et al., 2002; Hamaguchi et al., 2006; Nimmerjahn and Ravetch, 2005; Nimmerjahn et al., 2007; Ravetch and Clynes, 1998; Sylvestre et al., 1996; Uchida et al., 2004). As will be discussed later, the proinflammatory activity of activated complement components such as C5a, however, does play an important role by upregulating activating FcRs.

As discussed before the majority of innate immune effector cells express more than one activating FcR. For example, monocytes and macrophages express all activating FcγRs, followed by neutrophils that predominantly express FcγRIII and IV. Therefore, ICs could bind to several activating FcRs and the triggered response might be mediated by all of these receptors. It is important to keep in mind, however, that the individual activating FcRs have a differential affinity for different antibody isotypes (Nimmerjahn and Ravetch, 2005; Nimmerjahn et al., 2005). FcγRIII, for example, can bind to IgG1, IgG2a, and IgG2b subclasses in vitro; FcγRIV shows a more restricted specificity for IgG2a and IgG2b (Hirano et al., 2007; Nimmerjahn et al., 2005). Importantly, the affinity of FcγRIV for these subclasses is more than one order of magnitude higher than for FcγRIII, which suggests that FcγRIV might be the dominant activating receptor for IgG2a and IgG2b in vivo. In addition, this also predicts that the FcγRIIB-imposed negative regulation of these IgG subclasses might be less than for IgG1, for example. Indeed, these predictions could be validated in a variety of in vivo model systems including passive models of antibody-mediated platelet and B-cell depletion, tumor cell destruction, and in more complex active autoimmune models such as glomerulonephritis

(Hamaguchi *et al.*, 2006; Kaneko *et al.*, 2006b; Nimmerjahn and Ravetch, 2005; Nimmerjahn *et al.*, 2005). Low-affinity binding of certain IgE alleles to FcγRIV has been described and it was suggested that allergic responses triggered by IgE might be at least partially dependent on IgE binding to FcγRIV on monocytes (Hirano *et al.*, 2007).

For IgG1, the situation is more straightforward as neither FcγRIV nor FcγRI binds to this antibody subclass (Nimmerjahn and Ravetch, 2006). Thus, FcγRIII deletion abrogated IgG1-mediated effector functions *in vivo* in mouse models of arthritis, glomerulonephritis, IgG-dependent anaphylaxis, IgG-mediated hemolytic anemia, and immunothrombocytopenia (ITP) (Bruhns *et al.*, 2003; Fossati-Jimack *et al.*, 2000; Fujii *et al.*, 2003; Hazenbos *et al.*, 1996; Ji *et al.*, 2002; Meyer *et al.*, 1998; Nimmerjahn and Ravetch, 2005). Despite the capacity of the high-affinity FcγRI to bind to IgG2a, the contribution of this FcR for mediating antibody activity in many of the aforementioned model systems was negligible, which might be due to the saturation of this receptor with monomeric IgG2a serum antibodies in the steady state. Depending on the model system and cytokine environment, FcγRI might participate and enhance antibody-mediated inflammation on *de novo* upregulation or in the presence of high amounts of ICs in peripheral tissues (Barnes *et al.*, 2002; Bevaart *et al.*, 2006; Ioan-Facsinay *et al.*, 2002). In humans, the same principles may apply as human FcγRIIIA has a higher affinity for IgG1 compared to FcγRIIA. In addition, the presence of allelic variants with different affinities for selective antibody isotypes further supports this concept (Dijstelbloem *et al.*, 2001). This is validated by clinical studies with lymphoma patients that were treated with a B cell depleting CD20-specific antibody (Rituximab), as patients with the high-affinity FcγRIIA and FcγRIIIA alleles for this antibody had better clinical responses (Cartron *et al.*, 2002; Weng and Levy, 2003; Weng *et al.*, 2004).

Considering these powerful and potentially dangerous effector responses, regulatory mechanisms must be in place to prevent nonspecific activation. As indicated before this function is mediated by the inhibitory FcγRIIB on innate immune cells (Ravetch and Lanier, 2000). This becomes apparent in animals deficient in FcγRIIB, which have enhanced IC-mediated inflammation and phagocytosis as demonstrated by a stronger Arthus reaction, systemic IgG and IgE-induced anaphylaxis, collagen-induced arthritis (CIA), anti-GBM glomerulonephritis, ITP, hemolytic anemia, and IgG-mediated clearance of pathogens and tumor cells (Clynes *et al.*, 1999, 2000; Nakamura *et al.*, 2000; Nandakumar *et al.*, 2003; Nimmerjahn and Ravetch, 2005; Takai *et al.*, 1996). As FcγRIIB also regulates autoantibody production in B cells, in some of these models both increased autoantibody production due to FcγRIIB deficiency on B cells and heightened effector cell responses are likely to contribute to the observed phenotype, which will be discussed later. Due to the differential affinity of various IgG subclasses for activating FcRs, the level of negative regulation by the inhibitory FcR differs. IgG1 is most strictly regulated due to the lower affinity of FcγRIII compared to FcγRIIB. In contrast, IgG2a and IgG2b are less regulated as FcγRIV has a much higher affinity for these subclasses than FcγRIIB (Nimmerjahn and Ravetch, 2005). The ratio of the affinities of different antibody subclasses for their specific activating and the inhibitory FcR has been termed A/I ratio and is a good predictor of antibody activity *in vivo* (Nimmerjahn and Ravetch, 2005, 2007a).

5. MODULATION OF ANTIBODY ACTIVITY

There are several factors that can influence balanced signaling by activating and inhibitory FcR pairs by changing either their relative expression level or the ligand affinity for the receptor. Activating FcR expression on innate immune cells can be strongly increased by proinflammatory stimuli (LPS), TH-1 cytokines (IFN-γ), and the complement component C5a (Guyre *et al.*, 1983; Nimmerjahn and Ravetch, 2006; Shushakova *et al.*, 2002); in contrast TH-2 cytokines such as IL-4, IL-10, or TGF-β downregulate activating FcR expression and increase the level of FcγRIIB (Fig. 5.2) (Nimmerjahn *et al.*, 2005; Okayama *et al.*, 2000; Pricop *et al.*, 2001; Radeke *et al.*, 2002; Tridandapani *et al.*, 2003). These effects can be cell type specific as IL-4, for example, while upregulating FcγRIIB expression on myeloid cells, downregulates FcγRIIB expression on activated B cells (Rudge *et al.*, 2002). Regarding the mechanism of C5a generation it has been suggested that an FcR dependent and complement independent pathway leads to the generation of this strong proinflammatory mediator (Kumar *et al.*, 2006; Shushakova *et al.*, 2002; Skokowa *et al.*, 2005).

Another factor that can greatly influence antibody binding to FcRs and therefore antibody activity is the sugar moiety attached to all IgG subclasses at the asparagine residue 297 (N297) in the CH2 region of the antibody constant region (Arnold *et al.*, 2007). Genetic or biochemical deletion of this sugar side chain abrogates FcR binding but does not affect the interaction with other proteins such as the FcRn (Arnold *et al.*, 2007; Shields *et al.*, 2001). It consists of a branched heptameric core sugar structure consisting of *N*-acetylglucosamine (GlcNac) and mannose. In addition, this core sugar structure contains variable amounts of branching and terminal sugar residues such as sialic acid, galactose, fucose, and GlcNac. Indeed, in normal human serum more than 30 different IgG glycovariants were identified (Arnold *et al.*, 2006). Considering that at least some of these variants have a differential activity this introduces a high level of complexity. An even greater variety is introduced by the fact that the sugar moieties of the two antibody Fc-fragments might differ with respect to their exact composition (Arnold et al., 2007). Depending on the variable region sequence of the antibody a significant percentage of Fab-associated N-linked sugar side chains can be found. In contrast to the Fc-portion, these sugar moieties are generally fully processed with high levels of terminal sialic acid and galactose residues. Antibody–FcR interactions can be significantly influenced by the presence or absence of these terminal or branching sugar residues. Antibodies without fucose, for example, bind with up to 50-fold higher affinity to mouse activating FcγRIV and human FcγRIIIA (Nimmerjahn and Ravetch, 2005; Shields et al., 2002; Shinkawa et al., 2003). Interestingly, only FcγRIIIA and not FcγRIIA or FcγRIIB binding is influenced by the presence or absence of fucose. It was suggested that the sugar moiety attached to the ASN-162 residue in FcγRIIIA might be responsible for interacting with the branching fucose residues, as an aglycosylated FcR was unable to detect these differences (Ferrara et al., 2006). This finding might have significant implications for the optimization of antibody activity in the therapy of human infectious or malignant disease, and efforts are under way to test these antibody glycovariants in human clinical trials.

Tipping the scale in the opposite direction, high levels of terminal sialic acid residues significantly impair antibody binding to mouse and human FcRs (Kaneko et al., 2006a; Scallon et al., 2007). Interestingly, it is well known that the addition of terminal sugar residues can differ depending on the activation status of the immune system. Autoimmune-prone mouse strains such as MRL/lpr, for example, or human arthritis patients have antibodies with reduced amounts of terminal sialic acid and galactose residues during acute phases of the disease (Bond et al., 1990; Malhotra et al., 1995; Mizuochi et al., 1990). Similarly, in models of accelerated nephrotoxic nephritis and arthritis murine serum IgG antibodies had reduced amounts of sialic acid (Kaneko et al., 2006a; Nimmerjahn et al., 2007). Based on *in vitro* studies, it was initially suggested that antibodies devoid of terminal sialic acid and galactose might be able to activate the lectin pathway of complement activation by means of *de novo* MBL binding to the exposed mannose rich core sugar structure (Arnold et al., 2007; Malhotra et al., 1995). However, more recent *in vivo* studies with mice deficient in MBL argue against this scenario. By using murine autoimmune model systems, such as ITP or arthritis, it was demonstrated that autoantibodies without galactose were not functionally impaired in mice deficient in both MBL subunits (Nimmerjahn et al., 2007). The activity of these antibody glycovariants was abrogated, however, in mouse strains deficient in all activating FcRs, favoring a model in which FcRs and not the complement pathway are responsible for the pathogenicity of agalactosyl antibodies *in vivo*. Taken together, it seems that sialic acid and not galactose is an important regulator of antibody activity *in vivo*.

Importantly, this is not simply achieved by reduced binding of sialic acid rich antibodies to cellular FcRs, but by actively promoting an anti-inflammatory environment. Supporting this model, it was shown that the anti-inflammatory activity of high doses of intravenously administered immunoglobulin G (IVIG) therapy can be potentiated by enriching these IgG preparation for the sialic acid rich fraction (Kaneko et al., 2006a; Nimmerjahn and Ravetch, 2007b). IVIG therapy is an effective treatment for a variety of human autoimmune diseases including SLE, arthritis, multiple sclerosis, and ITP (Bayary et al., 2006). Emphasizing the important role of the inhibitory FcγRIIB in setting a threshold for innate immune effector cell activation, IVIG therapy is critically dependent on the presence of this negative regulator. FcγRIIB deficient animals are no longer protected by IVIG treatment in models of ITP, arthritis, and nephrotoxic nephritis (Bruhns et al., 2003; Kaneko et al., 2006a,b). However, this effect on FcγRIIB is not a direct one. Sialylated IgG upregulates FcRIIB expression on effector macrophages but only in response to another macrophage population, the regulatory macrophage, that is necessary for the anti-inflammatory activity of sialylated IgG. Therefore, despite the proposal of several mechanisms for IVIG activity in humans, it seems likely that here also FcγRIIB will turn out to be the crucial mediator of its anti-inflammatory action.

6. ACTIVATING AND INHIBITORY FcR EXPRESSION ON DCs

DCs are the key cell type for the initiation of cellular and humoral adaptive immune responses (Steinman and Hemmi, 2006). This is achieved by their extraordinary capacity to sample the body for invading pathogens, to phagocytose them, and to present antigenic peptides in the context of major histocompatibility (MHC) molecules to T cells. Besides this well-established function, DCs can also tolerize T cells, depending on the state of maturation of the DC (Steinman et al., 2003). If T cells recognize antigenic peptides on activated DCs, T-cell activation and expansion follows, whereas if peptides are presented on resting DCs, T cells become inactivated or turn into regulatory T cells (Hawiger et al., 2001, 2004; Kretschmer et al., 2005; Yamazaki et al., 2006). Thus, DCs are actively involved in the maintenance of peripheral T-cell tolerance during the steady state. The family of FcR proteins has a dual function on DCs: first, they will bind to ICs, which are the predominant form of an antigen during an immune response, thereby facilitating their phagocytosis and processing for presentation of antigenic peptides on MHC molecules; importantly, FcR-mediated uptake of ICs leads to the presentation of antigenic peptides in the context of MHC class I and MHC class II molecules thus priming CD4 as well as CD8 T-cell responses (Dhodapkar *et al.*, 2002; Groh *et al.*, 2005; Kalergis and Ravetch, 2002; Rafiq et al., 2002; Regnault *et al.*, 1999). Second, ICs trigger activating and inhibitory signaling pathways that, depending on the individual strength of these opposing signals, will determine whether DCs become activated or remain in a resting state. More recent evidence suggests that DCs are also important for the B-cell response. The capacity of DCs to retain antigens for prolonged times in an intact form might allow antigen transport from the periphery to lymphoid organs and presentation to B cells. Consistent with this it was shown that ICs taken up via FcγRIIB are degraded inefficiently and are recycled to the cell–cell surface where they can interact with B cells (Bergtold *et al.*, 2005).

The most important function of FcγRIIB on DCs is to control IC-mediated DC maturation; supporting this notion, DCs derived from FcγRIIB-deficient mice showed an enhanced potential to generate antigen-specific T-cell responses *in vitro* and *in vivo* (Kalergis and Ravetch, 2002). More importantly, FcγRIIB-deficient DCs or DCs incubated with a monoclonal antibody that blocks IC binding to FcγRIIB showed spontaneous maturation, evidenced by the upregulation of costimulatory molecules such as CD80, CD86, and MHC class II (Boruchov *et al.*, 2005; Dhodapkar *et al.*, 2005). This suggests that the inhibitory FcR does not only regulate DC activation but is also actively involved in preventing spontaneous DC maturation under noninflammatory steady-state conditions. Indeed, low levels of ICs can be identified in the serum of healthy individuals, emphasizing the importance of regulatory mechanisms that prevent unwanted DC activation (Dhodapkar *et al.*, 2005).

FcRs play a vital role in the *in vivo* mechanisms by which therapeutic antibodies mediate their activity. In addition to these FcR-mediated effector properties mediated by macrophages and NK cells, antibody therapy of malignant and infectious diseases might be enhanced by transient blockade of FcγRIIB activity on DCs as a novel strategy to enhance antigen-specific immune responses during immunotherapy (Nimmerjahn and Ravetch, 2007a). Considering the regulatory role of FcγRIIB in controlling DC activation it will be important to monitor if this systemic block of FcγRIIB activity initiates unwanted autoimmune responses. Moreover, antibody variants with enhanced binding to activating FcRs might have improved activities *in vivo* by circumventing the concomitant triggering of FcγRIIB (Lazar *et al.*, 2006; Shields *et al.*, 2001). It is important to consider, however, that mice and humans differ in expression of specific FcRs on DCs. Therefore, the development of animals carrying the human FcRs in place of their mouse counterparts will be important preclinical tools for assessing the *in vivo* activity of blocking antibodies for human FcRs.

7. FcγRIIB AS A MASTER REGULATOR OF HUMORAL TOLERANCE AND PLASMA CELL SURVIVAL

The BCR is generated by the random rearrangement of antibody genes in the bone marrow. This also leads to the generation of autoreactive receptors necessitating the presence of checkpoints such as receptor editing, deletion, and anergy of self-reactive BCR species, which ensure that autoreactive B cells are deleted or incapitated from the repertoire (Fig. 5.3) (Goodnow et al., 2005; Grimaldi et al., 2005; Meffre et al., 2000). This process, however, is incomplete and self-reactive cells can leave the bone marrow. Interestingly, there are differences between mouse strains in the efficiency of these checkpoints; Balb/c mice, for example, are more efficient in receptor editing than C57BL/6 mice, making the latter strain more permissive for the development of autoimmunity (Fukuyama et al., 2005). In the periphery, autoreactive B cells can be generated *de novo* during the process of somatic hypermutation in the germinal center (Bona and Stevenson, 2004; Ray et al., 1996). In particular, an expanded

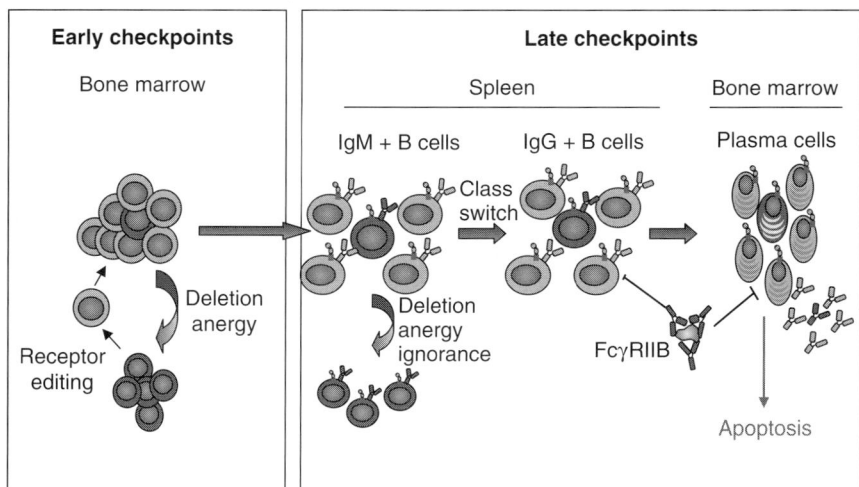

FIGURE 5.3 Regulation of B-cell responses by the inhibitory FcγRIIB. Shown are regulatory mechanisms that are in place to prevent the expansion of self-reactive B cells. During early B-cell development in the bone marrow, anergy, deletion, and receptor editing delete or inactivate the majority of B cells with a self-reactive receptor. In the periphery, anergy, deletion, and the inhibitory FcγRIIB represent major pathways to delete or control autoreactive cells. FcγRIIB is a very late checkpoint preventing the expansion of class switched IgG positive plasma blasts and plasma cells. In addition, FcγRIIB triggering on plasma cells induces apoptosis, thereby regulating plasma cell homeostasis. (See Plate 10 in Color Plate Section.)

repertoire of class switched self-reactive antibodies can trigger a wide variety of inflammatory effector functions (Dijstelbloem et al., 2001; Ravetch and Bolland, 2001). As indicated before, FcγRIIB is the only FcR on B cells. Together with other negative regulatory proteins such as CD22, CD5, and CD72, it regulates activating signals triggered by the BCR (Nitschke and Tsubata, 2004). The outstanding importance of the inhibitory FcR in this family has been demonstrated by the generation of the FcγRIIB knockout mouse that spontaneously develops an SLE-like disease characterized by the production of autoantibodies and a shortened life span (Bolland and Ravetch, 2000; Takai et al., 1996). This autoimmune phenotype is strain dependent as only C57BL/6 but not Balb/c mice develop spontaneous symptomatic disease, arguing for the involvement of other epistatic modifiers (Bolland et al., 2002; Nguyen et al., 2002). Indeed, studies showed that Balb/c mice double deficient in programmed death 1 (PD-1) and FcγRIIB developed severe autoimmune hydronephrosis and that Balb/c-Fcgr2b−/− mice had enhanced disease phenotypes in an inducible SLE model (Clynes et al., 2005; Okazaki et al., 2005). In addition, a polymorphism in the FcγRIIB promoter has been identified in autoimmune-prone mouse strains such as NZB, NOD, BXSB, and MRL, which leads to a reduced expression level of FcγRIIB on activated and germinal center B cells (Jiang et al., 1999, 2000; Pritchard et al., 2000; Xiu et al., 2002). Similar associations between FcγRIIB and the development or severity of autoimmune disease have been obtained in human autoimmune patients (Nimmerjahn, 2006). For instance, polymorphisms in the human FcγRIIB promoter have been linked to the development of lupus (Blank et al., 2005; Su et al., 2004a,b). A polymorphism that leads to decreased binding of the transcription factor AP-1 resulted in reduced surface expression of FcγRIIB on activated B cells of human SLE patients. Similarly, in a group of African-American SLE patients, memory B cells failed to upregulate FcγRIIB expression and this lower expression level correlated with a reduced threshold for B-cell activation (Mackay et al., 2006); this is consistent with another study describing that B cells from lupus patients showed enhanced triggering of activating signaling pathways after BCR stimulation. Besides polymorphisms in the FcγRIIB promoter, an allelic variant (I232T) of the inhibitory FcR that impairs its association with lipid rafts and thereby excludes it from active signaling complexes in the cell membrane has been associated with human SLE and arthritis (Floto et al., 2005; Kono et al., 2005). This represents yet another mechanism that aberrant FcγRIIB function could be involved in the initiation of autoimmune phenotypes. Nonetheless, it is important to note that there are disparities between different human populations and ethnicities, suggesting that as described before for mice the genetic background and other susceptibility factors are important for the development of autoimmune disease (Nimmerjahn, 2006).

Taken together, these results suggest that one therapeutic avenue to restore tolerance in autoimmune conditions might be to restore functional FcγRIIB expression on B cells, thereby regaining balanced immune responses. The therapeutic potential of this approach was demonstrated in a murine study using several autoimmune-prone mouse strains including NZM, BXSB, and FcγRIIB knockout animals (McGaha et al., 2005). After restoration of FcgRIIB expression by retroviral gene transfer, these animals had strongly reduced levels of autoantibodies and did not develop severe autoimmune symptoms. Importantly, this study highlights the threshold nature of autoimmunity as restoration of FcγRIIB expression on 40% of peripheral B cells was sufficient to interfere with the development of autoimmune disease. This clearly demonstrates that despite the complex nature of autoimmune diseases, therapeutic effects might be achievable by targeting a limited number of key regulatory proteins.

As autoreactive B cells can be generated at different points during central and peripheral B-cell development, it was important to define at which stage(s) FcγRIIB is preventing the generation of autoreactive B cells (Fig. 5.3). Results obtained in human autoimmune patients suggested that FcγRIIB might be a checkpoint during late phases of B-cell development. This has been validated in a mouse model containing an autoreactive prearranged VDJ region knocked into the immunoglobulin locus. It was shown that the absence of FcγRIIB did not impact early checkpoints in the bone marrow or prevent the development of IgM positive autoreactive B cells. FcγRIIB was crucial, however, to prevent the generation and expansion of IgG positive plasma cells secreting autoreactive antibody species (Fukuyama et al., 2005).

One other long-known outcome of isolated FcγRIIB triggering on B cells has recently received new attention: the induction of apoptosis. It has been suggested that this function is important to delete B cells that generate low-affinity BCRs during somatic hypermutation and therefore loose BCR interactions with their cognate antigen retained on FDCs (Pearse et al., 1999; Ravetch and Bolland, 2001).

While the importance of this mechanism remains to be established *in vivo*, there is evidence that these proapoptotic signals are important for plasma cell homeostasis.

Plasma cells express FcγRIIB and only very low levels or no BCR. They reside predominantly in niches in the bone marrow, where they have to receive survival signals from stromal cells (Radbruch et al., 2006). It is largely unclear how the limited amount of niches can accommodate new antigen-specific plasma cells generated with every new immune response when the body becomes challenged by various types of pathogens or after consecutive vaccinations with different antigens. With the demonstration that FcγRIIB cross-linking on plasma cells induces apoptosis, at least a partial solution for this problem might have been found (Ravetch and Nussenzweig, 2007; Xiang et al., 2007). ICs generated during an immune response could bind to plasma cells in the bone marrow and induce apoptosis on a fraction of cells, thus making space for cells with novel specificities (Fig. 5.3). Further supporting this model, secondary immunizations with a new antigen result in reduced levels of bone marrow plasma cells specific for the original antigen (Xiang et al., 2007). This finding is also of great importance for the role of FcγRIIB as a tolerance checkpoint, as plasma cells from autoimmune-prone mouse strains were shown to have absent or strongly reduced expression of FcγRIIB and are resistant to the induction of apoptosis. Restoring or overexpressing the inhibitory receptor could correct this defect, suggesting that the failure to control plasma cell persistence resulting from impaired FcγRIIB expression levels might account for the large number of these cells in autoimmune mouse strains, and ultimately be involved in the development of chronic autoimmune disease (Holmes and Burnet, 1963; Hoyer et al., 2004).

Taken together these studies emphasize the crucial role of FcγRIIB as a tolerance checkpoint during late stages of B-cell development and that correction of FcγRIIB expression levels might be a promising approach to interfere with autoimmune processes and to restore a balanced immune response, deletion of autoreactive IgG positive B cells, and ultimately tolerance.

8. SUMMARY AND OUTLOOK

An immune response is the result of complex interactions between a variety of innate and adaptive immune cells, and is tightly regulated at each step to adopt the magnitude of the response to the level of danger imposed by an infection with pathogenic microorganisms. Due to their broad expression level, studying FcR biology has lead to invaluable general insights into how immune responses are regulated and how minor changes in this regulation affect their outcome. The threshold set by paired expression of activating and inhibitory FcRs on innate and adaptive immune cells is crucial for a balanced response and aberrant expression of either component influences an immune response at several stages. Reduced FcγRIIB expression, for example, will lead to the expansion of IgG positive autoreactive B cells, a reduced amount of plasma cell apoptosis, stronger innate effector responses, and a lower threshold for DC activation, which in turn will influence the specificity and magnitude of the cellular response. This highlights the central importance of this negative regulator. Decreased levels of all activating FcRs will result in impaired antibody-mediated proinflammatory reactions. Due to the differential affinity of activating FcRs to individual IgG subclasses, the role of individual activating FcRs for each subclass will depend on the actual affinity for each IgG subclasses and the additional negative regulation by the inhibitory FcR. Moreover, the cytokine milieu will not only change the relative expression levels of activating and inhibitory FcRs but also influence the subclass of antibody that is generated, thus further enhancing or dampening FcR-dependent effector functions (Nimmerjahn and Ravetch, 2006). In addition, the changes in antibody glycosylation during proinflammatory and steady-state situations impact FcR binding and provide an environment that will determine which type of immune response will be triggered.

The wealth of data available from human autoimmune patients and from the corresponding mouse models clearly demonstrates that deregulated immune responses due to aberrantly expressed or nonfunctional FcR variants are greatly involved in the initiation and magnitude of these autoimmune diseases. One of the future challenges will be to translate this knowledge into novel therapeutic approaches, which will require the use of novel model systems that reflect the human clinical situation in more detail.

ACKNOWLEDGMENTS

This work was supported by grants from the German research foundation (DFG) and the Bayerisches Genomforschungsnetzwerk (BayGene) to F.N. and by grants from the NIH and E. Ludwig to J.V.R. We apologize to all colleagues whose important work was not directly cited due to limitations of space. These references can be found in the numerous review articles referred to in this chapter.

REFERENCES

Allen, J. M., and Seed, B. (1989). Isolation and expression of functional high-affinity Fc receptor complementary DNAs. *Science* **243**, 378–381.

Arnold, J. N., Dwek, R. A., Rudd, P. M., and Sim, R. B. (2006). Mannan binding lectin and its interaction with immunoglobulins in health and in disease. *Immunol. Lett.* **106**, 103–110.

Arnold, J. N., Wormald, M. R., Sim, R. B., Rudd, P. M., and Dwek, R. A. (2007). The impact of glycosylation on the biological function and structure of human immunoglobulins. *Annu. Rev. Immunol.* **25**, 21–50.

Azeredo da Silveira, S., Kikuchi, S., Fossati-Jimack, L., Moll, T., Saito, T., Verbeek, J. S., Botto, M., Walport, M. J., Carroll, M., and Izui, S. (2002). Complement activation selectively potentiates the pathogenicity of the IgG2b and IgG3 isotypes of a high affinity antierythrocyte autoantibody. *J. Exp. Med.* **195**, 665–672.

Barnes, N., Gavin, A. L., Tan, P. S., Mottram, P., Koentgen, F., and Hogarth, P. M. (2002). FcgammaRI-deficient mice show multiple alterations to inflammatory and immune responses. *Immunity* **16**, 379–389.

Bayary, J., Dasgupta, S., Misra, N., Ephrem, A., Van Huyen, J. P., Delignat, S., Hassan, G., Caligiuri, G., Nicoletti, A., Lacroix-Desmazes, S., Kazatchkine, M. D., and Kaveri, S. (2006).

Intravenous immunoglobulin in autoimmune disorders: An insight into the immunoregulatory mechanisms. *Int. Immunopharmacol.* **6,** 528–534.

Bergtold, A., Desai, D. D., Gavhane, A., and Clynes, R. (2005). Cell surface recycling of internalized antigen permits dendritic cell priming of B cells. *Immunity* **23,** 503–514.

Bevaart, L., Goldstein, J., Vitale, L., Russoniello, C., Treml, J., Zhang, J., Graziano, R. F., Leusen, J. H., van de Winkel, J. G., and Keler, T. (2006). Direct targeting of genetically modified tumour cells to Fc gammaRI triggers potent tumour cytotoxicity. *Br. J. Haematol.* **132,** 317–325.

Blank, M. C., Stefanescu, R. N., Masuda, E., Marti, F., King, P. D., Redecha, P. B., Wurzburger, R. J., Peterson, M. G., Tanaka, S., and Pricop, L. (2005). Decreased transcription of the human FCGR2B gene mediated by the -343 G/C promoter polymorphism and association with systemic lupus erythematosus. *Hum. Genet.* **117,** 220–227. Epub 2005 May 2014.

Bolland, S., and Ravetch, J. V. (1999). Inhibitory pathways triggered by ITIM-containing receptors. *Adv. Immunol.* **72,** 149–177.

Bolland, S., and Ravetch, J. V. (2000). Spontaneous autoimmune disease in Fc(gamma)RIIB-deficient mice results from strain-specific epistasis. *Immunity* **13,** 277–285.

Bolland, S., Yim, Y. S., Tus, K., Wakeland, E. K., and Ravetch, J. V. (2002). Genetic modifiers of systemic lupus erythematosus in FcgammaRIIB($-/-$) mice. *J. Exp. Med.* **195,** 1167–1174.

Bona, C. A., and Stevenson, F. K. (2004). *In* "Molecular Biology of B Cells" (T. Honjo, F. W. Alt, and M. S. Neuberger, Eds.), pp. 381–402. Elsevier, Boston.

Bond, A., Cooke, A., and Hay, F. C. (1990). Glycosylation of IgG, immune complexes and IgG subclasses in the MRL-lpr/lpr mouse model of rheumatoid arthritis. *Eur. J. Immunol.* **20,** 2229–2233.

Boruchov, A. M., Heller, G., Veri, M. C., Bonvini, E., Ravetch, J. V., and Young, J. W. (2005). Activating and inhibitory IgG Fc receptors on human DCs mediate opposing functions. *J. Clin. Invest.* **115,** 2914–2923.

Bruhns, P., Samuelsson, A., Pollard, J. W., and Ravetch, J. V. (2003). Colony-stimulating factor-1-dependent macrophages are responsible for IVIG protection in antibody-induced autoimmune disease. *Immunity* **18,** 573–581.

Cartron, G., Dacheux, L., Salles, G., Solal-Celigny, P., Bardos, P., Colombat, P., and Watier, H. (2002). Therapeutic activity of humanized anti-CD20 monoclonal antibody and polymorphism in IgG Fc receptor FcgammaRIIIa gene. *Blood* **99,** 754–758.

Clynes, R., and Ravetch, J. V. (1995). Cytotoxic antibodies trigger inflammation through Fc receptors. *Immunity* **3,** 21–26.

Clynes, R., Takechi, Y., Moroi, Y., Houghton, A., and Ravetch, J. V. (1998). Fc receptors are required in passive and active immunity to melanoma. *Proc. Natl. Acad. Sci. USA* **95,** 652–656.

Clynes, R., Maizes, J. S., Guinamard, R., Ono, M., Takai, T., and Ravetch, J. V. (1999). Modulation of immune complex-induced inflammation in vivo by the coordinate expression of activation and inhibitory Fc receptors. *J. Exp. Med.* **189,** 179–185.

Clynes, R., Calvani, N., Croker, B. P., and Richards, H. B. (2005). Modulation of the immune response in pristane-induced lupus by expression of activation and inhibitory Fc receptors. *Clin. Exp. Immunol.* **141,** 230–237.

Clynes, R. A., Towers, T. L., Presta, L. G., and Ravetch, J. V. (2000). Inhibitory Fc receptors modulate in vivo cytoxicity against tumor targets. *Nat. Med.* **6,** 443–446.

Daeron, M. (1997). Fc receptor biology. *Annu. Rev. Immunol.* **15,** 203–234.

Davis, R. S., Stephan, R. P., Chen, C. C., Dennis, G., Jr., Cooper, M. D., Ehrhardt, G. R., Hsu, J. T., Leu, C. M., and Ehrhardt, A. (2004). Differential B cell expression of mouse Fc receptor homologs. The inhibitory potential of Fc receptor homolog 4 on memory B cells. *Int. Immunol.* **16,** 1343–1353. Epub 2004 Aug 1349.

Davis, R. S., Ehrhardt, G. R., Leu, C. M., Hirano, M., and Cooper, M. D. (2005). An extended family of Fc receptor relatives. *Eur. J. Immunol.* **35,** 674–680.

Dhodapkar, K. M., Krasovsky, J., Williamson, B., and Dhodapkar, M. V. (2002). Antitumor monoclonal antibodies enhance cross-presentation of cellular antigens and the generation of myeloma-specific killer T cells by dendritic cells. *J. Exp. Med.* **195,** 125–133.

Dhodapkar, K. M., Kaufman, J. L., Ehlers, M., Banerjee, D. K., Bonvini, E., Koenig, S., Steinman, R. M., Ravetch, J. V., and Dhodapkar, M. V. (2005). Selective blockade of inhibitory Fcgamma receptor enables human dendritic cell maturation with IL-12p70 production and immunity to antibody-coated tumor cells. *Proc. Natl. Acad. Sci. USA* **102,** 2910–2915.

Dijstelbloem, H. M., van de Winkel, J. G., and Kallenberg, C. G. (2001). Inflammation in autoimmunity: Receptors for IgG revisited. *Trends Immunol.* **22,** 510–516.

Ehrhardt, G. R., Davis, R. S., Hsu, J. T., Leu, C. M., Ehrhardt, A., and Cooper, M. D. (2003). The inhibitory potential of Fc receptor homolog 4 on memory B cells. *Proc. Natl. Acad. Sci. USA* **100,** 13489–13494. Epub 12003 Nov 13483.

Ehrhardt, G. R., Hsu, J. T., Gartland, L., Leu, C. M., Zhang, S., Davis, R. S., and Cooper, M. D. (2005). Expression of the immunoregulatory molecule FcRH4 defines a distinctive tissue-based population of memory B cells. *J. Exp. Med.* **202,** 783–791. Epub 2005 Sep 2012.

Facchetti, F., Cella, M., Festa, S., Fremont, D. H., and Colonna, M. (2002). An unusual Fc receptor-related protein expressed in human centroblasts. *Proc. Natl. Acad. Sci. USA* **99,** 3776–3781. Epub 2002 Mar 3712.

Ferrara, C., Stuart, F., Sondermann, P., Brunker, P., and Umana, P. (2006). The carbohydrate at FcgammaRIIIa Asn-162. An element required for high affinity binding to non-fucosylated IgG glycoforms. *J. Biol. Chem.* **281,** 5032–5036.

Floto, R. A., Clatworthy, M. R., Heilbronn, K. R., Rosner, D. R., MacAry, P. A., Rankin, A., Lehner, P. J., Ouwehand, W. H., Allen, J. M., Watkins, N. A., and Smith, K. G. (2005). Loss of function of a lupus-associated FcgammaRIIb polymorphism through exclusion from lipid rafts. *Nat. Med.* **11,** 1056–1058.

Fossati-Jimack, L., Ioan-Facsinay, A., Reininger, L., Chicheportiche, Y., Watanabe, N., Saito, T., Hofhuis, F. M., Gessner, J. E., Schiller, C., Schmidt, R. E., Honjo, T., Verbeek, J. S., *et al.* (2000). Markedly different pathogenicity of four immunoglobulin G isotype-switch variants of an antierythrocyte autoantibody is based on their capacity to interact in vivo with the low-affinity Fcgamma receptor III. *J. Exp. Med.* **191,** 1293–1302.

Fujii, T., Hamano, Y., Ueda, S., Akikusa, B., Yamasaki, S., Ogawa, M., Saisho, H., Verbeek, J. S., Taki, S., and Saito, T. (2003). Predominant role of FcgammaRIII in the induction of accelerated nephrotoxic glomerulonephritis. *Kidney Int.* **64,** 1406–1416.

Fukuyama, H., Nimmerjahn, F., and Ravetch, J. V. (2005). The inhibitory Fcgamma receptor modulates autoimmunity by limiting the accumulation of immunoglobulin G+ anti-DNA plasma cells. *Nat. Immunol.* **6,** 99–106. Epub 2004 Dec 2012.

Goodnow, C. C., Sprent, J., de St Groth, B. F., and Vinuesa, C. G. (2005). Cellular and genetic mechanisms of self tolerance and autoimmunity. *Nature* **435,** 590–597.

Grimaldi, C. M., Hicks, R., and Diamond, B. (2005). B cell selection and susceptibility to autoimmunity. *J. Immunol.* **174,** 1775–1781.

Groh, V., Li, Y. Q., Cioca, D., Hunder, N. N., Wang, W., Riddell, S. R., Yee, C., and Spies, T. (2005). Efficient cross-priming of tumor antigen-specific T cells by dendritic cells sensitized with diverse anti-MICA opsonized tumor cells. *Proc. Natl. Acad. Sci. USA* **102,** 6461–6466. Epub 2005 Apr 6411.

Guyre, P. M., Morganelli, P. M., and Miller, R. (1983). Recombinant immune interferon increases immunoglobulin G Fc receptors on cultured human mononuclear phagocytes. *J. Clin. Invest.* **72,** 393–397.

Hamaguchi, Y., Xiu, Y., Komura, K., Nimmerjahn, F., and Tedder, T. F. (2006). Antibody isotype-specific engagement of Fcgamma receptors regulates B lymphocyte depletion during CD20 immunotherapy. *J. Exp. Med.* **203,** 743–753. Epub 2006 Mar 2006.

Hawiger, D., Inaba, K., Dorsett, Y., Guo, M., Mahnke, K., Rivera, M., Ravetch, J. V., Steinman, R. M., and Nussenzweig, M. C. (2001). Dendritic cells induce peripheral T cell unresponsiveness under steady state conditions in vivo. *J. Exp. Med.* **194,** 769–779.

Hawiger, D., Masilamani, R. F., Bettelli, E., Kuchroo, V. K., and Nussenzweig, M. C. (2004). Immunological unresponsiveness characterized by increased expression of CD5 on peripheral T cells induced by dendritic cells *in vivo*. *Immunity* **20,** 695–705.

Hazenbos, W. L., Gessner, J. E., Hofhuis, F. M., Kuipers, H., Meyer, D., Heijnen, I. A., Schmidt, R. E., Sandor, M., Capel, P. J., Daeron, M., van de Winkel, J. G., and Verbeek, J. S. (1996). Impaired IgG-dependent anaphylaxis and Arthus reaction in Fc gamma RIII (CD16) deficient mice. *Immunity* **5,** 181–188.

Hirano, M., Davis, R. S., Fine, W. D., Nakamura, S., Shimizu, K., Yagi, H., Kato, K., Stephan, R. P., and Cooper, M. D. (2007). IgE(b) immune complexes activate macrophages through FcgammaRIV binding. *Nat. Immunol.* **8,** 762–771.

Holmes, M. C., and Burnet, F. M. (1963). The natural history of autoimmune disease in Nzb mice. A comparison with the pattern of human autoimmune manifestations. *Ann. Intern. Med.* **59,** 265–276.

Hoyer, B. F., Moser, K., Hauser, A. E., Peddinghaus, A., Voigt, C., Eilat, D., Radbruch, A., Hiepe, F., and Manz, R. A. (2004). Short-lived plasmablasts and long-lived plasma cells contribute to chronic humoral autoimmunity in NZB/W mice. *J. Exp. Med.* **199,** 1577–1584.

Hulett, M. D., and Hogarth, P. M. (1994). Molecular basis of Fc receptor function. *Adv. Immunol.* **57,** 1–127.

Ioan-Facsinay, A., de Kimpe, S. J., Hellwig, S. M., van Lent, P. L., Hofhuis, F. M., van Ojik, H. H., Sedlik, C., da Silveira, S. A., Gerber, J., de Jong, Y. F., Roozendaal, R., Aarden, L. A., *et al*. (2002). FcgammaRI (CD64) contributes substantially to severity of arthritis, hypersensitivity responses, and protection from bacterial infection. *Immunity* **16,** 391–402.

Jefferis, R., and Lund, J. (2002). Interaction sites on human IgG-Fc for FcgammaR: Current models. *Immunol. Lett.* **82,** 57–65.

Ji, H., Ohmura, K., Mahmood, U., Lee, D. M., Hofhuis, F. M., Boackle, S. A., Takahashi, K., Holers, V. M., Walport, M., Gerard, C., Ezekowitz, A., Carroll, M. C., *et al.* (2002). Arthritis critically dependent on innate immune system players. *Immunity* **16,** 157–168.

Jiang, Y., Hirose, S., Sanokawa-Akakura, R., Abe, M., Mi, X., Li, N., Miura, Y., Shirai, J., Zhang, D., Hamano, Y., and Shirai, T. (1999). Genetically determined aberrant down-regulation of FcgammaRIIB1 in germinal center B cells associated with hyper-IgG and IgG autoantibodies in murine systemic lupus erythematosus. *Int. Immunol.* **11,** 1685–1691.

Jiang, Y., Hirose, S., Abe, M., Sanokawa-Akakura, R., Ohtsuji, M., Mi, X., Li, N., Xiu, Y., Zhang, D., Shirai, J., Hamano, Y., Fujii, H., *et al.* (2000). Polymorphisms in IgG Fc receptor IIB regulatory regions associated with autoimmune susceptibility. *Immunogenetics* **51,** 429–435.

Kalergis, A. M., and Ravetch, J. V. (2002). Inducing tumor immunity through the selective engagement of activating Fcgamma receptors on dendritic cells. *J. Exp. Med.* **195,** 1653–1659.

Kaneko, Y., Nimmerjahn, F., and Ravetch, J. V. (2006a). Anti-inflammatory activity of immunoglobulin G resulting from Fc sialylation. *Science* **313,** 670–673.

Kaneko, Y., Nimmerjahn, F., Madaio, M. P., and Ravetch, J. V. (2006b). Pathology and protection in nephrotoxic nephritis is determined by selective engagement of specific Fc receptors. *J. Exp. Med.* **203,** 789–797.

Kato, K., Fridman, W. H., Arata, Y., and Sautes-Fridman, C. (2000). A conformational change in the Fc precludes the binding of two Fcgamma receptor molecules to one IgG. *Immunol. Today* **21**, 310–312.

Kinet, J. P. (1999). The high-affinity IgE receptor (Fc epsilon RI): From physiology to pathology. *Annu. Rev. Immunol.* **17**, 931–972.

Kono, H., Kyogoku, C., Suzuki, T., Tsuchiya, N., Honda, H., Yamamoto, K., Tokunaga, K., and Honda, Z. (2005). FcgammaRIIB Ile232Thr transmembrane polymorphism associated with human systemic lupus erythematosus decreases affinity to lipid rafts and attenuates inhibitory effects on B cell receptor signaling. *Hum. Mol. Genet.* **14**, 2881–2892.

Kraft, S., and Novak, N. (2006). Fc receptors as determinants of allergic reactions. *Trends Immunol.* **27**, 88–95.

Kretschmer, K., Apostolou, I., Hawiger, D., Khazaie, K., Nussenzweig, M. C., and von Boehmer, H. (2005). Inducing and expanding regulatory T cell populations by foreign antigen. *Nat. Immunol.* **6**, 1219–1227.

Kumar, V., Ali, S. R., Konrad, S., Zwirner, J., Verbeek, J. S., Schmidt, R. E., and Gessner, J. E. (2006). Cell-derived anaphylatoxins as key mediators of antibody-dependent type II autoimmunity in mice. *J. Clin. Invest.* **116**, 512–520.

Lazar, G. A., Dang, W., Karki, S., Vafa, O., Peng, J. S., Hyun, L., Chan, C., Chung, H. S., Eivazi, A., Yoder, S. C., Vielmetter, J., Carmichael, D. F., *et al.* (2006). Engineered antibody Fc variants with enhanced effector function. *Proc. Natl. Acad. Sci. USA* **103**, 4005–4010.

Mackay, M., Stanevsky, A., Wang, T., Aranow, C., Li, M., Koenig, S., Ravetch, J. V., and Diamond, B. (2006). Selective dysregulation of the FcgammaIIB receptor on memory B cells in SLE. *J. Exp. Med.* **203**, 2157–2164.

Malhotra, R., Wormald, M. R., Rudd, P. M., Fischer, P. B., Dwek, R. A., and Sim, R. B. (1995). Glycosylation changes of IgG associated with rheumatoid arthritis can activate complement via the mannose-binding protein. *Nat. Med.* **1**, 237–243.

McGaha, T. L., Sorrentino, B., and Ravetch, J. V. (2005). Restoration of tolerance in lupus by targeted inhibitory receptor expression. *Science* **307**, 590–593.

Meffre, E., Casellas, R., and Nussenzweig, M. C. (2000). Antibody regulation of B cell development. *Nat. Immunol.* **1**, 379–385.

Meyer, D., Schiller, C., Westermann, J., Izui, S., Hazenbos, W. L., Verbeek, J. S., Schmidt, R. E., and Gessner, J. E. (1998). FcgammaRIII (CD16)-deficient mice show IgG isotype-dependent protection to experimental autoimmune hemolytic anemia. *Blood* **92**, 3997–4002.

Mizuochi, T., Hamako, J., Nose, M., and Titani, K. (1990). Structural changes in the oligosaccharide chains of IgG in autoimmune MRL/Mp-lpr/lpr mice. *J. Immunol.* **145**, 1794–1798.

Nakamura, A., Yuasa, T., Ujike, A., Ono, M., Nukiwa, T., Ravetch, J. V., and Takai, T. (2000). Fcgamma receptor IIB-deficient mice develop Goodpasture's syndrome upon immunization with type IV collagen: A novel murine model for autoimmune glomerular basement membrane disease. *J. Exp. Med.* **191**, 899–906.

Nandakumar, K. S., Andren, M., Martinsson, P., Bajtner, E., Hellstrom, S., Holmdahl, R., Kleinau, S., and Heyman, B. (2003). Induction of arthritis by single monoclonal IgG anti-collagen type II antibodies and enhancement of arthritis in mice lacking inhibitory FcgammaRIIB. Induction and suppression of collagen-induced arthritis is dependent on distinct fcgamma receptors. *Eur. J. Immunol.* **33**, 2269–2277.

Nguyen, C., Limaye, N., and Wakeland, E. K. (2002). Susceptibility genes in the pathogenesis of murine lupus. *Arthritis Res.* **4**, S255–S263. Epub 2002 May 2009.

Nimmerjahn, F. (2006). Activating and inhibitory FcgammaRs in autoimmune disorders. *Springer Semin. Immunopathol.* **28**, 305–319.

Nimmerjahn, F., and Ravetch, J. V. (2005). Divergent immunoglobulin G subclass activity through selective Fc receptor binding. *Science* **310**, 1510–1512.

Nimmerjahn, F., and Ravetch, J. V. (2006). Fcgamma receptors: Old friends and new family members. *Immunity* **24,** 19–28.

Nimmerjahn, F., and Ravetch, J. V. (2007a). Antibodies, Fc receptors and cancer. *Curr. Opin. Immunol.* **19,** 239–245.

Nimmerjahn, F., and Ravetch, J. V. (2007b). The anti-inflammatory activity of IgG: The intravenous IgG paradox. *J. Exp. Med.* **204,** 11–15.

Nimmerjahn, F., Bruhns, P., Horiuchi, K., and Ravetch, J. V. (2005). FcgammaRIV: A novel FcR with distinct IgG subclass specificity. *Immunity* **23,** 41–51.

Nimmerjahn, F., Anthony, R. M., and Ravetch, J. V. (2007). Agalactosylated IgG antibodies depend on cellular Fc receptors for in vivo activity. *Proc. Natl. Acad. Sci. USA* **104,** 8433–8437.

Nitschke, L., and Tsubata, T. (2004). Molecular interactions regulate BCR signal inhibition by CD22 and CD72. *Trends Immunol.* **25,** 543–550.

Okayama, Y., Kirshenbaum, A. S., and Metcalfe, D. D. (2000). Expression of a functional high-affinity IgG receptor, Fc gamma RI, on human mast cells: Up-regulation by IFN-gamma. *J. Immunol.* **164,** 4332–4339.

Okazaki, T., Otaka, Y., Wang, J., Hiai, H., Takai, T., Ravetch, J. V., and Honjo, T. (2005). Hydronephrosis associated with antiurothelial and antinuclear autoantibodies in BALB/c-Fcgr2b−/−Pdcd1−/− mice. *J. Exp. Med.* **202,** 1643–1648.

Park, S. Y., Ueda, S., Ohno, H., Hamano, Y., Tanaka, M., Shiratori, T., Yamazaki, T., Arase, H., Arase, N., Karasawa, A., Sato, S., Ledermann, B., et al. (1998). Resistance of Fc receptor-deficient mice to fatal glomerulonephritis. *J. Clin. Invest.* **102,** 1229–1238.

Pearse, R. N., Kawabe, T., Bolland, S., Guinamard, R., Kurosaki, T., and Ravetch, J. V. (1999). SHIP recruitment attenuates Fc gamma RIIB-induced B cell apoptosis. *Immunity* **10,** 753–760.

Pricop, L., Redecha, P., Teillaud, J. L., Frey, J., Fridman, W. H., Sautes-Fridman, C., and Salmon, J. E. (2001). Differential modulation of stimulatory and inhibitory Fc gamma receptors on human monocytes by Th1 and Th2 cytokines. *J. Immunol.* **166,** 531–537.

Pritchard, N. R., Cutler, A. J., Uribe, S., Chadban, S. J., Morley, B. J., and Smith, K. G. (2000). Autoimmune-prone mice share a promoter haplotype associated with reduced expression and function of the Fc receptor FcgammaRII. *Curr. Biol.* **10,** 227–230.

Qiu, W. Q., de Bruin, D., Brownstein, B. H., Pearse, R., and Ravetch, J. V. (1990). Organization of the human and mouse low-affinity Fc gamma R genes: Duplication and recombination. *Science* **248,** 732–735.

Radaev, S., and Sun, P. (2002). Recognition of immunoglobulins by Fcgamma receptors. *Mol. Immunol.* **38,** 1073–1083.

Radaev, S., Motyka, S., Fridman, W. H., Sautes-Fridman, C., and Sun, P. D. (2001). The structure of a human type III Fcgamma receptor in complex with Fc. *J. Biol. Chem.* **276,** 16469–16477.

Radbruch, A., Muehlinghaus, G., Luger, E. O., Inamine, A., Smith, K. G., Dorner, T., and Hiepe, F. (2006). Competence and competition: The challenge of becoming a long-lived plasma cell. *Nat. Rev. Immunol.* **6,** 741–750.

Radeke, H. H., Janssen-Graalfs, I., Sowa, E. N., Chouchakova, N., Skokowa, J., Loscher, F., Schmidt, R. E., Heeringa, P., and Gessner, J. E. (2002). Opposite regulation of type II and III receptors for immunoglobulin G in mouse glomerular mesangial cells and in the induction of anti-glomerular basement membrane (GBM) nephritis. *J. Biol. Chem.* **277,** 27535–27544. Epub 22002 Apr 27530.

Rafiq, K., Bergtold, A., and Clynes, R. (2002). Immune complex-mediated antigen presentation induces tumor immunity. *J. Clin. Invest.* **110,** 71–79.

Ravetch, J. V. (2003). Fc receptors. In "Fundamental Immunology" (W. E. Paul, Ed.), pp. 685–700. Lippincott-Raven, Philadelphia, PA.

Ravetch, J. V., and Bolland, S. (2001). IgG Fc receptors. *Annu. Rev. Immunol.* **19,** 275–290.

Ravetch, J. V., and Clynes, R. A. (1998). Divergent roles for Fc receptors and complement in vivo. *Annu. Rev. Immunol.* **16,** 421–432.
Ravetch, J. V., and Lanier, L. L. (2000). Immune inhibitory receptors. *Science* **290,** 84–89.
Ravetch, J. V., and Nussenzweig, M. (2007). Killing some to make way for others. *Nat. Immunol.* **8,** 337–339.
Ray, S. K., Putterman, C., and Diamond, B. (1996). Pathogenic autoantibodies are routinely generated during the response to foreign antigen: A paradigm for autoimmune disease. *Proc. Natl. Acad. Sci. USA* **93,** 2019–2024.
Regnault, A., Lankar, D., Lacabanne, V., Rodriguez, A., Thery, C., Rescigno, M., Saito, T., Verbeek, S., Bonnerot, C., Ricciardi-Castagnoli, P., and Amigorena, S. (1999). Fcgamma receptor-mediated induction of dendritic cell maturation and major histocompatibility complex class I-restricted antigen presentation after immune complex internalization. *J. Exp. Med.* **189,** 371–380.
Rudge, E. U., Cutler, A. J., Pritchard, N. R., and Smith, K. G. (2002). Interleukin 4 reduces expression of inhibitory receptors on B cells and abolishes CD22 and Fc gamma RII-mediated B cell suppression. *J. Exp. Med.* **195,** 1079–1085.
Scallon, B. J., Tam, S. H., McCarthy, S. G., Cai, A. N., and Raju, T. S. (2007). Higher levels of sialylated Fc glycans in immunoglobulin G molecules can adversely impact functionality. *Mol. Immunol.* **44,** 1524–1534.
Shields, R. L., Namenuk, A. K., Hong, K., Meng, Y. G., Rae, J., Briggs, J., Xie, D., Lai, J., Stadlen, A., Li, B., Fox, J. A., and Presta, L. G. (2001). High resolution mapping of the binding site on human IgG1 for Fc gamma RI, Fc gamma RII, Fc gamma RIII, and FcRn and design of IgG1 variants with improved binding to the Fc gamma R. *J. Biol. Chem.* **276,** 6591–6604.
Shields, R. L., Lai, J., Keck, R., O'Connell, L. Y., Hong, K., Meng, Y. G., Weikert, S. H., Presta, L. G., Namenuk, A. K., Rae, J., Briggs, J., Xie, D., *et al.* (2002). Lack of fucose on human IgG1 N-linked oligosaccharide improves binding to human Fcgamma RIII and antibody-dependent cellular toxicity. *J. Biol. Chem.* **277,** 26733–26740.
Shinkawa, T., Nakamura, K., Yamane, N., Shoji-Hosaka, E., Kanda, Y., Sakurada, M., Uchida, K., Anazawa, H., Satoh, M., Yamasaki, M., Hanai, N., and Shitara, K. (2003). The absence of fucose but not the presence of galactose or bisecting N-acetylglucosamine of human IgG1 complex-type oligosaccharides shows the critical role of enhancing antibody-dependent cellular cytotoxicity. *J. Biol. Chem.* **278,** 3466–3473. Epub 2002 Nov 3468.
Shushakova, N., Skokowa, J., Schulman, J., Baumann, U., Zwirner, J., Schmidt, R. E., and Gessner, J. E. (2002). C5a anaphylatoxin is a major regulator of activating versus inhibitory FcgammaRs in immune complex-induced lung disease. *J. Clin. Invest.* **110,** 1823–1830.
Skokowa, J., Ali, S. R., Felda, O., Kumar, V., Konrad, S., Shushakova, N., Schmidt, R. E., Piekorz, R. P., Nurnberg, B., Spicher, K., Birnbaumer, L., Zwirner, J., *et al.* (2005). Macrophages induce the inflammatory response in the pulmonary Arthus reaction through G alpha i2 activation that controls C5aR and Fc receptor cooperation. *J. Immunol.* **174,** 3041–3050.
Sondermann, P., Huber, R., Oosthuizen, V., and Jacob, U. (2000). The 3.2-A crystal structure of the human IgG1 Fc fragment-Fc gammaRIII complex. *Nature* **406,** 267–273.
Steinman, R. M., and Hemmi, H. (2006). Dendritic cells: Translating innate to adaptive immunity. *Curr. Top Microbiol. Immunol.* **311,** 17–58.
Steinman, R. M., Hawiger, D., Liu, K., Bonifaz, L., Bonnay, D., Mahnke, K., Iyoda, T., Ravetch, J., Dhodapkar, M., Inaba, K., and Nussenzweig, M. (2003). Dendritic cell function *in vivo* during the steady state: A role in peripheral tolerance. *Ann. NY Acad. Sci.* **987,** 15–25.

Su, K., Li, X., Edberg, J. C., Wu, J., Ferguson, P., and Kimberly, R. P. (2004a). A promoter haplotype of the immunoreceptor tyrosine-based inhibitory motif-bearing FcgammaRIIb alters receptor expression and associates with autoimmunity. II. Differential binding of GATA4 and Yin-Yang1 transcription factors and correlated receptor expression and function. *J. Immunol.* **172,** 7192–7199.

Su, K., Wu, J., Edberg, J. C., Li, X., Ferguson, P., Cooper, G. S., Langefeld, C. D., and Kimberly, R. P. (2004b). A promoter haplotype of the immunoreceptor tyrosine-based inhibitory motif-bearing FcgammaRIIb alters receptor expression and associates with autoimmunity. I. Regulatory FCGR2B polymorphisms and their association with systemic lupus erythematosus. *J. Immunol.* **172,** 7186–7191.

Sylvestre, D., Clynes, R., Ma, M., Warren, H., Carroll, M. C., and Ravetch, J. V. (1996). Immunoglobulin G-mediated inflammatory responses develop normally in complement-deficient mice. *J. Exp. Med.* **184,** 2385–2392.

Sylvestre, D. L., and Ravetch, J. V. (1994). Fc receptors initiate the Arthus reaction: Redefining the inflammatory cascade. *Science* **265,** 1095–1098.

Takai, T. (2002). Roles of Fc receptors in autoimmunity. *Nat. Rev. Immunol.* **2,** 580–592.

Takai, T., Li, M., Sylvestre, D., Clynes, R., and Ravetch, J. V. (1994). FcR gamma chain deletion results in pleiotrophic effector cell defects. *Cell* **76,** 519–529.

Takai, T., Ono, M., Hikida, M., Ohmori, H., and Ravetch, J. V. (1996). Augmented humoral and anaphylactic responses in Fc gamma RII-deficient mice. *Nature* **379,** 346–349.

Torigoe, C., Inman, J. K., and Metzger, H. (1998). An unusual mechanism for ligand antagonism. *Science* **281,** 568–572.

Tridandapani, S., Wardrop, R., Baran, C. P., Wang, Y., Opalek, J. M., Caligiuri, M. A., and Marsh, C. B. (2003). TGF-beta 1 suppresses [correction of supresses] myeloid Fc gamma receptor function by regulating the expression and function of the common gamma-subunit. *J. Immunol.* **170,** 4572–4577.

Tzeng, S. J., Bolland, S., Inabe, K., Kurosaki, T., and Pierce, S. K. (2005). The B cell inhibitory Fc receptor triggers apoptosis by a novel c-Abl-family kinase dependent pathway. *J. Biol. Chem.* **22,** 22.

Uchida, J., Hamaguchi, Y., Oliver, J. A., Ravetch, J. V., Poe, J. C., Haas, K. M., and Tedder, T. F. (2004). The innate mononuclear phagocyte network depletes B lymphocytes through Fc receptor-dependent mechanisms during anti-CD20 antibody immunotherapy. *J. Exp. Med.* **199,** 1659–1669.

Weng, W. K., and Levy, R. (2003). Two immunoglobulin G fragment C receptor polymorphisms independently predict response to rituximab in patients with follicular lymphoma. *J. Clin. Oncol.* **21,** 3940–3947. Epub 2003 Sep 3915.

Weng, W. K., Czerwinski, D., Timmerman, J., Hsu, F. J., and Levy, R. (2004). Clinical outcome of lymphoma patients after idiotype vaccination is correlated with humoral immune response and immunoglobulin G Fc receptor genotype. *J. Clin. Oncol.* **22,** 4717–4724. Epub 2004 Oct 4713.

Wines, B. D., and Hogarth, P. M. (2006). IgA receptors in health and disease. *Tissue Antigens* **68,** 103–114.

Woof, J. M., and Kerr, M. A. (2006). The function of immunoglobulin A in immunity. *J. Pathol.* **208,** 270–282.

Xiang, Z., Cutler, A. J., Brownlie, R. J., Fairfax, K., Lawlor, K. E., Severinson, E., Walker, E. U., Manz, R. A., Tarlinton, D. M., and Smith, K. G. (2007). FcgammaRIIb controls bone marrow plasma cell persistence and apoptosis. *Nat. Immunol.* **8,** 419–429.

Xiu, Y., Nakamura, K., Abe, M., Li, N., Wen, X. S., Jiang, Y., Zhang, D., Tsurui, H., Matsuoka, S., Hamano, Y., Fujii, H., Ono, M., *et al.* (2002). Transcriptional regulation of Fcgr2b gene by polymorphic promoter region and its contribution to humoral immune responses. *J. Immunol.* **169,** 4340–4346.

Yamazaki, S., Inaba, K., Tarbell, K. V., and Steinman, R. M. (2006). Dendritic cells expand antigen-specific Foxp3+ CD25+ CD4+ regulatory T cells including suppressors of alloreactivity. *Immunol. Rev.* **212,** 314–329.

Zhang, M., Zhang, Z., Garmestani, K., Goldman, C. K., Ravetch, J. V., Brechbiel, M. W., Carrasquillo, J. A., and Waldmann, T. A. (2004). Activating Fc receptors are required for antitumor efficacy of the antibodies directed toward CD25 in a murine model of adult t-cell leukemia. *Cancer Res.* **64,** 5825–5829.

Zhang, Y., Boesen, C. C., Radaev, S., Brooks, A. G., Fridman, W. H., Sautes-Fridman, C., and Sun, P. D. (2000). Crystal structure of the extracellular domain of a human Fc gamma RIII. *Immunity* **13,** 387–395.

SUBJECT INDEX

A

N-Acetylglucosamine (GlcNac), 188
Activating transcription factor (ATF) proteins, 61
Activation-induced cell death, 12, 116, 118
Acute disseminated encephalomyelitis, 3
Acute injury, 143
ADCC. *See* Antibody-dependent cellular cytotoxicity
ADEM. *See* Acute disseminated encephalomyelitis
Adoptive CD8+ T cell therapy
  TCR/CD27 pathway, 127–129
  in TGF-β and IL-2 role, 129
Age-related macular degeneration (AMD), 144, 160, 166
Age-related macular degeneration, immunopathogenesis and aHUS
  atypical HUS, 163–164
    binding of FH to human cells, 165–166
    complement activation and endothelial cells, 165
    excessive complement activation, 164
    function of FH and MCP, 165
  treatment options for aHUS
    liver/renal transplantation, 167–168
    plasma infusion/exchange, 167
    renal transplantation, 167
  treatment options for AMD
    antiangiogenic treatments and complement inhibitors, 168
AICD. *See* Activation-induced cell death
Altered self triggers innate immunity
  acute injury, 143–144
  debris accumulation, 144–145
Alternative complement pathway
  activation, 145–148
    initial activation of C3, 147
    positive feedback amplification loop, 146, 147
    regulation of, 148
  Factor H (CFH), 148–150
  Factor I, 152–153

membrane cofactor protein, 150–152
Alternative pathway (AP), 149
Antibody-dependent cellular cytotoxicity, 183
Antigen-presenting cells (APCs), 4, 21, 47
Antigen-selected CD8$^+$ T cells, subsets of, 106, 107
Antigen-specific T cells monitoring, 6
Antigen-stimulated CD8$^+$ T cells, models for development, 109–111
Apoptosis, 120, 121, 124, 143, 144, 149, 161, 165, 185, 194, 195
Atypical hemolytic uremic syndrome (aHUS), 144
  disease penetrance, 160
  Factor B, 159–160
  Factor H, 155–157
  Factor I, 159
  FH-related genes, 157
  hemolytic uremic syndrome (HUS), 154
  membrane cofactor protein, CD46, 157–159
Autoimmune disease models, 5, 12
Autoimmune disorders, 13
Autoimmune encephalomyelitis (EAE), 2
Autoimmune thyroid disease, 28
Autoreactive T cells, 12

B

4–1BB, deficiency of, 118. *See also* Coreceptors, in IL-2-independent CD8+ T cell clonal
B-Cell receptor (BCR), 182, 185, 192–194
Blood–brain barrier, 9, 24
Bone marrow-derived macrophages, 56
Bone marrow stromal cells, 51
Bone marrow transplantation, 15
Brain-derived neurotrophic factor (BDNF), 22
Bystander humoral response, 18

C

CCAAT/enhancer binding proteins (C/EBP), 61
CD70-dependent immunodeficiency syndrome, 119
$CD8^+$ T cell(s)
  central and peripheral clonal expansion, phases of, 128
  central memory, 107–108
  clonal expansion, model for CD27 and IL-2-dependent differentiation, 126–127
  models, development of antigen-stimulated cells, 108–111
  in persistent viral infections, 111–113
    cellular senescence, 113–114
    clonal persistence *vs.* clonal succession, 114
    inflationary epitopes, 111–113
    latent phase of mCMV infection, TCM and TEFF/EM phenotype, 112
    molecular requirements for clonal persistence, 114–115
    stimulation and expansion, inflationary epitopes, 111–113
  subsets of antigen-selected cell, 106
$CD8^+$ T cells, IL-2 in clonal expansion and effector differentiation, 115–116
  clonal expansion without IL-2R signaling, 116–118
  IL-2-independent $CD8^+$ T cell clonal expansion
    CD27, 118–120
    coreceptors in, 120
Cell–cell interactions, 53
Cell degradation process, 12
Central memory, $CD8^+$ T cells, 107–111
Central nervous system (CNS)
  cerebrospinal fluid, 7
  chronic autoimmunity in, 19
  immune response in, 2
  infiltrating T cells, 21
  inflammation by apoptosis, 12–13
Chemokines, 9–12
  driven migration, 12
  encoding genes, 11
Chinese Hamster Ovary (CHO), 156
Chromatin immunoprecipitation (ChIP), 51, 65
Chronic autoimmune disorders, 19, 29
Chronic neuroinflammation, 4
  apoptotic mechanism involved in, 13
  aspects of, 4
  B cells and antibody-mediated immune responses in, 16–19
  $CD4^+$ T helper cells in, 5–6
  induction and perseveration of, 5
  neurodegenerative features of, 20
  Th cells for controlling, 13–15
  Th1 *vs.* Th17, 6–7
Chronic neuronal damage, 20
Chronic tophaceous gout, 145
Clinical symptomatology, 10
Clonal expansion
  $CD8^+$ T cells for multiple viruses, 129
  cellular senescence in, 113
  coreceptors mediating, 118–120
  IL-2-independent pathway for $CD8^+$ T cell, 109
  mediated by IL2, 108–109
  model for $CD8^+$ T cell, 126–127
  IL-2 role in, 115–118
  type I IFN, enhancement in, 121
Cofactor activity (CA), 148–150, 154, 158, 159, 169
Complement control protein (CCP), 148, 151
Coreceptors, in IL-2-independent $CD8^+$ T cell clonal expansion
  CD27 and coreceptors, 118–120
Cortical disseminated lesions, 2
CREB. *See* Cyclic AMP-response element binding protein
CREB binding protein (CBP), 61
Cycle-independent cytotoxic agent, 23
Cyclic AMP-response element binding protein, 61
Cyclin-dependent kinase (CDK), 27
Cytokines (IFN-γ), 188
Cytosolic recognition systems, 54

D

Decay accelerating activity (DAA), 148
Delta proteins expression, 58
Dementia, 143
Dendritic cells (DCs), 47
Devic's syndrome. *See* Neuromyelitis optica
Diacylglycerol (DAG), 185
Diarrheal-associated/epidemic HUS (D+HUS), 154
DNA binding affinity, 66
DNA methylation, 74

Subject Index

E

Elispot technique, 5, 6. *See also* Chronic neuroinflammation
Em-myc transgenic mice, 124
Encephalitogenic T cell clones, 14
Eomesodermin (Eomes), 54, 71, 115
Eosinophil infiltration, 19
Escherichia coli, Shiga toxin, 154
Experimental autoimmune encephalomyelitis (EAE), 3, 5, 47
Extracellular bacteria, 47
Extracellular pathogens, 47

F

Fas-mediated apoptosis, 120
Fcγ-receptors (FcγRs), 180
Fc-receptors (FcRs)
    antibody activity, modulation of, 188–190
        FcgRIIB, in anti-inflammatory action, 190
        IVIG therapy, 190
        LPS, IFN-g, TH-2 cytokines and C5a in, 188
        sialic acid residues in antibody binding, 189–190
        sugar moiety role in, 188–189
    expression on DCs
        FcR proteins, dual function on, 190–191
        function of FcγRIIB on DCs, 191
        in vivo activity of blocking antibodies, 191
    FCγRIIB as regulator
        in autoreactive B cells, 194
        ICs, for induction of apoptosis, 195
        regulation of B-cell responses, 192
        in SLE patients and polymorphisms in FcγRIIB, 193
    FcR genes, 182
    human FcγRIIA/B/C and FcγRIIIA/B, 180–181
    IgG molecules, family of, 181
    in rodents and human proteins, 181
    role on innate immune effector cells, 185–188
    signaling pathways
        Abl-family kinase-dependent pathway, 185
        Btk and *PLCγ*, in ITAM signaling, 185
        coregulation of FcR signaling, 184
        phosphorylation of ITAM sequences, 184–185
        signaling adaptor molecules, 183
        Syk kinases and Tec family kinases, 185
Fetal hematopoiesis, 11
Follicular dendritic cells (FDCs), 185, 194

G

Gadolinium-enhancing MRI lesions, 25
GITR. *See* Glucocorticoid-induced tumor necrosis factor receptor
Glatiramer acetate (GA), 21
Glucocorticoid-induced tumor necrosis factor receptor, 14, 50
GM-CSF. *See* Granulocyte-macrophage colony-stimulating factor
G-protein-coupled receptors, 24
Granulocyte-macrophage colony-stimulating factor, 46
Gray matter and axonal pathology, 20

H

Helix-loop-helix transcription factors, 66
Hematopoietic stem cell transplantation, 19
Hemocyanin protein, 57
Herpes simplex virus, 16, 112
Histocompatibility complex, 44
Homozygous complement regulatory protein deficiencies
    membrane proteins MCP and Crry, 153–154
    plasma proteins FH and FI, 153
HSV. *See* Herpes simplex virus
Human immunodeficiency virus (HIV), 111
Human plasmacytoid dendritic cells (pDCs), 50
Humoral autoimmune CNS disease, 19
Humoral immunity, 46

I

IFN, types I and II, antiproliferative effects of $CD8^+$ cell, 120–123
    $CD8^+$ T cell proliferation in, 120–121
    effects on $CD8^+$ T cells
        IFN-γR signaling in apoptosis and IFNAR signaling, 121
        regulating IFN-γR expression, 121–122
        IFN-γR1 and IFN-γR2 in regulation, 122
        redistribution of IFN-γR1, to synapse and T-bet expression, 122
        switch determining effects of, 122–123

IFN, types I and II, antiproliferative effects of CD8$^+$ cell (*cont.*)
  E3 ubiquitin ligase and kinases, for Stat1, 123
IL-2, in clonal expansion and effector differentiation
  as mediator in CD8$^+$ T cell, 115–116
  without IL-2R signaling, 116–118
IL-2-independent CD8$^+$ T cell clonal expansion, coreceptors mediating, 118–120
IL-7Rα expression, 118
Immune complexes (ICs), 182
Immune modulatory effects, 22
Immunoglobulin G (IVIG) therapy, 180, 190
Immunological therapy of tumors, 129
Immunoreceptor tyrosine-based activating motifs, 48
Immunoreceptor tyrosine-based inhibitory motif, 48, 183
Immunothrombocytopenia, 187
Inflammatory bowel disease, 44
Inflammatory cells recruitment, 10
INK4b-Arf-INK4a tumor suppressor locus, 124
Inositol triphosphate (IP3), 185
Intercellular adhesion molecule-1, 9
Interferon (IFN-γ), 42, 106, 108, 109, 118, 122, 125, 128. *See also* Cytokines (IFN-γ) 186
  producing cells, 71
  transcription, 55
  types I and II, 120–121, 123, 127
Interferon response factor (IRF), 67
Interleukin (IL)
  IL-2, 109, 113, 115, 116–121, 123, 125–129
  IL-4, 188
  IL-10, 110, 152, 188
  IL-15, 115
Intracellular pathogens, 42, 47
Intrathecal antibody synthesis, 19
Intrathecal immunoglobulin (Ig) synthesis, 16
Intrathecal polyvalent antibody production, 18
Ischemia reperfusion injury (IRI), 143, 144
ITAM. *See* Immunoreceptor tyrosine-based activating motifs

ITIM. *See* Immunoreceptor tyrosine-based inhibitory motif
ITP. *See* Immunothrombocytopenia

J

JC virus, 24

K

Killer cell lectin-like receptor G1 (KLRG1), 113
Killer immunoglobulin receptors (KIR), 48

L

Lectin pathway (LP), 146
Leukocyte function-associated molecule-1, 9
Lineage-specific transcription factors, 47
Listeria infection, 122
Listeria monocytogenes, 60
Lyme arthritis, 7
Lymphocytic choriomeningitis virus (LCMV), 107
Lymphocytic systemic migration, 24
Lymphoid tissues, 51
Lytic granules cytoplasmic storage, 48

M

Macrophage-dominated inflammation, 4
Macular degeneration
  age-related, 160–161
  Factor B/C2, 163
  Factor H, 160–163
Major histocompatibility, 190
Mammalian genes, expression of, 77
Mannan-binding lectin (MBL), 181, 186
Mast cells, activation of, 47
MBP. *See* Myelin basic protein
Measles/Rubella/Herpes Zoster (MRZ) reaction, 16
Membrane cofactor protein (MCP), 150–151
Membranoproliferative glomerulonephritis type II, 153
Metalloproteinase, 110
MHC. *See* Major histocompatibility
Mitogen-activated protein kinase (MAPK), 48
Molecules induce differentiation, 53
Monoclonal antibody, 6
MPGN II. *See* Membranoproliferative glomerulonephritis type II
Multicellular helminths, 47

Multiorgan lymphoproliferative autoimmune disease, 13
Multiple sclerosis (MS)
  EAE mouse model of, 58
  in patients, 2
Murine cytomegalovirus (mCMV), 108
Murine influenza infection, 11
Mycobacterium tuberculosis, 3
Myelin basic protein, 3
Myelin oligodendrocyte glycoprotein (MOG), 4
  antibodies, 18
  B cell and T cell receptor, 4

N

Natalizumab treatment, 24
Natural killer cells (NK), 120, 183, 185, 191
  cell effector mechanisms, 48
  cytokines influence on, 50
  cytotoxicity of, 50
  effect of TGF-β on, 51–52
  IFN-γ production in, 49–51
  Ifng transcription in, 61
  IL-15 and IL-2 for developing, 51
  induction of, 44
Natural killer T (NKT) cells, 120, 121
  cytokine signals in, 54
  IFN-γ production by, 52–53
Neonatal Fc-receptor (FcRn), 181
Neurodegeneration, 2
Neuromyelitis optica (NMO), 19
Nonlymphoid tissues, 59

O

Oligoclonal bands (OCB), 16
Oligodendrocytes, 12

P

Pathogenic mycobacteria, 43
PDGF. *See* Platelet-derived growth factor
Peripheral blood mononuclear cells (PBMCs), 158
Phytohemagglutinin, 113
Plasma cell development, 18
Plasmacytoid dendritic cells, 120
Plasma serine protease, 148
Platelet-derived growth factor, 123
Pleckstrin homology (PH), 185
Pleiotropic cytokines, 7
PLP. *See* Proteolipid protein
PML. *See* Progressive multifocal leukoencephalopathy
Progressive multifocal leukoencephalopathy, 24
Proinflammatory cytokines, 7
Proinflammatory gene expression, 21
Promoter and intronic regulatory elements, 77
Proteases, 146, 147
Protein kinase C (PKC), 185
Protein tyrosine kinases (PTK), 48
Proteolipid protein, 4
Pyogenic bacterial infection, 145

R

Replicative senescence, transcriptional control of
  BMI-1, BLIMP-1, and BCL6/BCL6 in, 123–126
  IL-2, IFN-γ and presence of AICD, 125
  TCR stimulation and c-Myc binding role, 124
Retinal pigmented epithelial, 161–163, 166
Rheumatoid arthritis (RA), 25
RNA polymerase-containing complexes, 60
RPE. *See* Retinal pigmented epithelial

S

Serum autoantibodies, 19
SH2-domain containing inositol 5 phosphatase, 185
SHIP. *See* SH2-domain containing inositol 5 phosphatase
Single nucleotide polymorphisms (SNPs), 157, 159
SLE. *See* Systemic lupus erythematodes
SOCS-1. *See* Suppressor of cytokine signaling-1
Streptococcal pyogenes, 163
Streptococcus pneumoniae, 154
Sugar/lectin complexes, 147
Superagonistic CD28 antibody, 28
Suppressor of cytokine signaling-1, 59
Synthetic amino acid copolymer, 22
Systemic lupus erythematodes, 19

T

T-bet, $CD8^+$ T cell for, 115, 121–122, 128
T cell
  cytokine receptor, 57
  depletion, 119

T cell (*cont.*)
  precursor frequencies, 59
T cell receptor (TCR), 45
TCR/CD27 pathway. *See* Adoptive CD8$^+$ T cell therapy, TCR/CD27 pathway
TGF-β signaling, 59
T helper cells (Th), 5
Th1 lineage, factors influencing, 57–58
TNF-α signaling, 13
TNF-related apoptosis-inducing ligand, 12
Toll-like receptor (TLR), 22
Toll-like receptor 9 (TLR9), 50
Toxin-induced demyelination model, 21
Toxoplasm gondii, 65
TRAIL. *See* TNF-related apoptosis-inducing ligand
Transferringmyelin-specific lymphocytes, 2
Transforming growth factor (TGF), 26
Transgenic reporter assays, 76
Treg cells, 29, 47
Tumor necrosis factor (TNF), 7, 46

U

Urate crystals, 145

V

Vascular endothelial-derived growth factor (VEGF), 168
Vascular hyalinization, 19
Very Late Antigen-4 (VLA-4), 9
Viral and intracellular bacterial infections, 84
Viral protein synthesis, 50

Contents of Recent Volumes

Volume 85

Cumulative Subject Index Volumes 66–82

Volume 86

Adenosine Deaminase Deficiency: Metabolic Basis of Immune Deficiency and Pulmonary Inflammation
Michael R. Blackburn and Rodney E. Kellems

Mechanism and Control of V(D)J Recombination Versus Class Switch Recombination: Similarities and Differences
Darryll D. Dudley, Jayanta Chaudhuri, Craig H. Bassing, and Frederick W. Alt

Isoforms of Terminal Deoxynucleotidyltransferase: Developmental Aspects and Function
To-Ha Thai and John F. Kearney

Innate Autoimmunity
Michael C. Carroll and V. Michael Holers

Formation of Bradykinin: A Major Contributor to the Innate Inflammatory Response
Kusumam Joseph and Allen P. Kaplan

Interleukin-2, Interleukin-15, and Their Roles in Human Natural Killer Cells
Brian Becknell and Michael A. Caligiuri

Regulation of Antigen Presentation and Cross-Presentation in the Dendritic Cell Network: Facts, Hypothesis, and Immunological Implications
Nicholas S. Wilson and Jose A. Villadangos

Index

Volume 87

Role of the LAT Adaptor in T-Cell Development and T_h2 Differentiation
Bernard Malissen, Enrique Aguado, and Marie Malissen

The Integration of Conventional and Unconventional T Cells that Characterizes Cell-Mediated Responses
Daniel J. Pennington, David Vermijlen, Emma L. Wise, Sarah L. Clarke, Robert E. Tigelaar, and Adrian C. Hayday

Negative Regulation of Cytokine and TLR Signalings by SOCS and Others
Tetsuji Naka, Minoru Fujimoto, Hiroko Tsutsui, and Akihiko Yoshimura

Pathogenic T-Cell Clones in Autoimmune Diabetes: More Lessons from the NOD Mouse
Kathryn Haskins

The Biology of Human Lymphoid Malignancies Revealed by Gene Expression Profiling
Louis M. Staudt and Sandeep Dave

New Insights into Alternative Mechanisms of Immune Receptor Diversification
Gary W. Litman, John P. Cannon, and Jonathan P. Rast

The Repair of DNA Damages/Modifications During the Maturation of the Immune System: Lessons from Human Primary Immunodeficiency Disorders and Animal Models
Patrick Revy, Dietke Buck, Françoise le Deist, and Jean-Pierre de Villartay

Antibody Class Switch Recombination: Roles for Switch Sequences and Mismatch Repair Proteins
Irene M. Min and Erik Selsing

Index

Volume 88

CD22: A Multifunctional Receptor That Regulates B Lymphocyte Survival and Signal Transduction
Thomas F. Tedder, Jonathan C. Poe, and Karen M. Haas

Tetramer Analysis of Human Autoreactive CD4-Positive T Cells
Gerald T. Nepom

Regulation of Phospholipase C-γ2 Networks in B Lymphocytes
Masaki Hikida and Tomohiro Kurosaki

Role of Human Mast Cells and Basophils in Bronchial Asthma
Gianni Marone, Massimo Triggiani, Arturo Genovese, and Amato De Paulis

A Novel Recognition System for MHC Class I Molecules Constituted by PIR
Toshiyuki Takai

Dendritic Cell Biology
Francesca Granucci, Maria Foti, and Paola Ricciardi-Castagnoli

The Murine Diabetogenic Class II Histocompatibility Molecule I-A^{g7}: Structural and Functional Properties and Specificity of Peptide Selection
Anish Suri and Emil R. Unanue

RNAi and RNA-Based Regulation of Immune System Function
Dipanjan Chowdhury and Carl D. Novina

Index

Volume 89

Posttranscriptional Mechanisms Regulating the Inflammatory Response
Georg Stoecklin Paul Anderson

Negative Signaling in Fc Receptor Complexes
Marc Daëron and Renaud Lesourne

The Surprising Diversity of Lipid Antigens for CD1-Restricted T Cells
D. Branch Moody

Lysophospholipids as Mediators of Immunity
Debby A. Lin and Joshua A. Boyce

Systemic Mastocytosis
Jamie Robyn and Dean D. Metcalfe

Regulation of Fibrosis by the Immune System
Mark L. Lupher, Jr. and W. Michael Gallatin

Immunity and Acquired Alterations in Cognition and Emotion: Lessons from SLE
Betty Diamond, Czeslawa Kowal, Patricio T. Huerta, Cynthia Aranow, Meggan Mackay, Lorraine A. DeGiorgio, Ji Lee, Antigone Triantafyllopoulou, Joel Cohen-Solal Bruce, and T. Volpe

Immunodeficiencies with Autoimmune Consequences
Luigi D. Notarangelo, Eleonora Gambineri, and Raffaele Badolato

Index

Volume 90

Cancer Immunosurveillance and Immunoediting: The Roles of Immunity in Suppressing Tumor Development and Shaping Tumor Immunogenicity
Mark J. Smyth, Gavin P. Dunn, and Robert D. Schreiber

Mechanisms of Immune Evasion by Tumors
Charles G. Drake, Elizabeth Jaffee, and Drew M. Pardoll

Development of Antibodies and Chimeric Molecules for Cancer Immunotherapy
Thomas A. Waldmann and John C. Morris

Induction of Tumor Immunity Following Allogeneic Stem Cell Transplantation
Catherine J. Wu and Jerome Ritz

Vaccination for Treatment and Prevention of Cancer in Animal Models

Federica Cavallo, Rienk Offringa,
 Sjoerd H. van der Burg, Guido Forni,
 and Cornelis J. M. Melief

Unraveling the Complex Relationship Between Cancer Immunity and Autoimmunity: Lessons from Melanoma and Vitiligo
 Hiroshi Uchi, Rodica Stan, Mary Jo Turk,
 Manuel E. Engelhorn,
 Gabrielle A. Rizzuto,
 Stacie M. Goldberg, Jedd D. Wolchok,
 and Alan N. Houghton

Immunity to Melanoma Antigens: From Self-Tolerance to Immunotherapy
 Craig L. Slingluff, Jr.,
 Kimberly A. Chianese-Bullock,
 Timothy N. J. Bullock,
 William W. Grosh, David W. Mullins,
 Lisa Nichols, Walter Olson,
 Gina Petroni, Mark Smolkin, and
 Victor H. Engelhard

Checkpoint Blockade in Cancer Immunotherapy
 Alan J. Korman, Karl S. Peggs, and
 James P. Allison

Combinatorial Cancer Immunotherapy
 F. Stephen Hodi and Glenn Dranoff

Index

Volume 91

A Reappraisal of Humoral Immunity Based on Mechanisms of Antibody-Mediated Protection Against Intracellular Pathogens
 Arturo Casadevall and
 Liise-anne Pirofski

Accessibility Control of V(D)J Recombination
 Robin Milley Cobb, Kenneth J. Oestreich,
 Oleg A. Osipovich, and
 Eugene M. Oltz

Targeting Integrin Structure and Function in Disease

Donald E. Staunton, Mark L. Lupher,
 Robert Liddington,
 and W. Michael Gallatin

Endogenous TLR Ligands and Autoimmunity
 Hermann Wagner

Genetic Analysis of Innate Immunity
 Kasper Hoebe, Zhengfan Jiang, Koichi
 Tabeta, Xin Du, Philippe Georgel,
 Karine Crozat, and Bruce Beutler

TIM Family of Genes in Immunity and Tolerance
 Vijay K. Kuchroo, Jennifer Hartt Meyers,
 Dale T. Umetsu, and
 Rosemarie H. DeKruyff

Inhibition of Inflammatory Responses by Leukocyte Ig-Like Receptors
 Howard R. Katz

Index

Volume 92

Systemic Lupus Erythematosus: Multiple Immunological Phenotypes in a Complex Genetic Disease
 Anna-Marie Fairhurst,
 Amy E. Wandstrat, and
 Edward K. Wakeland

Avian Models with Spontaneous Autoimmune Diseases
 Georg Wick, Leif Andersson, Karel
 Hala, M. Eric Gershwin, Carlo Selmi,
 Gisela F. Erf, Susan J. Lamont, and
 Roswitha Sgonc

Functional Dynamics of Naturally Occurring Regulatory T Cells in Health and Autoimmunity
 Megan K. Levings, Sarah Allan, Eva
 d'Hennezel, and Ciriaco A. Piccirillo

BTLA and HVEM Cross Talk Regulates Inhibition and Costimulation

Maya Gavrieli, John Sedy, Christopher A. Nelson, and Kenneth M. Murphy

The Human T Cell Response to Melanoma Antigens
Pedro Romero, Jean-Charles Cerottini, and Daniel E. Speiser

Antigen Presentation and the Ubiquitin-Proteasome System in Host–Pathogen Interactions
Joana Loureiro and Hidde L. Ploegh

Index

Volume 93

Class Switch Recombination: A Comparison Between Mouse and Human
Qiang Pan-Hammarström, Yaofeng Zhao, and Lennart Hammarström

Anti-IgE Antibodies for the Treatment of IgE-Mediated Allergic Diseases
Tse Wen Chang, Pheidias C. Wu, C. Long Hsu, and Alfur F. Hung

Immune Semaphorins: Increasing Members and Their Diverse Roles
Hitoshi Kikutani, Kazuhiro Suzuki, and Atsushi Kumanogoh

Tec Kinases in T Cell and Mast Cell Signaling
Martin Felices, Markus Falk, Yoko Kosaka, and Leslie J. Berg

Integrin Regulation of Lymphocyte Trafficking: Lessons from Structural and Signaling Studies
Tatsuo Kinashi

Regulation of Immune Responses and Hematopoiesis by the Rap1 Signal
Nagahiro Minato, Kohei Kometani, and Masakazu Hattori

Lung Dendritic Cell Migration
Hamida Hammad and Bart N. Lambrecht

Index

Volume 94

Discovery of Activation-Induced Cytidine Deaminase, the Engraver of Antibody Memory
Masamichi Muramatsu, Hitoshi Nagaoka, Reiko Shinkura, Nasim A. Begum, and Tasuku Honjo

DNA Deamination in Immunity: AID in the Context of Its APOBEC Relatives
Silvestro G. Conticello, Marc-Andre Langlois, Zizhen Yang, and Michael S. Neuberger

The Role of Activation-Induced Deaminase in Antibody Diversification and Chromosome Translocations
Almudena Ramiro, Bernardo Reina San-Martin, Kevin McBride, Mila Jankovic, Vasco Barreto, André Nussenzweig, and Michel C. Nussenzweig

Targeting of AID-Mediated Sequence Diversification by *cis*-Acting Determinants
Shu Yuan Yang and David G. Schatz

AID-Initiated Purposeful Mutations in Immunoglobulin Genes
Myron F. Goodman, Matthew D. Scharff, and Floyd E. Romesberg

Evolution of the Immunoglobulin Heavy Chain Class Switch Recombination Mechanism
Jayanta Chaudhuri, Uttiya Basu, Ali Zarrin, Catherine Yan, Sonia Franco, Thomas Perlot, Bao Vuong, Jing Wang, Ryan T. Phan, Abhishek Datta, John Manis, and Frederick W. Alt

Beyond SHM and CSR: AID and Related Cytidine Deaminases in the Host Response to Viral Infection
Brad R. Rosenberg and F. Nina Papavasiliou

Role of AID in Tumorigenesis
Il-mi Okazaki, Ai Kotani, and Tasuku Honjo

Pathophysiology of B-Cell Intrinsic
  Immunoglobulin Class Switch
  Recombination Deficiencies
  *Anne Durandy, Nadine Taubenheim,
  Sophie Peron, and Alain Fischer*

Index

Volume 95

Fate Decisions Regulating Bone Marrow
  and Peripheral B Lymphocyte
  Development
  John G. Monroe and Kenneth Dorshkind

Tolerance and Autoimmunity:
  Lessons at the Bedside of Primary
  Immunodeficiencies
  *Magda Carneiro-Sampaio and Antonio
  Coutinho*

B-Cell Self-Tolerance in Humans
  *Hedda Wardemann and Michel
  C. Nussenzweig*

Manipulation of Regulatory T-Cell
  Number and Function with CD28-
  Specific Monoclonal Antibodies
  Thomas Hünig

Osteoimmunology: A View from the Bone
  Jean-Pierre David

Mast Cell Proteases
  *Gunnar Pejler, Magnus Åbrink,
  Maria Ringvall, and Sara Wernersson*

Index